The Manchu *Mirrors* and the Knowledge
of Plants and Animals in High Qing China

HARVARD-YENCHING INSTITUTE MONOGRAPH SERIES 143

The Manchu *Mirrors*
and the Knowledge of Plants and Animals
in High Qing China

He Bian and Mårten Söderblom Saarela

Published by the Harvard University Asia Center
Distributed by Harvard University Press
Cambridge (Massachusetts) and London 2025

© 2025 by the President and Fellows of Harvard College
All rights reserved. No part of this publication may be reproduced, translated, stored in a retrieval system, or transmitted in any form or by any means, electronic, mechanical, photocopying, recording or otherwise, without prior written permission from the publisher.

Published by the Harvard University Asia Center, Cambridge, MA 02138

The Harvard University Asia Center publishes a monograph series and, in coordination with the Fairbank Center for Chinese Studies, the Korea Institute, the Reischauer Institute of Japanese Studies, and other faculties and institutes, administers research projects designed to further scholarly understanding of China, Japan, Korea, Vietnam, and other Asian countries. The Center also sponsors projects addressing multidisciplinary and regional issues in Asia.

The Harvard University Asia Center gratefully acknowledges the generous support of the Harvard-Yenching Institute, whose funding contributed to the publication of this book.

The Harvard-Yenching Institute, founded in 1928, is an independent foundation dedicated to the advancement of higher education in the humanities and social sciences in Asia. Headquartered on the campus of Harvard University, the Institute provides fellowships for advanced research, training, and graduate studies at Harvard by competitively selected faculty and graduate students from Asia. The Institute also supports a range of academic activities at its fifty partner universities and research institutes across Asia. At Harvard, the Institute promotes East Asian studies through annual contributions to the Harvard-Yenching Library and publication of the *Harvard Journal of Asiatic Studies* and the Harvard-Yenching Institute Monograph Series.

Library of Congress Cataloging-in-Publication Data

Names: Bian, He, author. | Söderblom Saarela, Mårten, author.
Title: The Manchu Mirrors and the knowledge of plants and animals in High Qing China / He Bian and Mårten Söderblom Saarela.
Other titles: Harvard-Yenching Institute monograph series ; 143.
Description: Cambridge : Harvard University Asia Center, 2025. | Series: Harvard-Yenching Institute monograph series ; 143 | Includes bibliographical references and index. | English; some text in Manchu and Chinese.
Identifiers: LCCN 2024051418 | ISBN 9780674297517 (hardcover) | ISBN 9780674301726 (epub)
Subjects: LCSH: Manchu language—History. | Manchu language—Lexicography—History. | Chinese language—Lexicography—History. | Learning and cholarship—China—History. | Natural history—Dictionaries—China—History. |Multilingualism—China—History. | China—History—Qing dynasty, 1644–1912.
Classification: LCC PL472 .B53 2025 | DDC 578.01/4—dc23/eng/20250511
LC record available at https://lccn.loc.gov/2024051418

Index by Cynthia Col

Printed by Books International, 22883 Quicksilver Drive, Dulles, VA 20166, USA (www.books intl.com). The manufacturer's authorized representative in the EU for product safety is LOGOS EUROPE, 9 rue Nicolas Poussin, 17000, La Rochelle, France (e-mail: Contact@logoseurope.eu).

To Benjamin A. Elman, teacher and mentor

Contents

List of Illustrations — ix

Acknowledgments — xi

Conventions — xiii

Introduction — 1

PART I: LEXICOGRAPHY AND ENCYCLOPEDISM AT THE KANGXI COURT

1. Flora and Fauna in the *Mirror of the Manchu Language* — 21
2. The Manchu *Mirror* as Chinese Encyclopedia — 58
3. *Guang Qunfang pu* and the Reform of Literati Botany — 85

PART II: PLURILINGUAL NOMENCLATURE IN MANCHURIA

4. Naming Manchurian Things in Early-Qing Chinese Writings — 123
5. Manchurian and Inner Asian Plants in *Guang Qunfang pu* and *Gujin tushu jicheng* — 147

PART III: NEOLOGISMS FOR PLANTS AND ANIMALS AT THE QIANLONG COURT

6. *Ode to Mukden* and the Manchu Homeland in Chinese Literary Form — 193
7. Qianlong's Manchu-Language Reform and Natural-Historical Philology — 230

8. Plants and Animals in the *Expanded and Emended Mirror of the Manchu Language*   274

Conclusion: Reception of the Manchu *Mirrors* in Qing China and Beyond   313

Bibliography   335

Index   377

Illustrations

Figures

1.1	First page on herbs (*orho*) in the Kangxi *Mirror*	30
1.2	The swordfish and the mermaid in Verbiest's *Kunyu tushuo*	54
2.1	A page with saltwater fish from the Kangxi *Mirror*	82
3.1	Opening of *juan* 1 in Wang Xiangjin's *Qunfang pu*, late Ming edition	90
3.2	Pages from *Qunfang pu* showing top panel glosses and bottom panel notes	98
3.3	A page from *Guang Qunfang pu*, lithographic edition	107
3.4	Layout in *Yuanjian leihan* showing "original" (*yuan* 原) and "added" (*zeng* 增) content	111
3.5	Layout in *Guang Qunfang pu* showing "original" (*yuan* 原) and "added" (*zeng* 增) content	111
4.1	A map of Manchuria in the 1684 *Shengjing tongzhi*	129
4.2	Local product list in the 1684 *Shengjing tongzhi*, showing *fafaha* and *misu usiha*	141
6.1	A miniature copy of the *Ode to Mukden* (title in Chinese) from the Qianlong emperor's curio collection	192
7.1	Album leaf depicting *ayan hiyan* (Ch. *yunxiang*) in Yu Xing, *Jiachan jianxin*	243
8.1	A page from the *Expanded and Emended Mirror of Manchu Language* with entries for kinds of gourd	285
8.2	A page from the *Expanded and Emended Mirror of Manchu Language*'s supplement with entry for giraffe	310

| 9.1 | A selection of Rémusat's natural-historical note cards with material from the Manchu *Mirrors* and Japanese sources | 326 |

Tables

2.1	Chinese translations relative to total entries of section in 1708 *Mirror*	78
2.2	Chinese translations within section relative to overall total in 1708 *Mirror*	79
3.1	Corresponding sections and entry counts in *Qunfang pu* and *Guang Qunfang pu*	100
3.2	Structure of entries in *Qunfang pu* and *Guang Qunfang pu*	106

Acknowledgments

This collaborative project was carried out in separate locations, largely during the COVID-19 pandemic, from Yilan and Iowa City to Taipei, Princeton, and Los Angeles. Looking back, we feel as if it all happened in a different world, but one full of support from many people.

The project began in pre-pandemic times at the Max Planck Institute for the History of Science in Berlin, where Masato Hasegawa was convening a reading group. Present for the discussion of our sources were also Mackenzie Cooley and Yubin Shen. We also thank Dagmar Schäfer for her support at the institute. Söderblom Saarela thanks the staff and faculty of the Institute of Modern History, Academia Sinica, where he worked when writing the bulk of the manuscript. Lai Yu-chih in particular helped with references. Bian thanks her colleagues in the departments of History and East Asian Studies at Princeton University. We thank two peer reviewers and Mark C. Elliott (as the third reader) for their critical and constructive feedback on the initial submission. Robert M. Graham, Kristen Wanner, and Qin Higley guided the manuscript through revision and publication. Helen Glenn Court expertly copyedited the manuscript, Tore Wallert patiently and beautifully designed the cover, and Cynthia Col painstakingly made the index. We are responsible for any remaining errors.

Other friends and colleagues who at one point or another offered us help when we were writing this book include Michael D. Gordin, Anna M. Shields, He Yingtian 何映天, Martin Heijdra, Marcia Yonemoto, Nathan Vedal, Young Oh, Leigh Jenco, Sue Naquin, Kai Jun Chen, Bina Arch, Julia Manchu Wu, Tsai Sung-ying 蔡松穎, Chen Kuan-chieh 陳冠傑, Lin Shih-hsuan 林士鉉, Arina Mikhalevskaya, Andreas Hoelzl, José Andrés Alonso de la Fuente, Noël Golvers, Alice Crowther, Matthew Mosca, Christopher Atwood, Eric Schluessel,

David Porter, Steve Miles, Peter Lavelle, Anna Grasskamp, Devin Fitzgerald, Soren Edgren, and all the participants of the online U.S.-based Manchu reading group. We also thank the librarians and archivists at Princeton University Library, Harvard-Yenching Library, the British Library, the First Historical Archives in Beijing, the Bibliothèque Nationale de France, the Staatsbibliothek zu Berlin, the University of Iowa Libraries, and the Library of the Institute of Modern History.

Parts of the book were presented at the Association for Asian Studies Annual Conference in 2021 (Bian and Söderblom Saarela) and 2022 (Söderblom Saarela), the History of Science Society Annual Meeting in 2022 (Bian and Söderblom Saarela), Davis Center Work-in-Progress talk and EAS Colloquium (Bian), the online conference Natural History Writings and Material Culture in East Asian Cultural Imagery 東亞文化意象的博物書寫與物質文化 at Academia Sinica (Bian), the Institute of Modern History colloquium (Söderblom Saarela), the conference L'Apport des missions chrétiennes: Échanges des connaissances avec l'Asie (Söderblom Saarela), the MPIWG Philology Workshop (Söderblom Saarela), the Manchu reading group organized by the Institute of Modern History and National Taiwan University (Söderblom Saarela), and the Tsinghua University History Department (Bian).

National Science and Technology Council (R.O.C.) grants "Manchu Language Standardization and Qing Institutions during the Eighteenth Century" (108-2410-H-001-078) and "Language and Governance in Qing Inner Asia" (110-2410-H-001-006-MY3) supported research for this book. Bian has conducted initial research during her ACLS China Studies Early Career Fellowship in 2021, and finished revision during her sabbatical leave in Spring 2024 funded by the Department of History at Princeton.

Finally, we thank our partners and family for their loving support, a rare constant in today's world of rapid changes. We dedicate this book to Benjamin A. Elman, without whom this collaboration would not have been possible, and whose ideas continue to shape our work to this day.

Conventions

- Qing reign dates are given using two-letter abbreviations for post-1644 reign periods, for example, KX = Kangxi, QL = Qianlong, and so on.
- English-language translations that immediately follow foreign-language words are given in single quotation marks. Double quotation marks are used for everything else. "Idem" is used when a gloss is identical to the preceding.
- "Manchu" refers to the Manchu language or Manchu people, whereas "Manchurian" refers to all the languages and peoples of Manchuria, including Tungusic languages other than Manchu and non-Tungusic languages such as Nivkh.
- Books are generally referred to by their title in the original language, with two exceptions, for which we use English. The first is books with plurilingual titles, including all editions of the Manchu *Mirror*. Because all later installments in this book series are plurilingual, we refer also to the first, monolingual edition in English. Note that books that contain both Manchu and Chinese but have only a Chinese title are referred to in Chinese. Qianlong's catalogs of birds and beasts are special cases. These books are referred to in Chinese in court documents, but the catalogs themselves do not carry titles. Because they are bilingual, we refer to them in English. The second exception is very well-known books, such as the Confucian classics.

INTRODUCTION

The seeds of this book were sown, so to speak, in a seminar room at the Max Planck Institute for the History of Science in Berlin in May 2018. At a Manchu reading group that our colleague Masato Hasegawa set up at the institute, one of us shared transcripts of a few eighteenth-century documents that originated in the Qing Grand Council and are now held at the First Historical Archives in Beijing. What we thought would be a fun and relaxing reading of standard Qing memorials quickly turned into a much more intriguing conversation.

First, we noticed that although these documents were found among the palace memorial reference copies kept in the Grand Council archive, not all of them were in fact palace memorials. Some certainly were, being signed, dated, and addressed to the emperor. Others were bare-bones notes to superiors within the Grand Council whose dates can be inferred only from their placement in the archive. Second, the documents were bilingual in a way rarely seen in official documents. The documents were written in Manchu. However, Chinese book titles, nouns, and even conjunctions in Chinese characters but without Manchu transcriptions were interspersed directly in the Manchu text (more in chapter 7). Documents in the palace memorial collection are usually in either one language or the other, and if they are bilingual, the Manchu and Chinese portions are written as separate texts.[1] These two characteristics—the truncated format of some documents and the mix of languages—made us think that expediency rather than concern for proper bureaucratic form had guided the composition of the documents we had before us.

Third and most important, the topic of the documents was the meaning of certain Manchu words referring to flora and fauna and the coining of new related words. For instance, to find the appropriate Manchu name for ostriches (or some other large bird of non-Chinese provenance), the authors of the documents referenced a body of scholarly literature that had been published in both Manchu and Chinese at the Qing court since the beginning of the eighteenth century. This concerted effort at naming and describing flora and fauna systematically evidenced a kind of bilingual natural-historical scholarship at the Qing court that was strikingly new to both of us. On closer scrutiny, it turned out that the documents drew from not only written sources produced at the Qing court, but also life drawings originally attached to these documents but now lost.

At the conclusion of the brief reading session, we resolved to further investigate this unexpected way that ostriches and other creatures and plants intersected with the sprawling plurilingual administrative apparatus of the Qing empire.[2] This book presents the result of our joint effort, which owes its inspiration to the Max Planck Institute and the colleagues there who encouraged us to pursue the project.

As its title suggests, this book aims at connecting two largely separate lines of historical inquiry. On the one hand, it is firmly rooted in the history of Manchu language as spoken, written, read, and taught among and beyond the ruling elite of the Qing dynasty, in which the set of texts known as the Manchu *Mirrors* have long held iconic stature as imperially sponsored dictionaries compiled during the so-called

---

1. For this kind of document, see Chuang, "Guoli Gugong"; Söderblom Saarela, "Linguistic Compartmentalization," 141–44.

2. In general, we prefer "plurilingual" (Lat. *plūs* "more" or *plūres* "several") to "multilingual" (Lat. *multus* "much, many") because the former allows for a constellation of only two languages, whereas the latter implies three or more. We reserve the term "multilingual" for cases in which three or more languages are involved.

High Qing period.[3] On the other hand, it addresses the history of codified knowledge about the natural world, or plants and animals in particular, in the Chinese learned tradition, which stretches far into antiquity but underwent important permutations under Qing rule.

Our previous work on the history of pharmacy and Manchu-language studies in late imperial China, respectively, overlaps methodologically in that it shares a focus on book history, print culture, and the history of knowledge and, chronologically, in its concern with the epistemic ramifications of the Qing conquest in the mid-seventeenth century. In the case of pharmacy, studied in Bian's *Know Your Remedies*, an earlier model of state-commissioned pharmacopeia gave way in the High Qing era to a much more fluid nexus of knowledge that flowed from the marketplace, the pharmacy shop, and vernacular testimonies derived from cosmopolitan consumption, in which the Qing ruling elite enthusiastically took part.[4] The language studies in Söderblom Saarela's *The Early Modern Travels of Manchu* similarly allow us to situate the Qing court within a flow of texts that extended far beyond the empire's borders, one that enabled scholars in different parts of Eurasia to work on the common problems of analyzing and presenting new linguistic material for ease of access and reference.[5] Söderblom Saarela's *The Manchu Language at Court and in the Bureaucracy under the Qianlong Emperor* focused on part of the period covered here but was concerned mainly with administrative documents and the editing of the pre-conquest Manchu archive.[6] In examining Qing plant and animal knowledge through the Manchu *Mirrors*, we tell a story that builds on our work on distinct aspects of Qing intellectual culture but goes beyond it in ways that we would have been hard-pressed to do each on our own.

The story told in this book begins and ends with two iterations of the court's *Mirror* series of Manchu dictionaries, consisting of eight chapters

---

3. Wakeman, "High Ch'ing" (this essay is outdated but interesting for its use of "High Qing"); Rowe, *China's Last*, 63–89.
4. Bian, *Know Your Remedies*, esp. chaps. 4, 6.
5. Söderblom Saarela, *Early Modern Travels*.
6. Söderblom Saarela, *Manchu Language at Court*.

grouped into three parts. Part 1 centers on the Kangxi *Mirror* (*Han-i araha manju gisun-i buleku bithe*, presented to the throne in 1708), as well as a series of encyclopedic and lexicographical texts in Chinese commissioned by the Kangxi court during the 1700s and 1710s. Part 2 takes an excursion to focus on Manchuria and Inner Asia as a privileged space for naming and investigating plants and animals that had not been documented in the Chinese tradition before the Qing but featured prominently in late Kangxi encyclopedism and a new vision for the "investigation of things" (*gewu* 格物). Part 3 traces the intensifying efforts of coining Manchu neologisms for plant and animal names during the Qianlong reign (1736–1796) across various sources, from the *Ode to Mukden* (*Han-i araha Mukden-i fu* [later *fujurun*] *bithe* | *Yuzhi Shengjing fu* 御製盛京賦) in the 1740s to the Manchu-Chinese revised and expanded *Mirror* (*Han-i araha nonggime toktobuha manju gisun-i buleku bithe* | *Yuzhi zengding Qingwen jian* 御製增訂清文鑑, completed in 1772). The conclusion examines the subsequent reception of the plant and animal chapters in the Manchu *Mirrors* in Qing China and beyond.

Using the two Manchu *Mirrors* as the chronological anchors of our inquiry, we present a new perspective—with plants and animals at the center—on Qing intellectual history and epistemic change. Bracketed by these efforts completed in 1708 and 1772, respectively, our narrative uses the two *Mirrors* as a reflective device to shed new light on the dynamic relationship between Chinese and Manchu knowledge-making efforts at the Qing court, as well as between court and private scholarship. Much is at stake in the process that made Kangxi's monolingual dictionary—one primarily concerned with "the names that exist in Manchu," as the compilers at one point put it—into Qianlong's bilingual dictionary, which no longer presented the world as seen through Manchu but instead stretched the Manchu language to cover the world in its entirety, including things formerly foreign to the Manchus.[7] The Manchu *Mirrors* and related Chinese texts allow us to detect widespread fascination with creatures not named or described in earlier sources, the elaborate procedures for naming them, and the manipulation of the lexicon to achieve nomenclatural orderliness. Overall, our

7. Noted also by Tsai, "Qing qianqi," 243.

journey between the Manchu *Mirrors* presents an evolving vision for a Qing order of things—in the sense of words and things—consisting of an eclectic mixture of the mundane and homely items with the wondrous and fantastical, of the vernacular with the classical, and of the familiar and the shockingly new.

Most sources we use are well known among experts in Qing history. The Manchu and Chinese texts, however, have rarely—if ever—been discussed together as products of court scholarship rather than as sources belonging to either linguistic tradition. We introduce the key sources in greater detail at the beginning of each chapter. The rest of this introduction provides an overview of two central themes in current historiography and stakes the main interventions of this book. Our intention is twofold: to challenge the implicit centrality of Chinese sources and the supposed purity of Manchu knowledge in Qing intellectual and cultural histories, and to carve out a new line of inquiry between the history of science (knowledge about plants and animals) and the history of humanities (lexicography and encyclopedism).[8]

Never Pure: Manchu Lexicography and Scholarly Practices at the High Qing Court

The eclectic selection of references seen in the Grand Council documents mentioned earlier testifies to the blurring of linguistic and ethnic boundaries when it came to finding a proper Manchu name for, say, flower parts, ostriches—known since the early Ming exploration of the Indian Ocean—and juvenile locusts. We are left, then, with a question. Without assuming that what we see in these documents necessarily implies a cultural or scholarly agenda that was distinctly Manchu—and eo ipso of limited interest to non-Manchu actors—how can we better make sense of knowledge claims made at the Qing court as both a result of imperial policies specific to its times and part of learned traditions that transcend them?

---

8. For a similar historiographical intervention, see Daston and Most, "History of Science."

Much of our understanding of Manchu lexicography has built on scholarship retrospectively labeled as the New Qing History, characterized by an emphasis on non-Chinese sources and its stress of the centrality of Manchu identity within Qing ideology and actual governance.[9] Despite significant disagreements over the formation and salience of Manchu identity, scholars generally agree that the Qianlong reign witnessed an intense campaign that emphasized what it meant to be Manchu. As well as court rituals that reenacted a mystified Manchu past and enforced bodily regimens that tested bannermen's mounted archery skills, the Manchu language and its proper use figured prominently in Qianlong's agenda. Indeed, the Manchu *Mirrors,* especially the four- and five-language versions of the dictionary, have frequently assumed a metonymical function in scholarly discourses with Qing rulership over a multilingual and therefore multiethnic empire.[10]

Yet an innate distinction between Manchuness and Chineseness can prove problematic when we look at the actual dynamics of cultural production. In chapters 1 and 2, we analyze sections on flora and fauna in the Kangxi emperor's monolingual *Mirror,* tracing the diverse sources of plant and animal knowledge to the Chinese encyclopedic tradition and literature introduced by Jesuit missionaries, in addition to words used in Manchu and Mongolian but also drawn from other minority languages in northeast Asia. Later, we follow the massive invention of Manchu neologisms to name plants and animals in Qianlong's *Ode to Mukden* (chapter 6), reforms of the Manchu language as reflected in pictorial albums and Grand Council discussions (chapter 7), and the *Expanded and Emended Mirror of the Manchu Language* (chapter 8). Contrary to our expectations, our findings found no stable natural knowledge that remained comfortably Manchu throughout the eighteenth century.

9. For discussions of this historiographical trend, see Waley-Cohen, "New Qing History"; Dunnell and Millward, introduction to *New Qing Imperial History*; Elliott, "Manwen dang'an yu Xin Qingshi"; Wu Guo, "New Qing History."

10. For a critical assessment of Qianlong's construction of universal rulership in projects including the *Pentaglot,* see Crossley, *Translucent Mirror,* 269.

By complicating the supposed purity of the Manchu lexicon, then, are we summoning the exorcized ghost of Sinicization through the back door? Asserting the importance of Manchu identity for the Qing formation and the utility of Manchu-language sources for its study contributed to a now irreversible reorientation of Qing studies, and the field is all the better for it. In its wake, the complicated relationship between learning at the Qing court and extra-palatial scholarly communities has been brought to the fore. Thanks to Matthew Mosca's and others' work on Qing frontier policy and geographical scholarship, we now have a far more sophisticated picture of how imperial expansion—and the court's privileged access to intelligence and other sensitive information—eventually leaked outside Beijing and brought lasting impact to the way in which historical geographical research was conducted, for example, by private scholars of a Chinese background.[11] Scholars of literature and translation have made further use of literary output by men and women of the Eight Banners to interrogate how Qing authors traversed linguistic divides.[12]

Delving further into the material mechanisms of cultural change during the Qing, scholars have studied the rich archives that documented the daily operations of the Qing court, as well as the dazzlingly inventive artifacts produced under the aegis of the High Qing emperors, to show how the Manchu elite in Qing China was really building (in Dorothy Ko's words) a "material empire" that required tremendous care and resources to run and maintain, not to mention sophisticated knowledge and a command of materials extracted from the natural world. As Kai Jun Chen, Yulian Wu, Roslyn Hammers, and others have noted, this effort eventually serves to "rework the classics" as the

---

11. Mosca, *Frontier Policy to Foreign Policy*; Xue Zhang, "Imperial Maps."
12. For more on the Qing Eight Banners as a key institution for military mobilization and unique social space for entrenched Manchu identity, see Elliott, *Manchu Way*. For a discussion of cultural hybridity in popular culture among bannermen, see Chiu, *Bannermen Tales*. For a growing literature on Manchu literary culture and translation, see, among others, Nappi, *Translating Early Modern China*, chaps. 4–5; Idema, *Two Centuries of Manchu Women Poets*; Zheng, "Experimenting with the National Language"; Bramao-Ramos, "Manchu-Language Books"; Vedal, "Manchu Reading of *Jinpingmei*."

Manchu court sought to leave a long-lasting legacy in the Chinese scholarly tradition.[13]

At the same time, archival materials, especially in Manchu and Mongolian, have shed light on how the state managed the material wealth of what Jonathan Schlesinger has called the "natural fringes of Qing rule." The close monitoring of production and extraction of precious commodities such as ginseng, fur, fresh water pearls, and mushrooms left a rich paper trail in the official archive as well as reshaped the consumption habits of millions of Chinese, who might have lived far away from these places but still became intimately familiar with this new "wilderness" and the native peoples and creatures living there.[14] Thus the Manchu court during the High Qing era has increasingly been treated as a key site for scholarly innovation and technical ingenuity, motivated by the conquest elite's need to tame and remold Chinese civilization in its own terms.

Like much recent scholarship, we do not see the Manchu rulers of the Qing dynasty as purely antagonistic to the agenda of the Chinese intellectual elite, or as affecting Chinese thought merely by diverting it away from politically sensitive topics.[15] Instead, we find it more productive to examine how the Manchu imperial regime provided and

---

13. Ko, *Social Life of Inkstones*, esp. chap. 1; Wu Yulian, *Luxurious Networks*; Siebert, Chen, and Ko, *Making the Palace Machine Work*; Hammers, "Assimilating the Classics"; Chen, *Porcelain for the Emperor*. Art historians based in East Asia and elsewhere have pioneered in creatively reinterpreting the material legacy of the Qing court. We cite some of that scholarship in chapter 7. Moreover, here we note the existence of rich literature on the Buddhist material culture at the Qing court but are unable to dwell on it further.

14. Schlesinger, *World Trimmed with Fur*; Bello, *Across Forest, Steppe and Mountain*. See chapter 1 for a discussion on how the Manchu *Mirrors* were used in this line of scholarship. All frontiers were not equal in Qing administrative policy; although certain natural products in Manchuria and Mongolia received direct supervision by Qing court institutions, regions in the southwest and southeast (Taiwan) were less centralized or documented.

15. The old view is most clearly expressed in Goodrich, *Literary Inquisition*. For a classical and nuanced study of court scholarship as seen from the compilation of the *Complete Books of the Four Repositories*, see Guy, *Emperor's Four Treasuries*. Although its subtitle might suggest the book as a prime example of the state versus scholars paradigm, Guy's analysis in fact reveals a much more reciprocal and mutually dependent relationship.

allocated manpower and resources to scholarly projects that resulted in the creation of new knowledge. In chapter 3, we show how the Kangxi court took up a late Ming scholar official's compendium on plants and made aggressive changes to its scope, design, and organizational principles in the resultant *Guang Qunfang pu* 廣群芳譜 (Expanded catalog of myriad flowers). Coincidentally, *Guang Qunfang pu* was completed at court in the same year (1708) as the monolingual Manchu *Mirror*, but the way *Guang Qunfang pu* described plants departed markedly from both the Ming original and the Manchu dictionary. The same political and cultural milieu of the late Kangxi court also gave rise to the first iterations of *Gujin tushu jicheng* 古今圖書集成 (Consummate collection of texts and illustrations from past and present), usually hailed as the most voluminous and the very last collection of "texts [sorted into] categories" or "classified writings" (*leishu* 類書) of encyclopedic scope in the Chinese tradition. Neither Sinicization nor cultural hybridity could quite capture the court-centered departure from Ming literati scholarship and the renewed significance attributed to plants and animals in such writings.

Nowhere is this criss-crossing process of court-driven cultural renewal more obvious than in the sustained attention paid to Manchurian and Inner Asian flora and fauna during the eighteenth century. In her recent book, Ruth Rogaski traces the rich history of codified knowledge about Manchuria as a place, offering careful analysis of Chinese exile writings in the early Qing and an extensive literature generated by the Kangxi emperor's visits back to the Manchu homeland and surveying of the White Mountain. Although Rogaski rightfully highlights the role of sensual experience to analyze natural knowledge in multilingual archives, our account in this book (especially part 2) complements her narrative by emphasizing the sustained interest in Manchu names for Manchurian things and the centrality of language in their investigation.[16] Chapter 4 notes the early efforts of Chinese literati exiles, courtiers accompanying Kangxi's tours in the 1690s, and local gazetteers to grasp and describe flora and fauna across different tongues. Chapter 5 goes back to *Guang Qunfang pu* to reveal the embedded accounts of "beyond-the-border" plants compiled by scholar

---

16. Rogaski, *Knowing Manchuria*.

officials in Kangxi's hunting entourages of the 1700s, linking their historical and literary accounts of Manchurian and Inner Asian species also to Kangxi's new vision for the "investigation of things" as articulated in his collected writings. Chapter 6 in part 3 notes again the exceptional status of Manchuria that triggered the first sustained invention of Manchu neologisms across editions of the *Ode to Mukden* early on in the Qianlong reign. In sum, our analysis reveals the privileged status assumed by Manchurian and Inner Asian flora and fauna as an essential feature of High Qing scholarship in both the Manchu and Chinese language sources.

Between Words and Things: The Manchu *Mirrors*, Lexicographical Encyclopedism, and Natural History

Having defined our approach to the cultural politics of the High Qing state, we now turn to the question about why study plants and animals in particular. The historiography of natural history in the early modern world has grown much in recent decades to enable a new look at neglected places, actors, and sources. As Londa Schiebinger asked in her 2007 study of colonial plant knowledge in the Atlantic world, "Who owns nature?" The answer is no longer automatically leading to an unquestioned acceptance of the centrality of imperial enterprises behind such scientific endeavors. Fa-ti Fan's pioneering study of British naturalists in Canton also came out in the early 2000s, anticipating Harold Cook's now classic study of Dutch trade in the Far East and the extension of medical and natural knowledge via "matters of exchange." A rich literature has since grown out of this conjuncture to shed light on the early modern transformation of natural history not only resulting from but also embellishing and abetting the ambitions of colonialism and imperialism.[17]

17. Schiebinger, *Plants and Empire*; Fan, *British Naturalists in Qing China*; Cook, *Matters of Exchange*. See also a synthesis of recent work in Batsaki, Cahalan, and Tchikine, *Botany of Empire*. For a broader history of colonial expansion and museum collection, see Delbourgo, *Collecting the World*. For a new history of colonial bioprospecting with an emphasis on medicinal uses, see Breen, *Age of Intoxication*. For

Crucially, the global scope of the historical inquiry pioneered in this literature has made it possible to look beyond Western European colonialism, extending inquiries to the dynamics of knowledge and state power in imperial Russia, the Ottoman Empire, Tokugawa Japan, and late imperial China, among others.[18] The East Asian context has in particular complicated the binary of colonizer and the Indigenous: The White visitors who frequented Asian trading entrepôts were not the only inquisitive actors with a stake in knowing and controlling local nature in this part of the world. Instead, we need to seriously examine the various regimes of politics and commerce already at work in the region prior to large-scale European incursion if we are to properly understand the dazzlingly new forms of knowledge and the truth claims presented there in many tongues, Chinese and Manchu among them.

So far, it has proved challenging to integrate Qing China into the historiography of natural history in the early modern world.[19] One of the main reasons for this impasse is no doubt that there was simply no term corresponding to the European notion of natural history in late imperial Chinese sources, nor were "plants," "animals," and "minerals" ontological categories with self-evident boundaries in the Chinese tradition of "erudition over things" (*bowu* 博物).[20] Facing a situation akin

---

more critical engagement with the role of indigenous knowledge in the making of colonial natural history, see Menon, "Making Useful Knowledge"; Cooley, "Animal Empires."

18. For recent work on pharmaceutical knowledge and state power in Imperial Russia, see Griffin, "Russia and the Medical Drug Trade"; Koroloff, "Business of Gardens." For a discussion of "practical naturalism" and the political economy of scholarship under the Ottoman Empire, see Küçük, *Science without Leisure*.

19. Most studies assume a modernist periodization that skipped over the High Qing to look at the inception of modern botany in the late nineteenth century; see, for example, Menzies, *Ordering the Myriad Things*. The Qing received scant treatment in Needham and Lu's survey of Chinese botany in *Biology and Biological Technology*. Botany and zoology hardly received any mention in Elman's *On Their Own Terms*. One of the first comprehensive surveys of botany in China in any language, however, did pay attention to the botanical knowledge held by "Coreans, Manchoos, Mongols, and Tibetans," including a passing mention of the quadrilingual *Mirror*. See Bretschneider, *Botanicon Sinicum*, 102–5. Bretschneider also notes how many Manchu names for Chinese plants were "frequently forged" (103).

20. For scholarship on *bowu* in early modern (mostly Ming) China, see, among others, Zhang, "Infrastructure of Science Making"; Elman, "Collecting and Classifying."

to what Federico Marcon discusses in his study of the knowledge of nature in Tokugawa Japan, scholars of Qing China are dealing at the same time with a plethora of terms denoting erudition and curiosity toward the myriad creatures of the world, yet have no ready terms that could easily match the rich semantic history of European ones such as "nature." Whereas Marcon's study illustrates how Japanese samurai-scholars creatively adopted and reinvented *honzōgaku* 本草学—the encyclopedic pharmacopeias from Ming China—for their own uses, *bencao* 本草, the corresponding term in Chinese, ceased to command comparable privilege and lost its encyclopedic scope after the 1700s.[21] In an age of global discovery and imperial expansion, Qing China, judging by much of the available research, appears curiously silent on the subject of the nonhuman creatures inhabiting its expanding horizons.

This stereotype of a stagnant Chinese empire indifferent to the world and the advancement of knowledge is, of course, not true. Although the new scholarship on the Qing court we reviewed earlier offers many productive paths away from this conceptual impasse, such works largely focused not on the core corpus of Qing learning but instead on material culture and environmental history. In this book, we propose a new look at the conjuncture between classical scholarship and Qing imperial expansion to show that interest in plants and animals was considerable on the part of Qing actors, albeit expressed through the vehicle of lexicography, or the discussion over words.

Another obstacle lies in the fact that plants and animals have not attracted much attention in the standard narrative over the rise of so-called evidential learning (*kaozhengxue* 考證學), tides of transformative intellectual change across the Ming-Qing transition that emphasized "learning truth from facts" (*shishi qiushi* 實事求是) in classical philology.[22] Georges Métailié's careful study of some Qing

21. Marcon, *Knowledge of Nature*; Bian, *Know Your Remedies*.
22. General histories of evidential learning often bypass or only briefly touch on the study of plants and animals or technical subjects. For a general sketch of the problem, see Elman, *On Their Own Terms,* especially the two chapters in part 3, "Evidential Research and Natural Studies" (223–80). Elman's discussion is mostly limited to general discourses of method and mathematics in particular, leaving much room for further research. For how the "Way of Heaven and Earth" fit into the ethos

philologists who did take an avid interest in classical names for plants and insects has led him to conclude that this branch of inquiry, though impressive in its own right, cannot be compared with contemporary European natural erudition and is better understood as a kind of ethnobotany bearing Chinese characteristics.[23] Thus the path toward treating Qing botanical learning as on an epistemological footing similar to that of contemporary Western developments becomes effectively blocked. As our discussion makes clear, we hesitate to follow him in confining Qing botanical scholarship to any ethnographically definable realm. Yet Métailié's work to us nevertheless provides a rare exception to the overall trend in the history of Qing scholarship in which plants and animals still occupy a relatively marginal place.

It is also true that some outstanding scholarship on topics adjacent to natural history has developed in new directions seemingly without needing to consider the historical specificity of the Qing empire. Notably, Zou Zhenhuan 鄒振環, in his recent book on exotic creatures in late imperial China, studies the influx of information on the non-Chinese natural world through Chinese and European maritime networks. Zou stresses the continuity of the late imperial period, when early Ming court-sponsored voyages and the later Portuguese-sponsored Jesuit mission appear as two moments in the globalization of the early modern world.

---

of evidential learning in Qing scholarly identity, see Sela's *China's Philological Turn* (135–62), which still focuses mostly on astronomy, mathematics, and calendrical studies. This stands in contrast to the deep-seated fascination with the moral agency of cosmic technology in Song-Yuan-Ming Neo-Confucianism, as discussed in Schäfer, *Crafting of the 10,000 Things*.

23. Métailié, *Traditional Botany*. Métailié's work remains in close conversation with a long-running tradition in Sinology that examines the depiction of plants and animals as expressions of the Chinese cultural tradition since antiquity. Even though the specific context and style of scholarship vary greatly, one might appreciate the various works as having a common critical interest in illuminating the heterogeneity of cultural traditions with a focus on China: Elvin, *Retreat of the Elephants*, especially the chapter "Nature as Revelation" (321–68); Siebert, "Animals as Text" and other chapters in the same volume; the seminal discussion of the Chinese emphasis on utility in Bray, "Essence and Utility"; and the highlighting of transformation in Chinese notions of natural objecthood in Nappi, *Monkey and the Inkpot*, and Allen, *Vanishing into Things*, especially the chapter on resonance (210–32). Others, including Métailié himself in *Traditional Botany*, have warned against relying too heavily on textual sources detached from nature. See also Ptak, *Birds and Beasts*, 11.

For him, it was these maritime connections—and, implicitly, not the Qing conquest of Inner Asia—that led to a second wave of exotic creatures entering China and the Chinese imagination following a first wave during the reach of the Han and Tang empires into Central Asia in the first millennium CE.[24]

Zou's fascinating panorama certainly has a logic and coherence to it and succeeds in highlighting a facet of late imperial learning that was missing in earlier histories of scholarship and science. The cost of this historiographical choice, however, is the erasure of the Qing court as a site for knowledge production, a site whose ties to Inner Asia enabled a transformation of Chinese scholarship that was parallel to and at times intertwined with that generated by knowledge originating overseas. As the chapters in this book demonstrate, Manchu lexicography emerges as a meeting point between the continental and the maritime together in the offices and book depositories of the court.

Learning about things through a systematic study of the names of things, to be sure, was in no way new to Qing China. Joseph Needham and Lu Gwei-djen pay close attention to lexicographical and encyclopedic sources in their survey of Chinese botany, starting from the *Erya* 爾雅 (Approaching perfection) and *Shuowen jiezi* 說文解字 (Explaining the graphs and unraveling the written words, ca. 100 CE) and ending with high, if only brief, praise for the late Kangxi dictionaries and encyclopedias (in Chinese).[25] Carla Nappi's analysis of the late Ming work of Li Shizhen's *Bencao gangmu* 本草綱目 (Systematic materia medica) highlights Li's close attention to names as an important scholarly approach to understanding *materia medica*.[26] Nathan Vedal's recent book fills in an important gap by restoring, in its full richness and complexity, Ming language scholarship as understood within the cosmic order of things.[27] Our analysis of the Manchu *Mirrors* is informed by this scholarly tradition just as the Qing court scholars were fully aware of such precedents in Chinese learning in Ming times and

---

24. Zou, *Zai jian yishou*, 1–3.
25. Needham and Lu, *Biology and Biological Technology*, 182–219, in particular *Gujin tushu jicheng*, 206–7, and *Kangxi zidian*, 218. The translation of the title of *Shuowen jiezi* follows Bottéro, "Ancient China," 59.
26. Nappi, *Monkey and the Inkpot*, 85–86.
27. Vedal, *Culture of Language*.

earlier. At the same time, however, we seek to highlight the surprising and unprecedented departures from such precedents in the Chinese study of names and things. We cannot pursue to their full extent the possible connections between Manchu lexicography and the patterns of language study in evidential learning among Qing scholars, but resonance is considerable between the Qianlong emperor's lexicon reform and contemporary (or later) philological investigations into the names of things (*mingwu* 名物), a capacious category of learning that encompasses, but is not limited to, the etic category of plants and animals.[28]

Previous research on the Manchu *Mirrors* has focused on their role as language-learning aids, or simply sought to facilitate and extend the use of them in this capacity for the benefit of present-day students of the Manchu language (see the conclusion). Hence the *Mirrors* have largely remained outside the historiography of science and scholarship in late imperial China. Yet, as the Grand Council documents revealed, the issue of finding appropriate plant and animal names in Manchu clearly involved research efforts from a general library of scholarly literature in both Chinese and Manchu, most of which had been recently produced under the same Qing imperial auspices, including the *Mirrors*. Furthermore, this library, including both Chinese-language encyclopedias and Manchu reference works, has direct ties to European natural history. A growing scholarship shows how Catholic missionaries in China engaged in natural historical scholarship before the Manchu conquest and continued throughout the eighteenth century. A few of these were published in Chinese, others were written in Manchu, and, as we show in subsequent chapters, information about plants and animals from these texts made its way into the *Mirrors* and other court-based texts.[29] Although we in no way intend to privilege European commentaries on Qing sources, we use them deliberately as an alternative viewpoint to counterbalance the narrow framing of the

28. See He's recent dissertation, "Well-Ordered Textures."
29. More books were planned than published. See Bocci, "Animal Section"; Csillag, "Natural History Illustrations"; Wu Huiyi and Zheng, "Transmission of Renaissance Herbal Images." For Manchu books, see Saunders and Lee, *Manchu Anatomy*; Talpe, "Manchu Text"; Walravens, "Medical Knowledge of the Manchus"; Hanson, "Manchu Medical Manuscripts"; Hanson, "Significance of Manchu Medical Sources"; Jami, "Imperial Science Written in Manchu."

Manchu *Mirrors* and other court texts in the Sinocentric bibliographical tradition.

Taking Manchu lexicography seriously—together with all its explicit or implicit connections with other scholarly traditions—allows us to detect a widespread interest in plants and animals that reached far beyond the confines of classical exegesis that Chinese scholars conducted away from the Manchu court. It is our claim in this book that this passion to name new things (and to rename old things anew that at times bordered on obsession) cannot be separated from the politically charged and wide-ranging efforts to promote, adapt, and edify the study of the Manchu language in the Qing imperium. Indeed, if we were to speak of a Qing order of things—the entangled fields of knowledge over words and things evolving from the Song-Yuan-Ming times—we should first pay attention to the emergent and expanding lexicographical order of the Manchu language and an increasing codification of the vernacular in other languages as well, Chinese included. As is true of other dictionaries, especially those that like the *Mirrors* are arranged by subject matter, it is often difficult to distinguish the lexicographic (discussing words) from the encyclopedic (discussing facts).[30] At the most fundamental level, therefore, the historical space between the monolingual Manchu *Mirror* of 1708 and the bilingual Manchu-Chinese *Mirror* of 1772 reveals how Qing actors grappled with the unstable epistemic premises that anchored words to things, and vice versa. This should prove of interest, we hope, to readers who may not have opened this book with an abiding interest in Manchu studies and inspire more to make use of Manchu sources for similar inquiries.

Part 1 takes a close look at how encyclopedism at the late Kangxi court started to take on a more pronounced lexicographical quality in the wake of the compilation of Manchu and Chinese dictionaries. Part 2 follows an excursion to Manchuria and Inner Asia to show how Kangxi used frontier species to articulate a new imperial vision for the investigation of things but in so doing also created new challenges in establishing correspondences between ancient and modern accounts. Part 3

---

30. Söderblom Saarela, *Early Modern Travels*, 122–26; Söderblom Saarela, "Lexicography," 230.

examines how lexicography was transformed by Qianlong's Manchu language reform with the addition of large numbers of neologisms, offering a new interpretation of the politics of *tongwen* 同文 as achieving radical lexicographical parity between Manchu and Chinese.[31] The question of how this body of court-based plurilingual lexicographical encyclopedism has been received in Qing China and beyond is a subject of more reflection in the conclusion.

31. For the use and transformation of the idea of *tongwen* under Qing rule, see Lin Shih-hsuan, *Qingdai Menggu yu Manzhou*, 345–63; Ma and Borjigidai Oyunbilig, "'Tongwen zhi zhi.'" The phrase *tongwen* appears in different contexts in this book. Because its several meanings are difficult to grasp in one translation, we translate it differently as the situation demands.

PART I

*Lexicography and
Encyclopedism at the
Kangxi Court*

CHAPTER ONE

Flora and Fauna in the Mirror of the Manchu Language

Qing court lexicography has at times been treated primarily as a source reflecting the state of the concerned languages at the time of their writing rather than as tendentious, creative, and uneven scholarship that profoundly shaped the languages they contained. Notably, on some of the most fascinating pages of *A World Trimmed with Fur: Wild Things, Pristine Places, and the Natural Fringes of Qing Rule*, Jonathan Schlesinger contrasts the *Mirrors* of the Kangxi court with a commercial Manchu-Chinese dictionary from the late seventeenth century. Schlesinger convincingly and eloquently shows that the integration of Chinese and Manchu discourses on the natural world was very slow in the making. Yet in this process, the dictionaries of the Qing court in Schlesinger's account appear as something approaching the gold standard of lexicography.

Schlesinger writes that in the Kangxi period "translation proved to be a constant problem, and in the generation after the Qing conquest, literati struggled to bridge the gap between Manchu and Chinese words and to standardize terminology." He suggestively points out that in the first published Manchu-Chinese dictionary, Shen Qiliang's 沈啟亮 (fl. 1645–1693) *Complete Book of the Great Qing* (*Da Qing quanshu* 大清全書 | *Daicing gurun-i yooni bithe*) from 1683 (it was based in part on existing material that had circulated in manuscript), hundreds of

Manchu lemmata were left untranslated, followed only by a blank space on the page. Many of the untranslated words were flora and fauna common to the non-Chinese northern periphery of the Manchu empire.[1]

Schlesinger contrasts the Chinese lexicographer Shen Qiliang's work with the Manchu-Mongol edition of Kangxi's *Mirror* that was published in 1717. In this dictionary of two Inner Asian languages whose lexicon had been brought close together by centuries of coexistence in a shared environment, Schlesinger writes, "all of the terms missing from texts like the *Da Qing quanshu*" received "sentence- or paragraph-length definitions." Thus, he argues, a textual description of the natural world known to the Manchus already existed in the early eighteenth century. Only its translation into Chinese was still a long way off.[2]

Schlesinger's magisterial book goes on to chart how a new Qing material culture emerged with the integration of the empire that straddled Inner Asia and the Chinese heartland. Yet the textual sources that constitute his point of departure (including the *Ode to Mukden*,[3] to which we return in chapter 6) have a different story to tell as well. The approach in this book differs both from its simple use as a lexicographical resource and its treatment as a reflection of the Manchu lifeworld.

It is tempting to assume, following the lines of Schlesinger's argument, that a book like the Manchu *Mirror* represented the codification of a Manchu body of knowledge that was complete in itself but still impossible to express in Chinese. An investigation of the chapters that cover flora and fauna in this book suggests, however, that the *Mirror* is best understood as a product of textual scholarship as practiced at the Kangxi court. Such scholarship was not limited to Manchu or Manchuria. Indeed, as Schesinger has helped us see, much of Manchuria was a largely unknown periphery in the early Qing period, and knowledge about nature in this region was a newly gained consequence of

---

1. Schlesinger, *World Trimmed with Fur*, 34.
2. Schlesinger, *World Trimmed with Fur*, 35.
3. Schlesinger, *World Trimmed with Fur*, 2.

the empire's expansion further into Inner Asia. The *Mirror* presented this knowledge to Manchu readers by including words in part borrowed from the native languages of the region. Many of these words were only Manchu in the sense that they were included in the *Mirror*. Even the editors were at times unfamiliar with what they referred to, and in consequence used whatever sources they had available to make sense of them. As we suggest in relation to sea creatures, those sources might have included texts of ultimately European provenance.

In this chapter, we first contextualize the making of the 1708 Manchu *Mirror* in the early Qing decades following the conquest, highlighting precedents in private and court scholarship that paid attention to naming plants and animals in Manchu. We then explore the chapters on plants and animals in Kangxi's *Mirror*, highlighting instances in which the *Mirror* describes the natural world of Inner Asia, especially Mongolia and greater Manchuria, both regions into which the Qing empire had expanded in the decades before the dictionary was compiled. Last, we consider information from outside the Qing empire, which the *Mirror*'s compilers in one instance appear to have used to corroborate information gathered on the Inner Asian periphery.

Manchu Scholarship from the Pre-Conquest Period to the Kangxi *Mirror*

Kangxi's *Mirror* stands out among the Manchu books the court published, but it was not the earliest product of Manchu court scholarship. Scholarly projects apparently began in the 1620s, when the Manchus ruled only a small state outside the Chinese northeastern border, with translations of books from Chinese into Manchu. In what follows, we situate the Kangxi *Mirror* within the history of official Qing scholarship.

In the pre-conquest period, certain officials petitioned to have the canonical Confucian *Four Books* translated into Manchu, but the leadership opted to translate an assortment of other texts instead. Three of the texts were ancient works "associated with a rising power on the eve of its conquest," which is the position that the Manchu rulers might have considered themselves to be in at this time.[4] Books translated in the pre-conquest period also included the collected statutes of the Ming, a blueprint for state-building.[5] Of particular interest is that one of the court scholars in this period translated a book titled *Wanbao quanshu* 萬寶全書 (Complete book of a myriad treasures). This title had been in use for centuries, and in the early seventeenth century was used for various editions of a popular encyclopedia or collection of "classified writings."[6] Several of the great scholarly publications of the eighteenth century that we discuss in this book likewise belonged in this genre. A translation of a historiographical compendium in the *tongjian* 通鑑 (Comprehensive mirror) style was also undertaken at this time. The word *mirror* later reappeared in the title of Kangxi's Manchu dictionary.

After the move of the imperial capital to occupied Beijing in 1644, several institutions carried out officially sponsored scholarship, among them the Imperial Household Department (Ch. Neiwu fu 內務府; Ma. *dorgi baita be uheri kadalara yamun*). This institution with its great financial resources and technical expertise printed the books. During the Shunzhi reign (1644–1661), when Qing control over China was by no means certain, several books were printed by the Imperial Household Department, including Chinese texts on morality, statecraft, and the calendar. Manchu texts, always less numerous, included translations of some of the more recent dynastic histories and foundational texts of the preceding Ming dynasty, the translation of which had begun already before the conquest.[7]

Notably, a Manchu translation of the *Poetry Classic* was published in 1655 (SZ 11/12th month).[8] It was printed in Manchu only and included the main text of the *Classic*, translated along with two layers of commentary on the basis of a Chinese edition from the early fifteenth

---

4. Durrant, "Sino-Manchu Translations," 655.
5. Keliher, "Administrative Law," 65.
6. Wu Huey-Fang, *Wanbao quanshu*, 37–39; Durrant, "Sino-Manchu Translations," 655.
7. Weng, *Qingdai Neifu keshu tulu*, 3–4. For a Manchu translation of the Liao, Jin, and Yuan dynastic histories, see Elliott, "Whose Empire Shall It Be?"
8. Pang and Volkova, *Descriptive Catalogue of Manchu Manuscripts*, 77 (item 165).

century.[9] The translation of the *Poetry Classic*—a book from which, according to Confucius, "we become largely acquainted with the names of birds, beasts, and plants" 多識於鳥獸草木之名—arguably represented the court scholars' first engagement with the rich literary Chinese vocabulary for the natural world.[10] Yet the translation of Chinese words for plants and animals in this book appears to have been an ad hoc affair without any obvious editorial strategy.

The Kangxi period marked a departure in terms of the scale and ambition of scholarly projects undertaken at court. The Manchu-Chinese Translation Office was set up in circa 1671 with both bureaucratic and scholarly translations in its purview.[11] The Hanlin Academy, which housed outstanding recent graduates of the civil service examinations, managed editorial projects, but over time ad hoc bureaus were established for particular titles.[12]

The *Imperially Commissioned Mirror of the Manchu Language* was compiled in this new institutional context.[13] In 1673—at a time when no Manchu dictionary had yet been printed and shortly after the establishment of the Manchu-Chinese Translation Office—Kangxi gave an order to compile a Manchu dictionary modeled on a common, late-Ming Chinese dictionary.[14] Work apparently progressed slowly, if at

9. Tu, "On the Source Text." Cf. Yeh, "Man-Han hebi," 10, 10n43.
10. Legge, *Chinese Classics*, vol. 1, 323.
11. Söderblom Saarela (*Early Modern Travels*, 132n52) cites more relevant scholarship.
12. Weng, *Qingdai Neifu keshu tulu*, 8; Monnet, "Le livre impérial," 434–38.
13. Jiang's *Kangxi Yuzhi Qingwen jian yanjiu* is the only monograph dedicated to Kangxi's *Mirror*. We find it difficult to engage with, however. Jiang recognizes that the *Mirror*'s "contents primarily synthesized the cultures of the Chinese, Manchu, and Mongolian peoples" 它的内容，主要綜合了汉、满、蒙民族的文化, but argues in the same breath that it "had an effect on the rapid progress of the Manchu people and their cultural advancement at that time that ought to have been no less important than the effect that the *Encyclopédie* had on the French enlightenment" 对当时的满民族的迅速进化和满族文化进步的作用应当不亚于《法国百科全书》对法国启蒙运动的作用 (32). This surprisingly positive attitude toward what is first and foremost a product of imperial scholarship seems to dissuade Jiang from taking a truly critical approach toward the book.
14. The imperial diary says that Kangxi wanted the new dictionary to imitate the Chinese *Zihui* 字彙 (The characters collected, referred to as *dzi hûi* in the Manchu version of the imperial diary) from 1615; see FHA, *Kangxi qiju zhu*, vol. 1, 93; *Ilire tere* [KX 12/4], 9a; Tsai, "Huacheng tianxia," 72–73. The *Mirror* that was eventually

all, but in 1699 and again in 1701 the Jesuit Joachim Bouvet (1656–1730) wrote to Europe about the project. He was under the impression that the dictionary would be bilingual and even described the project as the "translation of the Chinese dictionary into Tartar, that the emperor is having carried out by very able individuals."[15] It is thus possible that the book was initially intended to be bilingual and that Chinese source material was used extensively.[16] It appears that most of the work was carried out during the few years before the book was published.[17] We know that, to provide a sense of what the dictionary might look like, five sheets were printed and presented to Kangxi on September 27, 1706 (KX 45/8/21). On August 17, 1707 (KX 46/7/20), *juan* 19 and 20— which contained the bulk of the natural-historical material—were presented for imperial perusal, though it is unclear whether in print or manuscript.[18] The book was completed on August 8, 1708 (KX 47/6/*dingmao*).[19]

---

published in 1708 was similar to *Zihui* primarily in its inclusion of a graphologically arranged index.

15. Söderblom Saarela, *Early Modern Travels*, 121, 126. This is an opportune moment to observe that Europe at this time had little knowledge of Manchu dictionaries. Spence (*Question of Hu*, 52–53), following Omont (*Missions archéologiques françaises*, 2:807), writes that the king of France in 1700 was presented with a Manchu dictionary by a returning Jesuit missionary. Omont's sources, however, do not say that it was a dictionary, only that it was twelve large volumes, some of which were in Manchu and some in Chinese (a format that is hard to imagine for a dictionary). See *Catalogue des livres imprimez*, xliv; Boivin, *Mémoires pour l'histoire*, 302r.

16. Relatedly, Imanishi, "*Shinbunkan*: Tantai kara gotai made," 128.

17. Söderblom Saarela, *Early Modern Travels*, 133.

18. Weng, *Qing Neifu keshu dang'an*, vols. 1, 7, 9.

19. *Shengzu Ren huangdi shilu*, 1986, vol. 6, 233:6b (329, top panel). According to Fudari's 傅達禮 (fl. 1671–1675) biography in Hūng Jeo, Maci, and Ortai (*Baqi tongzhi* [*chuji*], 236:19b; *Han-i araha jakûn gûsai*, 236:31a), where it says that the book was finished "over twenty years after" (Ch. 後二十餘年; Ma. *orin funcere aniya amala*) Kangxi first gave the order. Meyer ("Das schamanistische Begriffsinventar," 200n15) assumes that it was finished sometime after 1693. Linke (*Zur Entwicklung des mandjurischen Khanats*, 319) does as well, but a small mistake leads him to provide a different date. *Baqi tongzhi* is not a very good source, however (Söderblom Saarela, *Early Modern Travels*, 39–40). In light of the other sources cited here, we decided not to ascribe to this version of events. Similarly, the claim in Haenisch ("Die Abteilung 'Jagd,'" 60) that the original edition from 1708 was bilingual (Manchu-Chinese)—with the monolingual Manchu edition dating from 1709—is a mistake.

The book's preface linked the Manchu written language to the great political enterprise initiated by Kangxi's forefathers in Manchuria in the early seventeenth century. The preface had an overall linguistic focus, stressing the need to preserve the ancestral language. Yet language and the world it described are not clearly distinguished in the preface, which after all introduced a book that—like a Chinese encyclopedia—arranged entries on the basis of their meaning, not their pronunciation or graphic form. The preface was apparently drafted in Manchu, given that a Manchu-only version with Kangxi's corrections remains in the archive. Its conceptualization of the relationship between written language and its referent, however, is consonant with what we find in contemporary Chinese sources.

"The meaning and principles of the world are gathered in the written characters (*wenzi*), and the written characters of the world are gathered in the six ways of writing down (words)" 天下之義理統于文字，天下之文字統于六書, wrote Gong Dingzi 龔鼎孳 (1616–1673), a poet and official in the central government under Kangxi.[20] Gong might not have been the originator of this general statement, but it is not impossible that Kangxi or his court scholars encountered it through him: Gong made this statement in the preface to an edition, published in 1685, of the most prominent Chinese dictionary of the second half of the seventeenth century. Gong was very clearly referring to the Chinese writing system, whose characters since antiquity had been conceptualized as making up six classes, or "six ways of writing down (words)."[21] These classes have no counterpart in the Manchu script, but that did not deter Kangxi from including Gong's words at the opening of the imperial preface to the *Mirror of the Manchu Language* (where it said *abkai fejergi jurgan giyan be, gemu šu hergen be bakdambuha, abkai fejergi šu hergen be, gemu ninggun hacin-i bithe de bakdambuha*).[22] The

20. Zhang Zilie, *Zhengzi tong*, 1685, *Gong xu*:5b–6a. On Gong, see Tu, "Kung Ting-Tzŭ."
21. For the phrase and its translation, see Bottéro, "Ancient China," 59.
22. Jiang, *Kangxi* Yuzhi Qingwen jian *yanjiu*, 180. In the later collection of Kangxi's literary writings (and in the *Veritable Records*), the *Mirror*'s preface was translated into Chinese, which produced a Chinese version of this sentence that differs somewhat from Gong's wording. See Jiang, *Kangxi* Yuzhi Qingwen jian *yanjiu*, 185; *Shengzu Ren huangdi shilu*, vol. 6, 233: 6b–8a (329, top panel; 330, top panel).

changes made to the draft version of the preface suggest that Kangxi and his anonymous amanuensis were choosing between Manchu translations of the various acceptations of the polysemous Chinese *wen*, which could mean both "writing, written" and "culture, cultured." The emperor changed the draft preface's *bithe hergen* 'the characters of books/documents' to *šu hergen* 'the refined characters', both obvious attempts to render the Chinese *wenzi* in Manchu.[23]

The imperial preface thus shared the general conception of written language as the aggregate of discrete meanings, which in turn had a direct relationship with the world itself. Kangxi accordingly wrote that the new book contained everything from big to small, from the patterns of the heavens to the lines of the earth to the discrimination of "names and things and appearances and numbers [of divinatory cracks and tossed yarrows stalks]" (*gebu jaka, arbun ton*).[24] Listing and describing plants and animals, the focus of this book, should be understood as part of the discrimination of names and things.

There is no indication that the Manchu *Mirror* at the time of its publication in 1708 was intended to be the first installment of a series. Tellingly, its Chinese counterpart *Kangxi zidian* 康熙字典 (Character standard of the Kangxi reign), published a few years later, only ever received a supplement, and not until the nineteenth century. Yet the place of the Manchu language at the center of a plurilingual government apparatus led to the *Mirror* expanding into a series of bilingual and multilingual works.

Locating Plants and Animals in the Kangxi *Mirror*

Here we first give an overview of the plants and animals included in the book. Words related to flora and fauna are found toward the end of the *Mirror*. They are not listed in continuous succession, however. Interspersed among nouns or noun phrases referring to kinds of plants

---

23. Draft preface to the *Imperially Commissioned Mirror of the Manchu Language*, Manchu endorsed palace memorial collection, 04-02-002-000080-0007, FHA.

24. Jiang, *Kangxi* Yuzhi Qingwen jian *yanjiu*, 183.

and animals are lemmata referring to appearance, parts, behavior, and so on.

The *Mirror* has a three-level arrangement of chapters (*šošohon*), sections (*hacin*), and segments (*meyen*).[25] Only the chapters and sections are listed in the table of contents. Their division reflects semantic fields. Of the book's thirty-six chapters, the last ten contain information on plants and animals. Moreover, some relevant information is found in the eleventh to last chapter (foodstuffs). The sections are subdivisions of chapters, reflecting a smaller semantic field (e.g., small birds, river fish). The segments, by contrast, have the primary function to divide up the lemmata lists of each section into smaller units (see fig. 1.1). They have no explicit semantic basis, but semantic considerations evidently played a part in the arrangement of words into segments.[26] In the chapters on flora and fauna, kinds of plants and animals are usually listed in the early sections (if the chapter is long) or early segments or early part of the single segment (if it is short), with descriptive words and actions relegated to the end.

The first relevant section is "Vegetables and dishes" (*sogi booha*). This is the second section in the chapter titled "Foodstuffs" (*jetere jaka*), in which it is preceded by the section "Cooked grains and meats" (*buda yali*).[27]

The section "Fruits" contains a definition of the general terms of the section title, thirty-six names of fruits, and a few words for parts of fruits (juice, peel, pit) and actions performed when handling fruit (de-seeding). Curiously, this set of words are then again followed by the names of a few more fruits, which ipso facto might have been added after the initial lemmata list had already been established.

The chapter cum section "Herbs" is long, including seventy-six names of plants (the section has been split into two segments for this

---

25. Mingdo, *Yin Han Qingwen jian*, a commercial translation of the *Mirror*, translated *meyen* as *duan* 段 'segment, part'. Later, Qianlong's Manchu-Chinese *Mirror* translated it as *ze* 則 'entry, item'.

26. See Jiang, *Kangxi* Yuzhi Qingwen jian *yanjiu*, chap. 2, sec. 1.

27. Jiang, *Kangxi* Yuzhi Qingwen jian *yanjiu*, 53. We have to a great extent translated single Manchu words according to Norman, *Comprehensive Manchu–English Dictionary*. We do not cite it in every instance or note when we have chosen to depart from it.

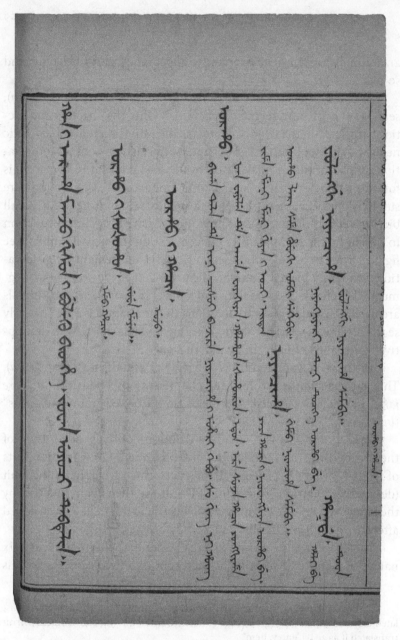

FIG. 1.1. First page of the section on grasses in the Kangxi *Mirror* (Beijing: Wuying Dian, 1708, 19:*orho-i hacin*:1a). Digital collections of the Staatsbibliothek zu Berlin – PK. http://resolver.staatsbibliothek-berlin.de/SBB0001F33E00000000.

reason). As in the other sections, words referring to plants processed in a certain fashion, as opposed to names of specific species, are occasionally interspersed within the section.

The chapter cum section "Trees" is similarly long and split up into five segments, the first two of which include the names of seventy-three plants. As in other sections, focus is strong on the use of the plant (e.g., for making arrows, boats, or herbal infusions). Segments three to five of the section contain words describing trees and their size and manner of growth.

The chapter cum section "Flowers" includes the names of fourteen plants, followed by words for the parts of flowers and actions involving flowers.

The chapters that follow include descriptions of fauna: birds, beasts, livestock, scaled animals, and bugs. The chapter "Birds large and small" (*gasha cecike*) is divided into four sections. The first two contain the names of birds (small, large), the second two words related to birds.

The chapter titled "Beasts" (*gurgu*) contains, in an eponymous section, seventy-five animal names by our count. In addition, names for the male, female, young, or old members of a certain species are listed as entries in their own right. The animals listed include macrofauna characteristic of the northeast, even creatures that were not commonly seen in the Qing ancestral land of southern Manchuria. One such animal was *kandahan* 'moose', which the Manchus referred to with a word that was Mongolic in origin and was widespread in the languages of Manchuria.[28] The *Mirror* called it "a kind of deer. It is big, with a hump on its back. There is skin under the throat that is like a bridle decoration. The neck is short, the horns on its head are flat and wide" (*buhū-i duwali. beye amba, nikde de bohoto bi, monggon-i fejile kandarhan-i adali sukū bi, meifen foholon, uju-i weihe halfiyan onco*).[29] The word used here for bridle decoration, *kandarhan*, was related to the name of the moose, as was *kanda* 'dewlap', which suggests the importance of

28. Doerfer, *Mongolo-Tungusica*, 81 (item 234).
29. *Han-i araha manju gisun*, 1708, 19:*gurgu-i hacin*:44b. Also translated in Schlesinger, *World Trimmed with Fur*, 35.

this animal in the Manchu imagination.[30] Other kinds of deer were listed as well. We get back to some of them and their treatment in various products of court scholarship later in this book.

The following chapter divides "Livestock and domestic animals" (*ulha ujima*) into several sections. The first, "Domestic animals" (*eiten ujima*), contains only the names of ten animals, by our count, as well as several words for different members of the same kind. For example, dogs of different colors and sizes are listed under different names, as are the male and female members of the same species.

The next section to include animal names is "Horses and livestock" (*morin ulha*). Here we find the names for the horse, camel, mule, and donkey, but that is it. After several sections with words relating to these animals, the section "Bovines" (*ihan*) then lists more animal names, including "cow" (likewise *ihan*), "water buffalo" (*mukei ihan*), and yak.

The next chapter featuring animal names is "Creatures with scales and shells" (*esihengge hurungge*). The sections including fauna are "Dragons and snakes" (*muduri meihe*), "River fish" (*birai nimaha*), and "Sea fish" (*mederi nimaha*). The first section, predictably, has only very few animal names: the dragon, python (*jabjan*), and (smaller) snake.

The section on river fish includes, in addition to the generic name for fish, fifty-one fish names (or fifty-two, if we count *nisiha* 'small fish' as the name of a species). In addition, the crab, shrimp, toad, frog, and several kinds of shellfish, turtles, and so on are found here as well. The definition of "sea fish," the word in the title of the next section, is quite interesting in that it delimits the coverage of the section to what is covered by the Manchu language. We return to this specification and its implications presently. We count forty-four names of animals here, including, again, kinds of shellfish and crab, sea cucumber, shrimp, whale, dolphin, and so on in addition to fish.

The last chapter of the dictionary includes the names of bugs, meaning insects, arachnids, and the like. A total of seventy-five names are listed in two sections, but that includes a few unclear cases such as *dondon*, defined as a "small butterfly" (*ajige gefehe*).[31]

---

30. Tsintsius, *Sravnitel'nyĭ slovar'*, 372.
31. *Han-i araha manju gisun*, 1708, 20:*umiyaha-i hacin*:44a.

In the preceding paragraphs, we have surveyed the chapters and sections that include information pertaining to the natural world. We return in the next chapter to analyze the relationship of this division to the Chinese scholarly tradition and earlier Manchu lexicography. Here, however, we discuss elements of these chapters' contents that do not reflect Chinese scholarship as much as knowledge gathered in Inner Asia and, to some extent, further afield.

Inner Asian Knowledge in the *Mirror*: Birds, Fossilized Trees, and Hibernating Snakes

The *Mirror* is a Manchu-language dictionary and evidences knowledge about the natural environment shared by the Manchus before the conquest of China. In addition, several entries inform the reader about the natural environment in Mongolia, which the Qing empire was still in the process of incorporating at the time the *Mirror* was compiled, and northern Manchuria, similarly recently incorporated. Including words that described the nature of these regions reflected the formation of an Inner Asian empire.[32]

The military conquest, penetration, and attendant economic integration of the region had begun before the conquest of China. Eventually, by the mid-Qing period, Beijing and Manchuria were no longer disconnected places, as Schlesinger has so eloquently shown.[33] Ties between center and periphery strengthened, and Manchuria grew closer as a region. In 1684, before work on the *Mirror* had yet to begin in earnest, the court dispatched a team to survey the lands up to the Amur, which brought new knowledge of the area to Beijing.[34] With time, population transfers and enrollment of native populations in the Eight Banners led the Manchu language to spread beyond its original homeland to areas where other Tungusic languages were natively spoken, and some communities (e.g., the Hūrha) "assimilated" to the

---

32. On the Qing as an Inner Asian empire, see, e.g., Di Cosmo, "The Qing and Inner Asia."
33. Schlesinger, *World Trimmed with Fur*.
34. Cams, *Companions in Geography*, 44.

Manchus.[35] Integration of Manchuria and Beijing notably included economic exchanges. These involved natural products submitted to the privy purse in the form of tribute to the Imperial Household Department, products from its estates (some of which were in Manchuria), and products gained through the monopoly trade that the institution managed.[36] Chinese civilians did not serve in this Manchu institution, but the transfer of objects from the northeast to the bilingual environment of the capital opened a conduit through which natural products moved into the Chinese language. With both the material objects and their Manchu names transmitted to Beijing, officials there could match them to Chinese words.

As suggested by the presented survey of flora and fauna in the dictionary, the unevenness of chapters and sections in terms of names for plants and animals reflects the relative abundance or paucity of Manchu terms in different areas. There are many names for trees and sweet-water fish, but few for flowers and snakes. The distribution of the words, of course, in itself informs us of the nature and extent of the compilers' knowledge about the natural world. In this section, however, we do not look at how Inner Asian vocabulary is distributed, but instead at how Inner Asian knowledge is presented within individual entries. Such knowledge takes several forms. It includes supplementary linguistic information, such as translations into Mongolian and sometimes vernacular Manchu, specifications that certain plants and animals are native to Mongolia or northern Manchuria, and information on the use of certain plants for certain purposes by the Mongols. Most interestingly, a few entries present what appears to be Manchurian—not necessarily Manchu as narrowly defined—folk knowledge about the natural world. We discuss these kinds of information in turn.

That some entries provide translations into Mongolian (recorded in Manchu transcription) is quite interesting given that the dictionary is nominally monolingual. These translations to some extent echo the

---

35. Matsuura, *Shinchō no Amūru seisaku*, 269; see also Schlesinger, *World Trimmed with Fur*, 64; Shirokogoroff, "Reading and Transliteration."

36. Chang, "Economic Role of the Imperial Household"; Torbert, *Ch'ing Imperial Household Department*, 84–92.

Chinese translations we also find in the dictionary and get back to in chapter 2.

The Mongolian translations are not particularly numerous. The Mongolian name is typically added to the end of the definition, for instance, in this description of a nettle-like plant:

> *gabtama* (cf. *gabtambi* 'to shoot [an arrow]') 'nettles'. The young plants are blanched and eaten. The stem and leaves have thorns. It cannot be plucked after it has grown tall, lest it stings the hand. The Mongols call it *halahai* [i.e., *qalaqai* or *qalayai* idem].[37]
>
> *gabtama, ajige de hethefi jembi. cikten abdaha de bula bi, den mutuha manggi jafaci ojorakū gala nukambi. gabtama be monggoso halahai sembi.*[38]

Other examples are similar. The entry for the Chinese wild peach tree (*hasuran* or *karkalan*) says that "the Mongols call it *kara hūna*" (*monggo kara hūna sembi*).[39] Furthermore, *paeonia albiflora*, or the Chinese peony (*šodan ilha*), "is called *cene*" (i.e., *čene*) in Mongolian (*monggo cene sembi*). In this case, the Mongolian word was a term that had been borrowed into Manchu.[40]

Sometimes, a different Manchu word is presented an alternative term. Such is the case of "Chinese toon tree" (*cūn moo*). The definition here presents the Manchu word *mooi wang* 'the king of trees' as a vernacular expression (*an-i gisun*), a notion that would be further elaborated in the Qianlong bilingual *Mirror* (see chapter 8).[41] The headword itself is a loan from Chinese (*chunmu* 椿木).[42] It might have been used as a learned or literary word in Manchu.

---

37. Lessing et al., *Mongolian-English Dictionary*, 916–17.
38. *Han-i araha manju gisun*, 1708, 19:*orho-i hacin*:6b. We discuss how *gabtama* entered Chinese-language encyclopedic texts during the same period in chapter 5.
39. *Han-i araha manju gisun*, 1708, 19:*moo-i hacin*:13b.
40. *Han-i araha manju gisun*, 1708, 19:*ilha-i hacin*:22a–b; Rozycki, *Mongol Elements in Manchu*, 46. The entry also gives the Chinese translation of the lemma.
41. To be sure, Manchu *wang* 'king' is a loan from Chinese as well, but it is old and attested already in Jurchen. See Schmidt, "Chinesische Elemente im Mandschu," 1933, 420.
42. *Han-i araha manju gisun*, 1708, 19:*orho-i hacin*:8a.

On rare occasions, the dictionary gives alternative Manchu dialect terms, or, arguably, words from other Tungusic languages. Among the verbal expressions relating to animals, we find the headword "snakes hibernate" (*meihe bulunambi*). The first part of the definition for this phrase reads as follows:

> When many snakes come together and enter a hole to spend the winter, it is said that the snakes hibernate. It is also called *eniyeniye*. The Kūyala [tribe] and the new Manchus call it *niyeniye*.
> *geren meihe isifi dung de dosifi tuweri heture be, meihe bulunambi sembi. geli eniyeniye sembi. kūyala, ice manju sa, niyeniye sembi.*[43]

Unlike other entries that offer synonyms for a headword, this entry explicitly links the alternative expressions to specific linguistic communities. Both Kūyala and "new Manchus" refer to Tungusic-speaking groups that were not among those incorporated in the Qing state when it was formed in the early seventeenth century.[44] The inclusion of expressions from their language here suggests that the dictionary's compilers conceived of it as including nonstandard expressions from Manchuria broadly defined, including its Inner Asian hinterland, which lay far from the Chinese frontier where the Qing polity was consolidated.

Some entries, although not including a Mongolian translation or alternative Manchu terms, do include information on Inner Asian plants or animals of varying detail. Some entries specify simply that an animal is native to Mongolia or northern Manchuria. For example, the dictionary lists a bird called "Mongolian swallow" (*monggo cibin*), which is "entirely black and somewhat smaller than the swallow" (*buljin yacin, cibin ci majige ajige*).[45] Furthermore, the marmot (*tarbahi*) "lives in the Mongolian steppe" (*monggo tala de banjimbi*) whereas the

---

43. *Han-i araha manju gisun*, 1708, 20:*muduri meihe-i hacin*:30a.
44. On the "new Manchus," see Uray-Kőhalmi, "Von woher kamen die Iče Manju?"; Kim, *Ethnic Chrysalis*, 101–4; for an identification of the Kūyala, see Hu, *Xin Man-Han*, 501.
45. *Han-i araha manju gisun*, 1708, 19:*cecike-i hacin*:35a.

snow rabbit (*cindahan*) "lives in cold places in the Khingan mountains" (*hing an-i* [sic] *šahūrun bade banjimbi*), which are located in northern Manchuria and had been explored earlier in the Kangxi reign.[46] Sometimes the importance of a plant for the Mongols is noted. We learn that the willow (*burga*) is used to "make the beams and pillars for Mongol yurts" (*monggo boo hana sun arambi*).[47]

In the entry for *isi*, referring to some kind of larch, the reader is presented with a relatively detailed description of the tree's distribution—which includes a holy place in Mongolian Buddhism, Mount Wutai in China—the culinary use, and fossilization in the soil of northern Manchuria:[48]

> *isi* 'larch'. Like the pine tree but heavier. The Mongols use the inner bark to boil tea. Its needles fall off in winter. It grows in very cold areas such as Mount Wutai, outside the border [i.e., in Mongolia], and in the Khingan mountains.[49] It does not rot naturally. If its needles sting the hand, the poison is very severe. Many years after this tree has fallen into water, it usually turns into a whetstone. When buried in the ground by the Amur river, it likewise turns into a whetstone.
>
> *isi, saksin de adali bime ujen, uriha be monggoso cai fuifumbi. erei sata tuweri sihambi. umesi šahūrun ba, u tai, jase-i tulergi hinggan-i jergi bade banjimbi. banjitai niyarakū. erei u gala de nukaci horon ambula ehe. ere moo muke de tuhefi aniya goidaha manggi, an wehe ubaliyambi. sahaliyan ulai šurdeme, na de umbubuhangge, inu an wehe ubaliyambi.*[50]

---

46. *Han-i araha manju gisun*, 1708, 19:*gurgu-i hacin*:48a, 49a. The spelling of the name for the Khingan in this entry is curious, and might suggest that the entry was translated from Chinese. A link to *Shengjing tongzhi* (a book we discuss in later chapters) cannot be excluded.

47. *Han-i araha manju gisun*, 1708, 19:*moo-i hacin*:10b.

48. On the difficulty of identifying the *isi* with an actual tree, see Wadley, "Preliminary Investigation," 115–16. On the mention of *isi* in Tulišen's 1723 travelogue and its Chinese translation there, see Schlesinger, *World Trimmed with Fur*, 35.

49. We are aware of the different translation in Wadley ("Preliminary Investigation," 116).

50. *Han-i araha manju gisun*, 1708, 19:*moo-i hacin*:9b; see also Corff et al., *Auf kaiserlichen Befehl*, vol. 2, 873 (item 4020.1).

The use of the fossilized wood of this tree to sharpen metal tools was a cultural practice of the northeast. It is possible that this knowledge was so common that it was shared by the dictionary's compilers. Alternatively, perhaps they obtained the information on the use of the *isi* larch through interviews or reports.

Knowledge with similar origins is found in other entries as well. In the entry for "snakes hibernate," quoted earlier, a sentence has been added to the alternative Manchu or Tungusic terms that we would, in fact, expect to end the entry. This last sentence might represent Inner Asian folk knowledge. It reads, "From the place where the snakes live, wintry air rises like smoke in a spiral out of the hole [in the ground]" (*meihe tehe baci tuweri sukdun šanggiyan-i adali burgašame tucime dung ni angga-i šurdeme sumbi*).[51] This description of snake hibernation was probably specialized knowledge not shared by every urban reader of the *Mirror* and might have reflected the knowledge of some of the Manchurian communities that were presented as having their own terms for this natural phenomenon.[52]

A Bird from Across the Sea: An Ostrich, Cassowary, or Peacock

The preceding section surveyed entries that described the natural environment of Manchuria and Mongolia. The information given in those entries probably to some extent reflected knowledge gleaned during Qing imperial expansion in the seventeenth century, and not knowledge shared by all competent speakers of the Manchu language. This section and the next continue to survey natural-historical knowledge that the *Mirror* contained by virtue of its being a book compiled at the

---

51. *Han-i araha manju gisun*, 1708, 20:*muduri meihe-i hacin*:30a.

52. Guo Pu's 郭璞 commentary to the *Erya* entry *teng* 螣, an obscure character attested primarily as the first part of the binome *tengshe* 螣蛇, says that it is "a kind of dragon. It can raise mist and roam within in" 龍類也，能興雲霧而遊其中. There was thus a connection between snakes and mist. But to our eyes, it looks too tenuous to be considered the source of the idea of smoke rising in a spiral above the winter den of snakes. Ruan, *Erya zhu shu*, 75 (2641 of series; middle panel; chap. 9).

center of an empire with far-reaching connections. We discuss first, in this section, one entry—for a bird—that reflects knowledge that the Manchus had acquired by succeeding the Ming empire in China. Second, in the section that follows, we discuss a few entries for sea creatures, focusing on the animal known as *edeng*.

First, though, is the entry for the bird. The entry in question is headed by a Chinese loanword, *to gi*, from early Mandarin $t^h\text{ɔ}'\ ki$ 駝雞 (> *tuoji*) 'camel chicken' or a related phonetic form.[53] The Chinese word had been used for birds brought as tribute from Southeast Asia to the Ming empire, perhaps referring to ostriches—which one Ming source claimed had hair like a camel and walked like a camel—or cassowaries.[54] These two birds are possible referents for *to gi* as defined in the *Mirror*. The description almost makes it sound like a peacock, however:

> *to gi*. It comes from south of the equator in the southern seas. It is very large, six feet tall, and incapable of flight. It exists with tail feathers in all five colors.
>
> *to gi, julergi mederi, fulgiyan jugūn-i julergi de tucimbi, umesi amba, ninggun ci den, deyeme bahanarakū, sunja boco-i funggala akūmbume banjihabi.*[55]

We return to this curious bird in in a later chapter because it featured in Qianlong's Manchu-language reform. Here we merely note that the bird's presence in Kangxi's *Mirror* shows the Manchus' inheritance of natural-historical discourse produced in the Ming period and evidencing contacts with Southeast Asia.

---

53. Reconstruction from Pulleyblank, *Lexicon of Reconstructed Pronunciation*.
54. Ptak, *Birds and Beasts*, 50; for ostriches, see 114, 125. Lai, "Images, Knowledge and Empire," 31. On large flightless birds in Chinese historical records up to the Qing period, see Wang Ting, *Xiyu, Nanhai shidi yanjiu*, 39–56. For an identification of *tuoji* with the ostrich, see Bocci, "Animal Section in Boym," 359.
55. *Han-i araha manju gisun*, 1708, 19:*gasha-i hacin*:29b.

Jesuit Sources and Inner Asian Knowledge: The Orca and the Mermaids

We turn now to a much less straightforward issue, the description of large sea creatures in the *Mirror*'s section "Sea fish." It appears that this section contains both knowledge obtained from communities living on the northern Manchurian coast as well as from European knowledge introduced by the Jesuits. The entries referring to aquatic creatures are a rich source for philological investigation and speculation. Indeed, Gerhard Doerfer (1920–2003), writing in reference to the later, pentaglot expansion of the *Mirror*, stated that the chapters on river fish and sea fish on their own "merit a monograph—one, however, that would run to several hundreds of pages" (*Diese Kapitel zu behandeln wäre eine Monographie wert—die aber mehrere hundert Seiten umfassen würde*).[56] If that is the case, then the corresponding sections in the original *Mirror*—which is not multilingual like the *Pentaglot*, but unlike it includes definitions—deserve at least an article, yet we will only be able to cursorily treat five of the relevant words.

Two of the five words we treat here are descriptive compounds, whereas the other three are words of probably non-Manchu origins. The Manchus were not a seafaring people, and anecdotal evidence suggests that marine animals were initially not well known at the Kangxi court. A new awareness of the variety of marine fauna might have been what prompted the *Mirror*'s compilers to preface the section "Sea fish" (*mederi nimaha*) with a note of a type that we do not see in the other chapters on plants and animals. The note reads as follows:

mederi nimaha. There are very many kinds and local names are not alike. [We] write all the names that exist in Manchu.

mederi nimaha, hacin umesi labdu bime, ba na-i gebulerengge adali akū. manju de gebu bisirengge be araha.[57]

---

56. Doerfer, *Mongolo-Tungusica*, 248.
57. *Han-i araha manju gisun*, 1708, 20:*mederi nimaha-i hacin*:36a.

Following this note, the section proceeds to list several ocean creatures, of which a particularly large number have extremely bare-bones definitions consisting of only a Chinese translation of the headword (an issue we return to in chapter 2). That is to say, even though the compilers restricted themselves to only list names that existed in Manchu—that is, by excluding foreign or Chinese words—their grasp of the identity of the creatures they listed appears to have been very weak. In this section, we unpack some of the context in which these words entered the dictionary.

One source for knowledge of exotic creatures, including sea creatures, at the Kangxi court were European natural history works interpreted with the help of the Jesuits. The Jesuits had already published Chinese-language texts with information from such natural histories in the late Ming period, including Giulio Aleni's (Ai Rulüe 艾儒畧, 1582–1649) *Zhifang waiji* 職方外紀 (Annals of foreign lands, 1623) and *Xifang dawen* 西方答問 (Questions and answers on the Western regions, 1637).[58] This kind of scholarship continued in the early Qing. The prolific scholar Ferdinand Verbiest (Nan Huairen 南懷仁, 1623–1688), who was close to the Kangxi emperor, with the help of Chinese collaborators added to Aleni's work, notably by publishing illustrations drawn from European sources. Verbiest finished a map of the world in 1674, onto which he added drawings of several animals. About the same time, Verbiest produced an accompaniment titled *Kunyu tushuo* 坤輿圖說 (Illustrated explanation of the world), likewise based on earlier Jesuit writings and copying animal images from European sources.[59] The Kangxi emperor was clearly aware of the existence of these European books, and other members of the court also remarked on the quality of European book illustrations.[60]

---

58. Ai, *Xifang dawen*, 2: 1:7a–8a (133, upper panel, lower panel); De Troia, "Real and Unreal Animals."

59. The date of *Kunyu tushuo* is disputed. We follow Lai ("Zhishi, xiangxiang yu jiaoliu," 144–47). Cf., e.g., Gang and Dematté, "Mapping an Acentric World," 79. On the animal illustrations, see Lai as well as Walravens, "Konrad Gessner"; Ianaccone, "Lo zoo dei Gesuiti"; Menegon, "New Knowledge of Strange Things"; Winter, "Conrad Gessner."

60. Söderblom Saarela, *Early Modern Travels*, 146.

Images of sea creatures drawn from European sources were thus available at court, both in the original, as books imported from Europe, and in Chinese books. Furthermore, an anonymous album titled *Haiguai tuji* 海怪圖記 (Illustrated record of sea monsters), Daniel Greenberg speculates, probably dates from 1688 and might have been produced with Jesuit participation to curry favor with the curious emperor.[61] Moreover, a contemporary illustrated encyclopedia of marine creatures, Nie Huang's 聶璜 (b. ca. 1640s) *Haicuo tu* 海錯圖 (Illustrations of the miscellaneous things from the sea) from 1698, was partially based on what Nie had witnessed or heard about during two decades living on the southeastern coast, but also drew on Jesuit writings and pictorial sources. It is possible that Nie had seen *Haiguai tuji* or something like it.[62] Pictorial sources of partially European origin thus enjoyed a certain degree of circulation in the period.

Interestingly, European natural histories were used to identify at least one marine mammal in the years during which the *Mirror* was being compiled. In 1682, Verbiest was on tour in Manchuria with the emperor. The European books came handy when a group of Koreans brought a sea creature—referred to by Verbiest as a "fish" (*piscis*)—to Kangxi. The emperor asked Verbiest, in the latter's words, "whether there was any mention of this fish in our European books." Verbiest said there was, and his brethren in Beijing sent two books, which Noël Golvers identifies as certain sixteenth- or seventeenth-century encyclopedias. With the help of the illustrations in these books, the "fish" was identified as a "sea calf" (*vitulus marinus*), a seal, which Kangxi then had moved to Beijing to be cared for.[63]

The story suggests two things. Most obviously, it shows Kangxi's use of European natural history of the identification of a marine mammal, but it also shows that the Qing court came into contact with these creatures through the integration of the empire's northeastern coastal

---

61. Greenberg, "Yuan cang *Haiguai tuji* chutan," 45–47; Greenberg, "Weird Science," 385–86.

62. Zou, "Jiaoliu yu hujian," 102. Nie's album only reached the Qing palace in 1726, however, and was likely not known by the compilers of the *Mirror*.

63. Golvers, *Libraries of Western Learning*, 3:318–19, 554–55; Golvers, *Letters of a Peking Jesuit*, 453.

periphery. Those who brought the seal were Koreans—thus hailing from this periphery—and the animal was delivered to Kangxi when he was on tour in Manchuria. We would thus expect that words for such previously unknown creatures, and information about them, had two sources: European books or non-Manchu peoples living on the northeastern littoral. At least in the case of one creature, the two sources of information appear to have been fused, as we will see.

We turn now to the section on marine creatures in the *Mirror*. The first five creatures listed in the section are *boo nimaha* 'whale' (lit. house fish), *edeng* 'water tiger' (sawfish? swordfish? orca?), *kenggin* 'walrus' (?) or 'seal' (?), *sahamha* 'sea bass' (?), and *niyalma nimaha* 'mermaid' (lit. human fish).[64] These words present an interesting quintet.

The whale is relatively straightforward. It is "very big, some are several tens of feet [*jang* < Ch. *zhang* 丈] long" (*umesi amba, ududu jang ningge bi*), and "when it squirts water it rises upward in the shape of a tower" (*muke fusumbihede subargan-i gese mukdembi*).[65] The *Mirror* gives the alternative name *kalimu*, of which, William Rozycki remarks, "the ultimate source may be Nivkh, a non-Tungus language spoken at the mouth of the Amur and on Sakhalin and which had a larger geographic spread in the past."[66] In the case of the whale, we see that the *Mirror* used one paraphrastic, descriptive name as well as a loanword

---

64. Li Yanji (*Qingwen huishu*, 4:4b [78, lower panel]), who translates the *Mirror*, defines *boo nimaha* as *jingni* 鯨鯢. This can arguably be translated as "whale." Using this word, the earlier writer Yang Shen 楊慎 (1488–1559) described a "great fish of the eastern ocean, a kind of *jingni*, of which the largest are the size of mountains and those second to them are the size of houses" 東海大魚，鯨鯢之屬，大則如山，其次如屋. Here, a large sea creature is also compared to a house, as in the *Mirror* (Yang, *Yiyu tu zan*, 847:3:7a [745, upper panel]; Zou, *Zai jian yishou*, 51). Note, however, that the word *jingni* has a long history and was not always used in reference to whales. See Hargett, "Whales in Ancient China," 97–98.

65. *Han-i araha manju gisun*, 1708, 20:*nimaha-i hacin*:36a. The entry ends with a reference to a section of *Hafu buleku bithe*, one of the Manchu abridgments of some version of *Zizhi tongjian* 資治通鑒, for which see Standaert ("Comprehensive Histories," 294–96).

66. Rozycki, *Mongol Elements in Manchu*, 130. See also Doerfer, *Mongolo-Tungusica*, 261. A people from Sakhalin whom the Manchus called Kuye (Ch. *kuye* 庫野), perhaps speakers of Nivkh, are listed as in a tributary relationship to the Qing in *Huang Qing zhigong tu* from 1761 (see chapter 6), but there is no reason to believe

from a language spoken in the coastal region they had acquired dominance over in the seventeenth century.

The second kind of sea creature listed has the name *edeng*. The *Mirror* gives the following definition:

> *edeng*. This fish is also called "water tiger." According to what old people say, the ridge bone of this fish is very sharp and dazzles the eyes like polished steel.[67] When a whale eats one of its offspring by mistake, it [sc. *edeng*] seeks revenge and kills the whale by cutting off piece after piece.
>
> *edeng, ere nimaha be muke tasha seme inu hūlambi. sakdasa-i gisurere be donjici, ere nimaha-i mulu haga umesi dacun bime, niorombuha selei*

---

that they were known to the Qing court only at that late date. See Chuang, *Xie Sui "Zhigong tu,"* 177.

67. Here we provide some background to the word *haga* 'bone', which is key to understanding the appearance of the creature described in this entry. The *Mirror* defines *haga* as "the bones of fish" (*haga, nimaha-i giranggi be, haga sembi*), which we have followed (*Han-i araha manju gisun*, 1708, 20:*nimaha-i hacin*:39a–b; see *haha* in Hayata and Teramura, *Daishin zensho*, vol. 3, 687 [item 0417a2]). That is to say, *haga* did not mean "fin." Two words were available to express the notion of "fin." One was *fethe* 'tail fin', which the *Mirror* defines as "the wing-like growths on the skin of fish are called *fethe* when speaking of the big ones at the back" (*nimaha-i dakūla de asha-i adali banjiha amargi amba ningge be fethe sembi*). See *Han-i araha manju gisun*, 1708, 20:*nimaha-i hacin*:39a; cf. *bethe* 'foot'). This is the only word listed in the *Mirror* referring to fins. Shen Qiliang, furthermore, lists *fethe* in the sense of *yuchi* 魚翅 'fins', with *yusai* 魚腮 'gills' as a second acceptation. The word *senggele* or *senggile* was also available in the sense of gills, however (Hayata and Teramura, *Daishin zensho*, vol. 3, 687: 1406b4, 0649b5, 0649a5). Li Yanji (for whom see the main text), writing in 1724, provides a second word: *ucika* "the front paddle[s] on a fish's belly" 魚肚子前划水. This is a metaphorical use of the word *ucika* 'rain-proof bow case', which is listed in the *Mirror* but only in its original sense (*Han-i araha manju gisun*, 1708, 4:*jebele dashūwan-i hacin*:34b: *ucika, aga muke de dashūwan beri be dalime etuburengge be, ucika sembi*). Amiot, *Dictionnaire Tartare-Mantchou François*, 236, s.v. *outchika*, writes that *ucika* is used in reference to dorsal fins.

We cannot exclude that the *Mirror*'s editors did not have a suitable word available to indicate a large, pointy dorsal fin, and that they thus chose to use *haga* 'fish bone' to refer to the wet, slick, black dorsal fin of the orca that in a certain light might glisten like metal. However, all translators, from the Qing period to the present, translate *haga* as a "sharp bone" or "needle," which to us seems far removed from bone-less dorsal fins. We find this detail significant, as we explain later in the main text.

Now we turn to the phrase *yasa jerkišembi* 'dazzling eyes' or 'dazzling the eyes' and, relatedly, the punctuation of this sentence. *Yasa* means "eye"; that part is straightforward. The second word is a verb that the *Mirror* defines thus: "When the eyes (*yasa*) sting from looking at the sun, that is called *jerkišembi*. Also, any gleaming light is called *jerkišembi*. Also, when the eyes [i.e., the vision] become unclear and indistinct when riding on a horse at high speed, that too is called *yasa jerkišembi*" (*šun-i išun tuwara de yasa nukajara be jerkišembi sembi, geli yaya elden-i giltaršara be, inu jerkišembi sembi, geli yaluha morin-i feksire hûdun de, yasa geri garilara be, inu yasa jerkišembi sembi*) (see *Han-i araha manju gisun*, 1708, 7:*tuwara šara hacin*:23a). Only the last of the three acceptations gives the exact phrase *yasa jerkišembi*. In all of these three expressions, it appears that *yasa* 'eye(s)' is the subject, hence unmarked by a case particle, which accords with the intransitive meaning of the phrases "the eyes sting" or "the eyes [or vision] become unclear and indistinct" in the first and third expressions. In the second expression, transitivity is arguably a bit unclear, but it seems to us that it should be understood that gleaming light is blinding or dazzling. If this reading is correct, the word is not used with a direct object in any of the three expressions.

It is worth noting, however, that at least in Sibe, which is closely related to literary Manchu, the accusative case particle *be* can be omitted when the direct object immediately precedes the verb, or when the object does not have a definite or specific reference. See Norman, "Sketch of Sibe Morphology," 166; Gorelova, *Manchu Grammar*, 171–72.

Hu Zengyi reproduces several examples of *jerkišembi* from Manchu translations of Chinese fiction, bilingual Qing phrase books, and bilingual imperial edicts. One of his examples contains a use of *yasa . . . jerkišembi* in which the eyes are dazzled but without the use of a case particle. In two other cases, this is expressed by the construction *yasa de jerkišembi*, in which the dative case is used. In several other examples, unmarked subjects are shining, flashing, and the like. This is inconclusive for how to interpret the sentence from the *Mirror*. See Hu, *Xin Man-Han*, 864, s.v. *zherkishembi*.

Li Yanji reproduces the three acceptations of the verb *jerkišembi* from the *Mirror* and adds a fourth: "Bright in general" (*fan guangming* 凡光明), which would arguably accord with some of the examples listed by Hu (Li Yanji, *Qingwen huishu*, 9:15a [169, top panel]). Li chooses to interpret the verb according to this fourth acceptation in his translation of the phrase *yasa jerkišembi* in the entry for *edeng* (Li Yanji, *Qingwen huishu*, 1:23a [14]). We have not followed this reading, but reading the sentence in accordance with Li does not change the bigger picture. See the main text for a full translation of the entry in Li's dictionary.

The punctuation of this sentence contributes to the uncertainty. It contains a single punctuation mark (represented in the transcription by a comma), which in the *Mirror* is used to close sentences. The double punctuation mark, by contrast, represented with a period, either ends entries or separates the definition proper from a quote that exemplifies the use of the headword. In the copy of the *Mirror* available to us, there is a single punctuation mark following *bime* (the "is" in "the bone at the ridge of the *adali yasa jerkišembi. boo nimaha tašarame erei deberen be jeke manggi kimulefi, boo nimaha be lasha lasha secime wambi.*[68]

The entry appears to describe a big sea creature that has some kind of sharp appendage and that attacks whales by "cutting" (*secime*), a word that in all its acceptations refers to the making of incisions with a sharp metal instrument.[69]

There is a lot to unpack in this entry, and it needs to be done on three levels. First is philology. How was this entry understood by Qing-period readers? Second is etymology. What is the history of the word that is described in this entry? We discuss these two levels in turn, and then try to reconcile them on the third level, history, by situating the entry in the context of the Qing court at the turn of the eighteenth century.

First we consider philology. The entry on the *edeng* has three translations from the Qing period.[70] The first is found in the Mongolian

---

back of this fish is very sharp") but not following *adali* (the "like" of "like polished steel"). In the *Mirror*, *adali* often ends clauses, both in the construction X *de adali* and X-*i adali*, which is what we have here. However, we have taken the absence of a punctuation mark to mean that *adali* does not end the sentence, as have other translators. By contrast, Li Yanji seems to assume that the lack of a punctuation mark following *adali* is a slip of the brush.

68. *Han-i araha manju gisun*, 1708, 20:*nimaha-i hacin*:36a.

69. This is a key word in the entry. The word *secimbi* that the *Mirror* uses here means either to "to cut" ("to break something open using a knife or sword is called *secimbi*" [*yaya jaka be, huwesi jeyen-i hūwalara be, secimbi sembi*], see *Han-i araha manju gisun*, 1708, 18:*juwere faitara hacin*:33a) or "to plow" ("to make a ditch in a field by hauling a plow is called *secimbi*" [*anja ušame usin be yohoron arara be, secimbi sembi*], see *Han-i araha manju gisun*, 1708, 13:*usin weilere hacin*:30b).

70. We count only translations of the whole entry, not simply of the headword *edeng*. The few translations of the headword that we know of rendered *edeng* as *shuihu* 水虎 'water tiger' in Chinese. This word is a translation of the alternative Manchu name *muke tasha* idem. The Chinese *shuihu* is evidently not the origin of the Manchu term *muke tasha*. The Chinese monster known before the Qing period as *shuihu* was not an ocean-living creature (see Li Shizhen, *Bencao gangmu*, vol. 773, 42:28b [249, top panel]). It is thus not the origin of the term *muke tasha*.

The Qing scholars who followed the *Mirror* and translated *edeng* as *shuihu* probably did not know what animal it referred to. Juntu's 屯圖 pedagogical collection *Yi xue san guan Qingwen jian* 一學三貫清文鑑 (Mirror of the Manchu language, which will direct you to three things when you consult only one) simply listed *edeng~shuihu* as translation of the *Mirror* published at court in 1717. In this book, *edeng* is simply transcribed into Mongolian and the entry is rendered word for word, particle for particle, into Mongolian. It is thus not especially instructive as an interpretation.[71]

The second translation is found in Li Yanji's 李延基 (fl. 1693–1724) influential *Qingwen huishu* 清文彙書 (Manchu collected), a Manchu-Chinese dictionary from 1724 that was based on Kangxi's *Mirror*.[72] Li's translation of the entry is at odds with ours, but his reading still confirms a key aspect of the *edeng* as described in the *Mirror*.

First, Li understands the entry to say that the creature's eyes are bright or dazzling, not that the eyes of the onlooker are dazzled by the steel-like bone. Second, and more significantly, Li mistakes the Manchu word for "whale" (*boo nimaha*) for what he calls in Chinese a *shenyu* 蜃魚. The character *shen* can refer to oysters of some sort, which it does in the classical text *Zhou li* 周禮 (Rites of Zhou).[73] The association of *shen* with oysters might suggest that Li might have taken the Manchu name *boo nimaha* for a half transcription, half calque of the Chinese *baoyu* 鮑魚 'abalone' (the Manchu syllable *boo* is pronounced *bao*, and *yu* means fish, as does *nimaha*). Indeed, it appears that the abalone was referred to as *boo nimaha* in colloquial Manchu.[74] Accordingly, Joseph Amiot (1718–1793), translating Li, rendered *shenyu* as "oyster" (*huître*).[75] Yet Li was probably thinking of a different definition of *shen*. In

---

an example of a sea creature "named after a beast" (Ma. *gurgu be jafafi hebulehengge*; Ch. 以獸名之者也). The description only makes sense in reference to *shuihu*, not *edeng*. Yet Juntu does not mention the alternative Manchu name *muke tasha*, which means that his gloss is nonsensical unless the reader refers to the Chinese text. See Juntu, *Yi xue san guan Qingwen jian*, 73a (153, bottom panel).

71. *Han-i araha manju gisun*, 1717, 20:*dalai-yin jiyasun-u jüil*:77b–78a. For instance, *yasa jerkišembi* is rendered as *nidü jirgelümüi*, similarly without a case particle following "eye." It is perhaps telling, however, that Lessing and colleagues, *Mongolian-English Dictionary*, 1046, s.v. *zergel-* specifies that *jergel-* (= *jirgel-*) is an intransitive verb. This appears to favor Li Yanji's reading but might just be the result of the will to provide an exact equivalent. Furthermore, *mulu haga* is translated as *niruyun-u qayadasu* 'back [or spine] [fish] bone' and *secime* is *qadur-un* 'cutting'.

72. Chunhua, *Qingdai Man-*, 300–303.

73. Biot, *Le Tcheou-li*, vol. 1, 382.

74. Shen Qiliang translates *boo nimaha* as *fuyu* 鰒魚, an alternative name of the abalone. See Hayata and Teramura, *Daishin zensho*, vol. 2, 80 (item 0249b5).

75. Amiot, *Dictionnaire Tartare-Mantchou François*, 101.

addition to meaning oyster, *shen* also referred to a legendary creature. This *shen* had the appearance of a big snake with horns like a dragon and could "exhale mist in the shape of towers and city walls" 能吁氣成樓臺城郭之狀.[76] Li, having never heard of a whale, identified *boo nimaha* with the monstrous *shen*.

Li's confusion suggests that the words *boo nimaha* and *edeng* were not words known among urban, eighteenth-century speakers of Manchu, but peripheral words that had perhaps been introduced into the language through this very entry in the *Mirror*. At the same time, his translation shows how he understood the physical characteristics of the *edeng* as the *Mirror* describes it. Li writes,

> *edeng*. The water-tiger fish of the ocean. Its spine needle (*ji ci*) is very sharp, like polished iron. Its eyes are bright. When the *shen* fish eats its fry (*yuzi*) by mistake, it takes revenge by cutting the *shen* fish apart piece by piece.[77]
>
> 海裡的水虎魚,脊刺很快如錚磨的鐵,眼睛光明。蜃魚錯吃了他的魚仔他結仇能畫開蜃魚一斷斷的。[78]

The misunderstanding notwithstanding, the translation is suggestive of how Li understood the reference to the *edeng*'s "spine needle." The phrase Li uses suggests that the *edeng* cut or sliced the *shen* fish into strips, which makes sense if Li understood the *edeng*'s appendage to resemble a pointy metal instrument. It is not clear from his translation whether the needle would have pointed out in the same direction as the spine (forward) or in a way perpendicular to it (upward).

The third Qing-period reading of this entry is found in the dictionary written by Ivan Zakharov (1817–1885), who studied Manchu and

---

76. Li Shizhen, *Bencao gangmu*, vol. 774, 43:7b (258, lower panel).

77. Amiot (*Dictionnaire Tartare-Mantchou François*, 101), translating Li Yanji, writes "eggs" where we write "fry."

78. Li Yanji, *Qingwen huishu*, 1:23a (14, bottom panel). The character *zi* 仔 is a conjecture; Li uses a variant character that we have not been able to identify with certainty.

Chinese in Beijing between 1839 and 1850.[79] Zakharov was evidently translating from the *Mirror* and not from Li Yanji, as he does not repeat the mistaken identification of *boo nimaha* with the *shen* fish. Zakharov writes that the *edeng* is

> A sea fish with a sharp bone on its spine, polished and shining like steel. If its offspring falls prey to a whale, then they cut through their innards and kill it.
>
> морской рыбы съ острою костью на хребтѣ, полированною и блстящею какъ сталь, детенышъ ее если попадетъ въ пасть кита, то, прорѣзывая ёго внутренности, умерщвляетъ его.[80]

First, Zakharov interprets the entry to mean that it is the "sharp bone" that is shining, not the eyes of the fish.[81] This accords with our translation, but it is secondary to his understanding of *edeng* as having a "sharp bone on its spine." In Zakharov's case, we know what kind of creature he believed was being described here, for he translated *edeng* as "sawfish" (пила-рыба).[82] Thus he clearly assumed that the "sharp

---

79. Walravens, *Ivan Il'ič Zacharov*, 1–4; Alexander Vovin's introduction in Zakharov, *Grammatika man'chzhurskogo iazyka*, v.

80. Zakharov, *Polnyĭ Man'chzhursko-Russkiĭ slovar'*, vol. 1, 74.

81. Zakharov apparently takes *yasa jerkišembi* to refer to the shining properties of the steel-like bone. This interpretation accords with Imanishi, Tamura, and Satō, *Gotai Shinbunkan yakukai*, vol. 1, 960 (item 16837), who give the following translation of the definition of *edeng*: "A sea monster. According to what old people say, the needle on the back of this fish is very sharp and furthermore dazzles the eye like polished steel" 海中の怪魚。古老の言に、この魚の背の刺頗る鋭くて磨きをかけた鉄の如く眼も眩む.

82. This translation is not seen in earlier European-language dictionaries. Gabelentz (*Sse-schu, Schu-king, Schi-king*, vol. 2, 50) simply writes "name of a sea fish." Zakharov's translation was followed in later dictionaries: Hauer, *Handwörterbuch der Mandschusprache*, 1952, vol. 1, 230; Norman, *Comprehensive Manchu–English Dictionary*, 89; Lessing et al., *Mongolian-English Dictionary*, 294.

Unrelatedly, Lessing and colleagues cite S. J. Gunzel et al.'s *Mongol-English Practical Dictionary* (1949–1953), place of publication uncertain. See Hartmut Walravens and colleagues (*Bibliographies of Mongolian*, 17), which gives the following definition: "Sea-beaver, sea-otter, Enhydra marina." This is obviously far from what is described in the *Mirror*.

bone" extended horizontally to the spine and out over the *edeng*'s face. To conclude, the two Qing readers of which we have any evidence (Li and Zakharov) understood *edeng* as having a needle or saw-like appendage, in one case at least extending forward from the creature's head.

We turn now to the etymology of the word *edeng*. Phonologically, the word *edeng* is quite curious and has been the subject of diverging interpretations. It is clearly not of Manchu origin. Rozycki suggested it might have been borrowed from (northern vernacular) Chinese, but does not propose any possible Chinese candidates.[83] Forms of the word, either as cognates or borrowings, are attested in several Manchurian languages, however, both Tungusic and otherwise.[84] In Uilta, a Tungusic language that is today spoken on Sakhalin, the word is ədə. In Eastern Sakhalin Nivkh, a non-Tungusic language, it is əzŋ. Both words mean "master" (cf. Ma. *ejen* idem), but have the secondary meaning of "orca," a creature associated with the "Master of the Waters" in the religion of the Nivkh people.[85]

We can now transition from etymology to history and ask how the word *edeng* ended up in Kangxi's *Mirror*, and why the description there differs so radically from the appearance of orcas. In response to the first question, we can assume that Qing agents either came into contact with this word along with a description of its referent, the orca, as the empire expanded into the Amur region, or sighted orcas themselves and were told by native informants that they were called ədə, əzŋ, *edeng*, or something similar.

The "old people" or "old men" (*sakdasa*) mentioned in the entry could, then, refer to these native informants. Indeed, the phrase used here evokes wording used in the dictionary's imperial preface. The preface mentions "asking among the aged old men" (*fe sakdasa de aname*

---

83. We disagree with Rozycki that in both Manchu and Mongolian, the word would have been "borrowed separately . . . from a third language, probably Chinese" (Rozycki, *Mongol Elements in Manchu*, 66). The word in Mongolian is clearly borrowed from the Manchu word and a product of Qing lexicography (that is, through the Mongolian translation of the *Mirror* that dates from 1717).

84. Doerfer, *Mongolo-Tungusica*, 143 (no. 645).

85. Ikegami, *Uiruta go jiten*, 55; Fortescue, *Comparative Nivkh Dictionary*, 167; see also Black, "The Nivkh (Gilyak) of Sakhalin," 93.

fonjime) as a method used when compiling the dictionary.[86] It is tempting to assume, given that the *Mirror* is a dictionary of the Manchu language, that these individuals were all speakers of Manchu. That would be mistaken, however. Manchu in this period was an umbrella term that comprised communities in Manchuria that until a few generations earlier had been distinct, such as the Yehe and Yeren.[87] Furthermore, as mentioned, linguistic material included in the *Mirror* also came from the "new Manchus" whose incorporation into the Qing state happened during the expansion into northern Manchuria in the second half of the seventeenth century. In the case of *edeng*, speakers of Manchurian languages who were only very recently incorporated into the empire's banner system appear to have been consulted.[88] In that sense, the entry on *edeng* is thus another example of Inner Asian folk knowledge making its way into the *Mirror* as a result of northward expansion.

We turn now to the question of why the description of *edeng* makes it appear so different from orcas, who do kill whales, but do not have sharp "ridge bones" extending from their spine and who do not "cut" their prey. If we are to follow Li Yanji's reading, the *edeng* in addition had "bright" or "dazzling" eyes, which cannot easily be reconciled with the appearance of the orca, whose small eyes are hard to distinguish against its black skin. Sea creatures with needle-like appendages

---

86. *Han-i araha manju gisun*, 1708, sioi:4a–b. Relatedly, see Söderblom Saarela, *Early Modern Travels*, 139.

87. Elliott, *Manchu Way*, 69–71.

88. The "old people" consulted regarding the *edeng* cannot have been the Fujianese informants who told Nie Huang about the *yuhu* 魚虎 'fish tiger' whose "skin on the back resembles the hedgehog and can sting people" 背皮似蝟，能刺人 (Nie, *Qinggong haicuo tu*, 51). The description is a poor fit and southern Chinese people were not considered part of the Manchu community. Furthermore, *sakdasa* 'old people' should not be taken as a reference to what in Chinese would be called *guren* 古人 'the ancients'. In the *Mirror*, the notion of "ancients" is expressed as *julgei niyalma* 'persons of antiquity', which is a direct calque of the Chinese phrase (see *Han-i araha manju gisun*, 1708, 19:*orho-i hacin*:7b). A great deal of material in the *Mirror* derives from Chinese antiquity, but not this entry. Moreover, we have also found nothing in the scholarship on huge sea creatures in Chinese texts (including Buddhist) that matches the appearance of the *edeng* (Hargett, "Whales in Ancient China"; Goode, "On the Sanbao," 85–107).

certainly exist, and descriptions of them were available at the Kangxi court. However, fish whose appendages look like steel and in addition kill whales are known only from accounts of European origin. The *Mirror*'s description of the *edeng*, though referring to a word of Manchurian (but not Manchu) origins, might have been influenced by such European accounts.

Orcas were known to European natural history. They were described as attacking whales since antiquity. In the seventeenth century, fantastic accounts of sea monsters merged elements from the description of orcas, present in the Atlantic, with that of swordfish, present in the Mediterranean. The orca was sometimes called "sword-fish" in seventeenth-century Dutch, and an influential sixteenth-century account described the (Mediterranean) swordfish as attacking "the great cetacean fish" (*les grands poissons cétacées*), much like the orca.[89] This conflation arguably contributed to the description of the sea monster known as the *ziphius* or *xiphias*, the latter of which was a word used in antiquity in reference to the swordfish (Lat. *xiphia(s)* < Gr. *xiphías*, in turn < *xíphos* 'sword').

A well-known and elaborate description of the legendary creature "called the xiphias" is found in Olaus Magnus's (1490–1557) *Historia de gentibus septentrionalibus* (A description of the northern peoples) from 1555. Magnus worked from several sources, textual as well as pictorial. His description contains elements (e.g., a head like that of an owl) not seen in the *Mirror*, but importantly includes the following characteristics. The xiphias has "a back that is tapering, or rather, raised into the form of a sword" (*dorsum cuneatum, vel ad gladii formam eleuatum*).[90] This is the closest parallel we have found to the phrasing used

---

89. Olaus Magnus, *Description of the Northern Peoples*, vol. 3, 1144, OM 21:14n1. Further on the ziphius, Van Duzer, *Sea Monsters*, 84–85. For the quote, see Rondelet, *Histoire entière des poissons*, bk. 8, chap. 14, 201. The descriptions of ziphius or xiphias in several Renaissance natural histories are tabulated in Zucker ("Zoologie et philologie," 145–47).

90. Olaus Magnus, *Historia de gentibus septentrionalibus*, bk. 21, chap. 14, 743. Translation from Magnus, *Description of the Northern Peoples*, vol. 3, 1096. In addition, Magnus's xiphias has "dreadful eyes" (*oculos horribiles*), which brings to mind Li Yanji's translation of the *Mirror*'s entry on *edeng*, in which the creature has bright eyes.

in the *Mirror* in reference to the *edeng*. Magnus accompanied his account with an illustration that depicted a xiphias quite different from what he described, a sawfish and a unicorn fish.

The xiphias made it into Chinese-language accounts authored by the Jesuits. A whale-eating xiphias somewhat different from what Magnus described is found in Aleni's *Zhifang waiji*. Aleni's book tells of "a fish called the swordfish (*jianyu* 劍魚 < *xiphias*)." The description, however, makes it sound more like a sawfish:

> Its mouth is about ten feet long. It has jagged teeth cut like a saw. Fierce and very strong, it can fight the *ba-le-ya* (< Portuguese *baleia* 'whale') fish.[91] When the water of the ocean has turned completely red, then this fish has won.
>
> 一名劍魚。其嘴長丈許,有齟刻如鋸。猛而多力,能與把勒亞魚戰。海水皆紅,此魚輒勝。[92]

Verbiest reproduced the account and added an illustration of the creature, which has a long, saw-like snout (see fig. 1.2a).[93] Aleni's and Verbiest's swordfish are similar to Zakharov's understanding of the *edeng*, and arguably compatible, at least, with Li Yanji's understanding of its physiology.

It is possible that European accounts of the xiphias, in the form of either the Chinese text on the *jianyu* or illustrated European encyclopedias, influenced the description of the *edeng* in the *Mirror*. We saw that Kangxi identified Manchurian marine creatures by consulting European books with Verbiest's help, so it is not inconceivable that a

---

91. Ptak, *Birds and Beasts*, 119.
92. Ai, *Zhifang waiji*, 1:5:4a (74, upper panel).
93. Verbiest, *Kunyu tushuo*, 157, 213–14. This swordfish has two blowholes (which the *ba-le-ya* also has in Verbiest's book). The image appears to have been copied from a European map unrelated to Aleni's text, which explains the mismatch between text and image in Verbiest's book (no mention of blowholes in Aleni's text). Furthermore, the saw-like jaws of the creature in this image are curious. Walravens shows that the image on which Verbiest probably relied showed the pointy snout but not the saw-like teeth ("Die Deutschland-Kenntnisse," 234–35; see also Cheng, "Mingmo-Qingchu," 93–94; Zou, *Zai jian yishou*, 55).

FIG. 1.2. The swordfish (a) and the mermaid (b) in Verbiest's *Kunyu tushuo*. Courtesy of the University of Iowa Libraries.

description of a sea monster with a sword-like growth on its back, taken directly from European-language sources, made its way into the *Mirror* through Jesuit mediation. Even though the word *edeng* was of Inner Asian origin, parts of its description clearly are not, and we do not hold it impossible that the "old people" mentioned in the entry referred to the Jesuits.[94] A consideration of the remaining three of the first five entries in the section on ocean creatures suggests that European influence was definitely manifest in some other entries.

European influence on the *Mirror*'s section on marine creatures is very strongly suggested by the entry on the "human fish," the section's fifth entry. Two entries stand between this creature and the *edeng*. First, the third entry presents a *kenggin*, which the *Mirror* says is "like a bear but with a fish's body" (*lefu de adali bime nimaha beye*). In a rendering that parallels the Chinese name of the *edeng*, this creature was later

---

94. We are grateful to Henning Klöter for suggesting this possibility.

translated into *shuixiong* 水熊 'water bear' in Qing dictionaries.[95] The word *kenggin* "ought to be [a] loanword from an eastern Tungusic language" (*Die Wörter mit k- [wie* keŋgin*] müssten Lehnwörter aus einer östlicheren tu. Sprache sein*).[96] It is tempting to assume that it refers to a walrus or seal of some kind—perhaps what the Koreans presented to Kangxi back in 1682.[97] However, the *Mirror* adds that "boats and ships are very wary of it" (*weihu cuwan ede ambula targambi*), which makes the identification uncertain.

Following *kenggin*, the *Mirror* lists *sahamha*, which probably refers to a sea bass or sea perch. The name is curious and Doerfer is unable to satisfactorily account for it. The description is, moreover, less than unambiguous.[98] Li Yanji does not give a Chinese name.[99]

There is more to say about the "human fish," the fifth entry and the last sea creature that we discuss. It is not, like the *kenggin*, referred to by a Tungusic loanword, nor by a name of unclear origin like *sahamha*. The human fish is called *niyalma nimaha* in Manchu, a descriptive

---

95. E.g., Corff et al., *Auf kaiserlichen Befehl*, vol. 2, 979 (item 4484.3).

96. Doerfer, *Mongolo-Tungusica*, 251, 260. Cf. Rozycki, *Mongol Elements in Manchu*, 137.

97. Imanishi, Tamura, and Satō (*Gotai Shinbunkan yakukai*, 960, item 16839) glosses it as a seal.

98. The *sahamha* is said to be as long as two arms extended, or five feet (*golmin emu da funcembi*), which could with allowance for some hyperbole describe an Asian sea bass. It is interesting that Doerfer—working from the later pentaglot *Mirror* and unaware of the chronology of Manchu-language reform—writes with confidence that *sahamha* is a neologism made from the first and last elements of *sahaliyan nimaha* 'black fish' but mistakenly has "white" (Doerfer, *Mongolo-Tungusica*, 250). We find this unlikely, but it is an indication of the marginality of this word within Tungusic. Regarding the measurement *da*, Hayata and Teramura (*Daishin zensho*, vol. 1, 128, 0813a2) define it as five feet (*wu chi* 五尺). The 1708 *Mirror* defines one *da* as "the measure of the two arms extended" (*juwe gala saniyafi emgeri kemnehengge be, emu da sembi*) (*Han-i araha manju gisun*, 1708, 2:baita sita-i hacin:17b). This is translated by Li Yanji (*Qingwen huishu*, 6:9b [115, top panel]) as *liang shou shenkai liang* 兩手伸開量. Li also gives the measurement of five feet.

The translation of *sahamha* as "bass" or "perch" goes back to *Yuzhi zengding* (vol. 233, 32:45a [299, bottom panel]). The Mongolian translation is a calque on the Manchu. See Lessing et al., *Mongolian-English Dictionary*, 657, s.v. SAΓAMX-A: "salt water fish; perch (?)"; Kowalewski, *Dictionnaire mongol-russe-français*, 1296, s.v. *sakhamkha*: "nom d'un poisson de mer."

99. Li Yanji, *Qingwen huishu*, 4:18b (85, lower panel).

phrase that mirrors the Chinese *renyu* 人 (alt. 魜) 魚 idem, which was used in later bilingual Manchu-Chinese dictionaries to translate it. Human fish of various sorts are known from multiple Chinese records, but none that we know of is a close match to that given in Kangxi's *Mirror*:

> *niyalma nimaha.* Lives in the sea. Above the navel (*du ci* < Ch. *duqi* 肚臍), it looks like a human, and below the navel it looks like a fish. They are obtained from places in different countries in the very great oceans.
>
> *niyalma nimaha, mederi de banjimbi, du ci wesihun niyalma adali, du ci fusihūn nimaha adali, umesi amba mederi encu gurun-i bade bahangge bi.*[100]

Like a human above the waist, like a fish below. This is a mermaid, much like the "sirens" (*xi-leng* 西楞) of Verbiest's *Kunyu tushuo*, where they are also illustrated to conform to this description (see fig. 1.2b).[101] The Chinese tradition includes many mentions of mermaids, but, as Zou Zhenhuan remarks, "the mermaids in European folklore and in Chinese and Japanese folklore are entirely different in appearance and character" 欧洲传说中的人鱼与中国、日本传说中的人鱼，在外形上和性质上是迥然不同的.[102] What the *Mirror* described was the European mermaid with a fish tail, not the creatures with legs described in most Chinese sources. Nevertheless, *Shanhai jing* 山海經 (Classic of mountains and seas) included a description of a mermaid with a "human face and fish body without legs" 人面魚身無足, which in an undated, but later than 1783, edition of the work looks quite a bit like Verbiest's sirens (which might for all we know have influenced it).[103] Yet the *Mirror*'s reference to foreign countries further supports the inference that Jesuit sources influenced the description.

---

100. *Han-i araha manju gisun*, 1708, 20:*nimaha-i hacin*:36b. The description is truncated in Li Yanji (*Qingwen huishu*, 2:24b [40, lower panel]): "The human fish. Lives in the great oceans. It is like a human above the hip and like a fish below the hip" 人魚，生於大海。從胯上似人，胯下似魚.

101. Verbiest, *Kunyu tushuo*, 209–10; Walravens, "Die Deutschland-Kenntnisse," 232; Magnani, "Searching for *Sirenes*," 97.

102. Zou, "*Haicuo tu*," 131; Zou, *Zai jian yishou*, 223.

103. Bi, *Shanhai jing xin jiaozheng*, 2: *yiyu* 異域.

We are inclined to conclude that European knowledge of sea creatures influenced the *Mirror*. As mentioned, the unusual note at the beginning of the section that the "local names" of sea creatures differ and that the section would limit itself to those names "that exist in Manchu" was probably prompted by the encounter with European accounts. Clearly, words derived from Manchurian languages other than Manchu—words such as *edeng, kalimu,* or *kenggin*—were acceptable in the section, but *ba-le-ya* or *xi-leng* were not. If European accounts were used, it ought to have been as a source of corroboration or supplementary material for things that at least had a name that the compilers could allow within the umbrella term of Manchu, even if, as in the case of the so-called human fish, the name might have been a translation of a Chinese expression that referred to another kind of legendary creature, or derive from a Tungusic language other than Manchu or even a Manchurian but non-Tungusic language, as was the case with *edeng*. If our reading of the *Mirror*'s description of the orca is correct, the *Mirror*'s editors might have matched the Northeast Asian *edeng* with the ziphius or xiphias of European natural history. If true, foreign knowledge and knowledge gathered during the empire's expansion into the northern Manchurian littoral were fused even within individual entries in the *Mirror*. If not, at least the section on sea creatures as a whole evidences such fusion, as the mermaid is most probably European.

In addition to the material considered here, the section on sea creatures evidently also drew heavily on information presented in Chinese. We return to the implications of the presence of such information in the following chapter.

CHAPTER TWO

The Manchu Mirror as Chinese Encyclopedia

In the previous chapter, we investigated the Inner Asian and European contents found in the Kangxi *Mirror*'s sections on flora and fauna. In this chapter, we turn to the aspects of the book best understood in the context of Chinese encyclopedic learning. Despite being dressed in nominally Manchu linguistic garb, the *Mirror* frequently used Chinese sources. As Tsai Ming-che 蔡名哲 recently put it, "it looks as if it was without influence from Chinese culture, yet many of its definitions are marked by such influence" 看似未受汉文化影响，然而其释义有不少已受汉文化影响.[1] In part, the Chinese influence took the form of numerous quotations from the Confucian classics, some of which had been translated into Manchu by the time of the dictionary's compilation. On occasion, the references were less specific reporting of information from "the ancients" (*julgei niyalma*), which were presented as neither Manchu nor Chinese but which were most probably the latter.

More interestingly, the *Mirror* contains references to recent or contemporary Chinese discourses on the natural world. Some of it was drawn from specialized Chinese literature that did not exist in Manchu translation at the time the dictionary was compiled or later. In the

1. Tsai, "Qing qianqi Manzhou," 244.

entries with such references, the compilers appear to work the way they would have on a Chinese lexicographical project. In addition to providing some of the information contained in individual entries, the Chinese scholarly tradition likewise affected the arrangement of entries and, certainly, the structure of the dictionary itself.

In this chapter, we first look at the principles behind the arrangement of Kangxi's *Mirror*, which is reminiscent of Chinese encyclopedias (collections of "classified writings"). We then consider the role of Chinese scholarship in the book and finally discuss the presence of vernacular Chinese translations hidden, as it were, in Manchu transcription within the entries.

Arrangement by Utility and Form: Gourds and Willows

When it comes to flora and fauna, the arrangement of headwords in the *Mirror* is to a great extent determined by their function. Classification according to function, or utility, was common in Chinese natural-historical scholarship and, indeed, in Chinese classification in general. The importance of function is a first indication that the *Mirror* to an important degree is a work of Chinese scholarship.

The criteria according to which headwords were arranged in the *Mirror* bring to mind other editorial projects undertaken at the Qing court in the eighteenth century. In the largest of these, the *Siku quanshu* 四庫全書 (Complete books of the four repositories), the socio-scholarly function of texts determined their placement within the hierarchically arranged library.[2] Natural-historical scholarship was similar in this regard. In Chinese pharmacopoeias, agricultural treatises, encyclopedias, and florilegia—all genres with some degree of similarity to the *Mirror*—classification was "based on widely known criteria of utility and not on abstracted physical characteristics."[3]

---

2. Gandolfo, "To Collect and to Order," 35.
3. Bray, "Essence and Utility," 13.

Another aspect of the *Mirror*'s arrangement suggests a similarity with scholarship on the world beyond language. Within the *Mirror*'s chapters, the names of flora and fauna are listed together on the basis of perceived real-world similarity rather than any similarity of linguistic form. Thus, in these chapters at least, the structure of the book is encyclopedic rather than linguistic. We use the example of gourds and melons to illustrate the two points that the botanical and zoological chapters are, first, largely encyclopedic and, second, arranged according to criteria used in Chinese encyclopedic works.[4] The word *hengke* 'melon, gourd, cucurbitaceous plants' (as in neighboring languages, ultimately < Ch. *xianggua* 香瓜) occurs in seven lemmata, but always with a modifier.[5] In addition, it is used to describe two gourds that do not have *hengke* in their name.

In the section "Vegetables and dishes," three entries are listed in succession whose headwords are all noun phrases with *hengke* as its head: *nasan hengke* 'salting gourd' (i.e., pickling gherkin), *lugiya hengke* 'pickling melon', and *cirku hengke* 'winter melon' (lit. pillow melon).[6] The culinary use is noted in all three cases ("can be eaten raw or cooked" [*eshun jeci ombi, urebufi jeci inu ombi*]; "eaten pressed and pickled" [*janggūwan gidafi jembi*], "eaten boiled" [*fuifufi jembi*]). At the end of the section, moreover, we find *guwalase*, which is said to be "like shepherd's purse and grows green buds that look like jujubes. Eating it is like eating cucumber" (*niyajiba sogi adali soro-i gese niowanggiyan bongko banjimbi jeci nasan hengke-i adali*).[7] This *guwalase* is thus compared to a *hengke* without being one.

---

4. See *Han-i araha manju gisun*, 1708, 18:*sogi booha-i hacin*:7a–b, 10a, 50a–b; 19:*orho-i hacin*:5b–6a.

5. Doerfer, *Mongolo-Tungusica*, 245.

6. *Han-i araha manju gisun*, 1708, 18:*sogi booha-i hacin*:7a–b; Corff et al., *Auf kaiserlichen Befehl*, vol. 2, 817 (lemma 3773.4) give the translation *länglicher, bitterer Kürbis, Momordica charantia*. The latter is the learned name of the bitter melon (*kugua*, for which see following). The later, plurilingual *Mirrors* (such as the one edited by Corff and colleagues) translate *lugiya hengke* as *shaogua* 稍瓜 'pickling melon' (*Cucumis melo, Conomon*, or *Makino*). We have followed this identification. See Read, *Chinese Medicinal Plants*, 15 (item 59).

7. *Han-i araha manju gisun*, 1708, 18:*sogi booha-i hacin*:10a. Cf. the translation in Imanishi, Tamura, and Satō, *Gotai Shinbunkan yakukai*, vol. 1, 803 (item 14188).

Later, in the section "Fruits" in the chapter "Various fruits," three more gourds are listed: *jancuhūn hengke* 'sweet melon', *solho hengke* 'Korean muskmelon', and *nikan hengke* 'Chinese melon'.[8] Unlike the *hengke* listed among the "Vegetables and dishes," these are not prepared in any way, but instead eaten raw. Thus the entries all comment on their taste ("fine flesh, sweet taste, fragrant, [and with] many seeds" [*amtan jancuhūn, wa sain, use labdu*], "in terms of taste, superior to the sweet melon" [*amtan, jancuhūn hengke ci wesihun*], and "in terms of taste, inferior to the Korean muskmelon" [*amtan solho hengke de isirakū*], respectively).

Finally, in part 2 of the section "Herbs" in the chapter "Herbs," we find another two gourds. They are *gaha hengke* 'small red snake gourd' (lit. crow gourd) and *gaha oton* 'bitter melon' (lit. crow tub).[9] The names both contain the element *gaha* 'crow' and the description is similar in both cases. Both plants are said to be "like the small gourd" (*ajige hengke-i adali*; a word that is not itself lemmatized in the book). The function that is highlighted in the descriptions of these two plants is their use as medical drugs. This use is referenced in a most interesting way, which we will get back to presently.

All the gourds listed, all but two of which are labeled a kind of *hengke* in Manchu, are thus distributed in the *Mirror* across three categories: vegetables and dishes (eaten cooked or otherwise prepared), fruits (sweet, eaten raw), and herbs (have medicinal properties). The use of the plants is the main criterion in arranging them. Different varieties of *hengke* 'gourds' are spread across several chapters. Yet nowhere is the word *hengke* itself defined; a gourd is a category of plants with many uses, not a specific type, so there is no obvious place to describe it. The result is that this word, which is used to characterize different gourds within entries, is not itself defined anywhere. The

---

8. *Han-i araha manju gisun*, 1708, 18:*tubihe-i hacin*:50a-b. See also Corff et al., *Auf kaiserlichen Befehl*, vol. 2, 862 (lemmata 3980.1-3).

9. *Han-i araha manju gisun*, 1708, 19:*orho-i hacin*:5b-6a. See also Corff et al., *Auf kaiserlichen Befehl*, vol. 2, 870 (items 4008.3, 4008.4). Corff and colleagues identify *gaha oton* as *Momordica charantia* 'bitter melon', the same gloss they give for *lugiya hengke*.

example of gourds shows that when it comes to flora and fauna, a classification of useful things rather than words was important for the arrangement.

The grouping of the different gourds within the chapters they occur is further evidence of the primacy of the referent over the word. Under both "Vegetables and dishes" and "Fruits," the gourds are listed in succession. True, in these chapters, the headwords all contain the element *hengke* in their Manchu names. Similarly, in the chapter "Herbs," the two gourds *gaha hengke* 'small red snake gourd' and *gaha oton* 'bitter melon' both carry *gaha* 'crow' in their name, even though the latter is not called a *hengke*.[10] (The word *gaha* is onomatopoetic in origin and has cognates in several Tungusic languages as well as Mongolian.)[11] Furthermore, in the arrangement of words, real-world physical similarity often trumps a similarity in name. The names for willow trees present an example. The willow is referred to by two words in Manchu, *fodoho* and *burga*. Both words are of Mongolian origin. The former is an old loan, whose cognate in written Mongol (*uda*) is a marginal word that would probably not have been recognized by Manchu speakers as related to *fodoho*.[12] The other word (*burga*) is a recent loan that is more obviously related to written Mongol *buryasu(n)*. In the *Mirror*'s chapter "Trees," a succession of four lemmata containing the element *fodoho* are listed, followed by six containing the element *burga*. These two groups of entries were probably listed together because of the similarity of their references, even though the words sounded nothing alike.

---

10. However, two other plants also contain *gaha* as a modifier in their name: *gaha yasa* 'prickly waterlily' (*Euryale ferox*), which had *gaha* 'crow' in its name because "its shape is somewhat similar to the head of a chicken" (*banin coko-i uju de adalikan*), and *gaha poo* 'puffball' (akin to a mushroom). These two are not listed along with the two gourds, nor with each other (one among the fruits, the other among the herbs). See *Han-i araha manju gisun*, 1708, 18:*tubihe-i hacin*:49b, 19:*orho-i hacin*:7a; Corff et al., *Auf kaiserlichen Befehl*, 861 (item 3977.3), 871 (4013.4).

11. Tsintsius, *Sravnitel'nyĭ slovar'*, 137, s.v. ГАКИ; Rozycki, *Mongol Elements in Manchu*, 85.

12. Rozycki, *Mongol Elements in Manchu*, 77–78, 39–40; *Han-i araha manju gisun*, 1708, 19:*moo-i hacin*:10b–11a.

In sum, neither the small red snake gourd and the bitter melon nor the different kinds of willow were listed next to each other because of their similar Manchu name. Most probably, they were brought together because of shared material characteristics, which in some cases were noted in the entries.[13]

The distribution of plants across sections and their placement within sections are, then, guided by both their function and their physical similarity. Arrangement on linguistic grounds, though perhaps present in some cases, is clearly secondary. In the chapters on flora and fauna, the *Mirror* looks more like an encyclopedia than a dictionary, and arguably a very Chinese encyclopedia at that.

Tacit Borrowings from Earlier Chinese Scholarship

Manchu lexicography was a relatively new pursuit when the *Mirror* was compiled, but a few works had already appeared by the time the court scholars began work in earnest.[14] Sangge's *Manchu and Chinese Classified Writings* (*Man-Han leishu* 滿漢類書 | *Man han lei šu bithe*) from 1700, which might very well have been consulted by the *Mirror*'s compilers, divided flora and fauna into the categories "Fruits and trees" (Ma. *tubihe moo*; Ch. *guo mu* 果木), "Flowers, herbs, and vegetables" (Ma. *ilga orho sogi*; Ch. *hua hui caishu* 花卉菜蔬), "Birds and beasts" (Ma. *gasha gurgu*; Ch. *niao shou* 鳥獸), "Horses" (Ma. *morin*; Ch. *ma* 馬), and "Fish and bugs" (Ma. *nimaha umiyaha*; Ch. *yu chong* 魚虫).[15] The other book, *Broadly Collected Complete Text in the Standard Script* (*Tongwen guanghui quanshu* 同文廣彙全書 | *Tung wen guwang lei ciowan šu*)

---

13. All of the gourds listed here, as well as all of the willows, are nouns or noun phrases with *gua* 'gourd' and *liu* 柳 'willow' as their head in Chinese. It cannot be excluded that the Chinese terminology used to refer to these plants influenced the arrangement.

14. Söderblom Saarela, *Early Modern Travels*, 133.

15. Sangge, *Man-Han leishu*; Söderblom Saarela, *Early Modern Travels*, 134, 139.

from 1693, used the categories "Rice and cereals" (*migu* 米穀), "Vegetables" (*caishu* 菜蔬), "Fruits" (*guopin* 菓品), and then, later in the book, "Flying fowl" (*feiqin* 飛禽), "Running beasts" (*zoushou* 走獸), "Fish and bugs" (*yu chong* 魚虫), "Trees" (*shumu* 樹木), and "Flowers and herbs" (*hua cao* 花草).[16]

It is probable that books with such an arrangement circulated already before their publication in the decades around the turn of the eighteenth century, as indicated in Shen Qiliang's dictionary, which was arranged by the spelling of the Manchu lemmata rather than by topic. In the entry for *lefu* 'bear', Shen included the following the example sentence:

> The lion, elephant, tiger, leopard, [black?] bear, brown bear, seal, roe deer and deer, red wolf and wolf, corsac fox, otter, lynx, zeren, boar, and sow are all running beasts.
>
> *arsalan, jufan* [read: *sufan*], *tasha, yarha, lefu, bi, fethi, sirhan* [read: *sirgan*, i.e., *sirga*], *buhū, jarhū, niohe, kirsa, hailun, silun, jerin* [read: *jeren*], *aidaha* [read: *aidagan*], *sakda be gemu faksime* [read: *feksime*], *gurgu kai,* 獅象虎豹熊羆海豹獐鹿豺狼沙狐水獺猞狸黃羊牙猪母猪皆走獸也。[17]

This example sentence restates the contents of the category "Running beasts," which is used in *Broadly Collected Complete Text in the Standard Script*. The number, order, and Manchu names for the animals it lists are not identical to the eponymous section in *Broadly Collected Complete Text in the Standard Script*. Yet it is not invented out of thin air either. The sequence "tiger, leopard, black bear, brown bear" at least has a long history, occurring in the ancient, layered *Shanhai jing* and having antecedents in even older literature.[18] We have not found the full sequence of animals listed in this order in an earlier work, however, which suggests that Shen's source might have been a manuscript that

16. A-dun, *Tongwen guanghui quanshu*.
17. Hayata and Teramura, *Daishin zensho*, vol. 1, 159.
18. Yuan, *Shanhai jing jiaozhu*, 343–44; for a translation, see Mathieu, *Étude sur la mythologie*, 2:528; on *Shanhai jing jiaozhu*, see Fracasso, "Shan hai ching."

was never published. It seems, then, that topically arranged vocabularies circulated in both manuscript and print in Beijing in the period when Kangxi's court scholars worked on the *Mirror*, and that they probably consulted such works.

Both of the earlier thematic dictionaries were much less voluminous than Kangxi's great reference work. These dictionaries, like the vocabularies for foreign languages that the Ming and Qing governments produced for the handling of foreign embassies, had a thematic arrangement ultimately based on ancient Chinese models.[19] The *Mirror*'s compilers were certainly aware of the existing Manchu-Chinese dictionaries, even though they did not mention them. Yet it is also clear that Kangxi's court scholars directly consulted the Chinese canonical sources when compiling the new Manchu lexicon.

As books that arranged lexical material according to subject matter, the *Mirror* and its immediate Manchu-Chinese predecessors were similar to the genre that had its earliest representative in the ancient work *Erya*. This and other ancient Chinese lexica did not attempt to unravel "nature as a biological texture." Rather, these books expressed "the belief that the whole world could be explained by means of graphs and categorized through lexicographic clarification and textual exegesis."[20] Yet in later times, Chinese natural-historical scholarship used them frequently. *Erya* and other ancient works were similarly used to forge the lexicographic encyclopedism that we see in Kangxi's *Mirror* and other scholarly projects in Chinese compiled around the same time, as we discuss in chapter 3.

The last seven of *Erya*'s nineteen chapters contain flora and fauna. They include "Herbs" (*cao* 草, covering grasses, herbs, and vegetables), "Trees" (*mu* 木, covering trees and bushes), "Invertebrates and creeping creatures" (*chong* 蟲, covering insects, spiders, reptiles, etc.), "Aquatic creatures" (*yu* 魚, covering fish, amphibians, and crustaceans), "Wildfowl" (*niao* 鳥, covering birds and flying creatures), "Wild animals"

---

19. Chunhua, "*Yuzhi Qingwen jian* leimu," 110–11.
20. Sterckx, "Animal Classification," 34.

(*shou* 獸, covering quadruped beasts), and "Domestic animals" (*chu* 畜, the coverage also includes poultry).[21]

Similarities with *Erya* itself are obvious in the arrangement of the flora and fauna chapters of the *Mirror*, but its division into chapters and sections is more refined and reflects later developments in the Chinese lexicographical and encyclopedic tradition. Indeed, Chunhua 春花 has shown that the *Mirror*'s compilers drew on the voluminous Chinese encyclopedia *Taiping yulan* 太平御覽 (Imperial digest [from the era of Rousing the State through] Great Tranquility) from 977 through 983 CE in the structuring of the new Manchu reference work.[22]

The entries in *Taiping yulan*, in accordance with the genre of "classified writings," were substantially longer than those in the *Mirror*. Its entries consisted of excerpts from earlier literature. It is probable that the *Mirror*'s compilers relied on these excerpts when writing some of the entries on flora and fauna, but it is also possible that they acquired those sources from the original texts themselves. Regardless, the earlier topical Manchu-Chinese dictionaries *Manchu and Chinese Classified Writings* and *Broadly Collected Complete Text in the Standard Script* could not have served as sources for the more discursive of the *Mirror*'s entries because the earlier books merely provided translations of single Manchu or Chinese words and phrases, not definitions and definitely not quotations.

Quotes from *Erya* are among the material that the *Mirror*'s compilers might have gotten secondhand through *Taiping yulan*. We find such quotes in the definitions of *gasha* 'bird' and *gurgu* 'beast'. At the end of the chapter "Wildfowl" in *Erya*, the text contrasts birds and beasts by how their bodies are covered and their number of feet: "[That which has] two feet and feathers is called a bird, [that which has] four feet and hair is called a beast" 二足而羽謂之禽，四足而毛謂之獸.[23]

---

21. Translations and descriptions of chapters based on Coblin, "Introductory Study," 7; Valenti, "Biological Classification," 30. For a discussion on animal classification in China from antiquity onward, see Siebert, "Klassen und Hierarchien."
22. Chunhua, "*Yuzhi Qingwen jian* leimu."
23. Ruan, *Erya zhu shu*, 84, middle panel (2650 in series pagination, chap. 10 in original). See also Carr, "Linguistic Study," 75.

The *Mirror* echoes these definitions by writing "[That which has] two feet and feathers is called a bird" (*juwe bethe, dethe funggala bisirengge be, gasha sembi*) under *gasha* and "[That which has] four feet and fur all over is called a beast" (*duin bethe, funiyehe nohongge be, gurgu sembi*) under *gurgu*.[24] *Taiping yulan* quotes these passages in places where someone researching "beasts" and "birds" was likely to look, so it is certainly possible that the *Mirror*'s compilers got them from there.[25]

In addition, the *Mirror* restates the *Erya*'s identification of "six domestic animals" (*liu chu* 六畜) as "horses, cattle, sheep, pigs, dogs, and chicken" (*morin, ihan, honin, ulgiyan, indahūn, coko*). However, the Manchu dictionary makes the addition that "domesticated geese, ducks, and other animals" (*boode ujiha niongniyaha, niyehe-i jergi ergengge jaka*) are also called "domestic animals" (*ujima*).[26] The compilers did thus not simply copy the ancient sources, but instead used them creatively. That is especially clear here because the term for "animal" used in this case, *ergengge jaka* 'thing endowed with breath', is a word with Buddhist overtones (cf. Ch. *shengling* 生靈, its translation in later official dictionaries of Manchu).[27] The phrase *ergengge jaka* does thus not appear to be a translation from a Chinese text of the classical period.

In these entries, the *Mirror*'s compilers tacitly used the definition given in *Erya*, with or without further elaboration. The entries for categories of plants of animals similarly presented information that ultimately derived from ancient Chinese texts. Such is the case with the definition of *nimaha* 'fish'. The entry first gives a basic description of fish—"scaled creatures living in water" (*muke de banjiha esihengge jergi jaka*)—and then provides further details on their appearance. Finally, it relates this interesting piece of information: "When the moon is full, their brain is full. When the moon is shrunken, their brain is shrunken" (*biya jaluci, fehi jalumbi, biya ekiyeci fehi ekiyembi*).

24. *Han-i araha manju gisun*, 1708, 19:*gasha-i hacin*:24b, *gurgu-i hacin*:42a.
25. *Taiping yulan*, vol. 901, 889:1b (24, bottom panel), 914:4a (179, top panel).
26. *Han-i araha manju gisun*, 1708, 20:*eiten ujima hacin*:1a, s.v. *ujima*.
27. Corff et al., *Auf kaiserlichen Befehl*, vol. 2, 945 (item 4316.3).

The *Mirror* does not say it, but the *locus classicus* for the idea that the brains of fish wax and wane with the phases of the moon is found in *Huainanzi* 淮南子. This book is "a collection of essays, resulting from the scholarly debates that took place under the patronage and at the court of Liu An 劉安 (?179–122 [BCE]) . . . sometime before 139 [BCE]."[28] In the astronomical chapter of *Huainanzi*, it says that "the moon is ancestor of the Yin, which is why when the moon wanes the brains of fishes diminish" 月者，陰之宗也，是以月虛而魚腦減.[29] The passage was quoted in *Taiping yulan*'s entry for "Fish, first part" (*yu shang* 魚上), but hardly in a prominent position among the many sources excerpted there.[30] Thus, though the route of transmission is unclear, the ancient idea that the brains of fish grow with the moon made it into the *Mirror*.

An influence of the Chinese scholarly tradition is seen elsewhere in the chapters on plants and animals as well. For example, the section "Dragons and snakes" begins with *muduri* 'dragon', which is said to be "first among the scaled creatures" (*esihengge jaka-i dorgi uju*). This phrasing goes back to *Da Dai liji* 大戴禮記 (Record of rites of the elder Dai), a compilation probably from the second century CE that, however, incorporates older material.[31] One passage in this text says that "the essence (or most representative example) . . . of the scaly [animals] is the dragon" 鱗蟲之精者曰龍, and another that "of scaly animals [there are] 360 kinds and the dragon is their headman (*zhang*)" 有鱗之蟲三百六十，而蛟龍為之長.[32] The second passage in particular is similar to the phrasing in the *Mirror*, with the original's *zhang* paralleling the Manchu *uju*. The compilers could have gleaned this from *Taiping yulan*.[33]

---

28. Le Blanc, "*Huai nan tzu* 淮南子," 189.

29. *Huainan honglie*, 531 (top panel, 3:2a). Translation from Graham, *Disputers of the Tao*, 333, italics removed. See also Sterckx, "Animal Classification," 44.

30. *Taiping yulan*, vol. 901, 935:9a (347, top panel).

31. Riegel, "*Ta Tai Li Chi*," 456.

32. First quote: *Da Dai liji*, 5:17a (458, bottom panel). Translation from Needham and Wang, *History of Scientific Thought*, 2:269. Second quote: *Da Dai liji*, 13:22a (539, bottom panel). Translation from Needham and Wang, *History of Scientific Thought*, 2:271.

33. Both passages were cited in *Taiping yulan*, the first under *chong* 蟲 and the second, tellingly, under *long* 龍: *Taiping yulan*, vol. 901, 929:2a (304, top panel), 944:1b (403, top panel).

Chinese Scholarship Quoted in the *Mirror*

These examples all involve tacit use of classical Chinese texts, perhaps mediated through *Taiping yulan*. In addition to such unacknowledged influences, the *Mirror* includes numerous direct quotations. In fact, one of the characteristics of its entries is the frequent inclusion of a quote at the end from a most often canonical Chinese text. Many of the quoted texts had already appeared in Manchu translation, but in some cases were evidently translated in order to include them in the dictionary. The reason for including them is somewhat elusive. Because the texts they are drawn from are all translations (either previously published or made expressly for inclusion in a particular entry), they do not exemplify earlier attestations of the Manchu word in question, as was the case in contemporary Chinese dictionaries. The quotes were probably intended to lend a literary air to the *Mirror* by making it similar to comparable Chinese books. Furthermore, as Tsai Ming-che points out, they "in themselves contributed to the standardization of a corresponding Manchu-Chinese terminology" 本身也将满汉文对译更加制式化, even if they in the process offered example sentences that inaccurately presented Manchu words as the doubles of their common Chinese translations, ignoring the particularities of Manchu concepts such as *mukūn* and *hala* (two degrees of consanguinity in the Manchu clan system).[34]

In the chapters for plants and animals, texts frequently cited include the *Poetry Classic*—a commonly cited book overall—and the chapter "Monthly ordinances" (*yueling* 月令) from *Liji* 禮記 (Record of rites).[35] The *Poetry Classic* was known, as mentioned, as a rich source for plant and animal names. The "Monthly ordinances," furthermore, illustrates the cosmography of five evolutionary phases that was current in ancient China.[36]

Despite their relevance for animal taxonomy, the quotes can be quite obscure and do not provide much natural-historical information. The entry on *nasan hengke* 'salting gourd', or pickling gherkin, ends with

---

34. Tsai, "Qing qianqi," 218–19.
35. Yu, "*Yuzhi Qingwen jian*," 58–59.
36. Carr, "Linguistic Study," 54.

such a quote: "In the 'Dong shan' [東山] chapter of 'Bin feng' [豳風] of the *Poetry Classic* it says: 'The salting gourd (*nasan hengke*), indeed it hung from the eaves'" (*ši ging ni bin fung ni dung šan fiyelen de, nasan hengke inu sihin de sirenehe sehebi*).[37]

The quote is a translation of the Chinese "The fruit of the kuo-lo gourds (viz. snake gourd) | reach the eaves" 果臝之實,亦施于宇.[38] Even though the Manchu lemma *nasan hengke* occurs in the Manchu translation of the *Poetry Classic*, this word is arguably a poor translation of the original Chinese. The obscure passage in the original Chinese probably did not refer to a pickling gherkin. Moreover, the language of the Manchu version reads as a kind of translationese and could hardly serve as a model for someone seeking to use the word *nasan hengke* in composition.

Quotes from other literary sources can be obscure as well and do not always properly speaking include the lemma that is being illustrated. Consider, for example, the quote that ends the definition of *jancuhūn hengke* 'sweet melon': "In the 'Shuo gua' [說卦] commentary [*juwan*, i.e., *zhuan* 傳] on the [trigram] *gen* [艮] of the *Change Classic*, it says that it 'is fruits and gourds'" (*i ging ni šo guwa juwan de, gen, tubihe hengke ohobi sehebi*).[39]

The short quote *tubihe hengke ohobi* 'is fruits and gourds' is a translation of the Chinese *wei guo luo* 為果蓏 'is fruits and seeds'.[40] The quote does not include the lemma *jancuhūn hengke* 'sweet melon', only *hengke* 'melon' or 'gourd'. Now, Kong Yingda 孔穎達 (574–648 CE), the

---

37. *Han-i araha manju gisun*, 1708, 18:*sogi booha-i hacin*:7a.

38. This is a quote from *Han-i araha Ši ging bithe* (vol. 8, n.p.), in the poem "Dung šan"; Karlgren, *Book of Odes*, 101. We have translated *guoluo* (kuo-lo) 果臝 as "snake gourd." This is based on the conventional identification of *guoluo* with *guolou* 栝樓 'Chinese snake gourd' (*Trichosanthes kirilowii*). See Ruan, *Mao Shi zhengyi*, 129, bottom panel (397 in series; chap. 8.2 in original). Shen Qiliang translates *nasan hengke* as *wanggua* 王瓜 (Hayata and Teramura, *Daishin zensho*, vol. 2, 296). This Chinese word was also used in later dictionaries in the tradition of the *Mirror of the Manchu Language*; for example, Corff and colleagues in *Auf kaiserlichen Befehl* (vol. 2, 817 [lemma 3773.2]) note that *wanggua* is probably a variant of *huanggua* 黃瓜 'cucumber'. They translate it as *Salzgurke*.

39. *Han-i araha manju gisun*, 1708, 18:*tubihe-i hacin*:50a. See also Corff et al., *Auf kaiserlichen Befehl*, vol. 2, 862 (lemma 3980.1).

40. Wilhelm, *I Ching or Book of Changes*, 299.

author of a canonical commentary on the *Change Classic*, says that "*guo* is the fruit of trees, *luo* is the fruit of vines" 木實為果,草實為蓏.[41] It is this commentary, which is not quoted in the Manchu dictionary entry, that the Manchu translation appears to follow: *luo* 'seeds' are produced by vines, as are gourds, thus *luo* might be taken as a metonymy for 'gourd' or 'melon'. But how does that help the reader of the *Mirror*?

This type of quote is not interesting when considering the *Mirror* as a piece of natural-historical scholarship. There is no connection between the source quoted and the thing described, other than the word in question appearing in the quote, if even that. On a few occasions, however, the *Mirror* quotes books that actually say something about the plant or animal that is the subject of the entry. The information can be literary or, more interestingly, natural historical, but it is always Chinese.

We find a few quotes from Chinese encyclopedias. Consider the entry for *modan* [< Ch. *mudan* 牡丹] *ilha* 'wooden peony'. After a description of its appearance and fragrant smell, the entry goes on to specify that "in Chinese it is called *mu dan*" (*nikan mu dan sembi*), using a transcription that is closer to the Beijing Mandarin of the time than the older transcription used in Manchu (*modan*). It follows this up with a quote:

> In *Qianque* [*ju*] *leishu*, it says,
> "The wooden peony is the king of flowers; the Chinese peony [*šodan* < Ch. *shao* 芍 + *-dan* from *modan*] is the prime minister of flowers."[42]
> *ciyan kiyo lei šu bithede, mudan be, ilha-i wang, šodan be, ilha-i dzaisiyang sehebi.*[43]

The saying that the wooden peony (*mudan*) is king and the Chinese peony (*shaoyao*) prime minister is well known, but the book quoted

---

41. Ruan, *Zhou Yi zhengyi*, 83, middle panel (chap. 9).
42. Schmidt, "Chinesische Elemente," 1933, 401.
43. *Han-i araha manju gisun*, 1708, 19:*ilha-i hacin*:22a.

here is a late-Ming encyclopedia.[44] However, the quote is literary rather than natural-historiographic.

Furthermore, in two instances—*kilin* 'unicorn' and *buhū* 'deer'—the *Mirror* quotes a text called *šui ing tu*, which might refer to Sun Rouzhi's 孫柔之 (fl. 6th c. CE) *Ruiying tu* 瑞應圖 (Illustrated compendium of auspicious responses), a text only known in fragments by the late imperial period. Although *Taiping yulan* quoted *Ruiying tu* under *lu* 鹿 'deer', the quote in the *Mirror* is not an exact match to it or to the reconstructed *Ruiying tu*.[45] The information related in the quotes concerns the political symbolism of the animals in question.

A few other quotes are more interesting for our purposes here. We mentioned earlier that uses of the "small red snake gourd" and the "bitter melon" were explained in the dictionary in an interesting way. In both cases, the use is explained using explicit quotes from Li Shizhen's 李時珍 (1518–1593) *Bencao gangmu* 本草綱目. Here, the source quoted is a work of the pharmacopoeia tradition, and it is precisely in that capacity that the *Mirror* quotes it while adapting it to its own purpose.

Following the descriptions of the appearance of the "small red snake gourd" and the "bitter melon," the *Mirror*'s compilers present the following quotations. Regarding the small red snake gourd,

---

44. Yu, "*Yuzhi Qingwen jian*," 58. The Manchu transcription cites a Chinese source *ciyan kiyo lei šu bithe*, which we have traced to a late Ming *leishu*. See Chen Renxi, *Qianque ju leishu*, 97:26b.

45. *Han-i araha manju gisun*, 1708, 19:*gurgu-i hacin*:43b; *Taiping yulan*, vol. 901, 906:9a (129, bottom panel). On the *Ruiying tu*, see Sturman, "Cranes above Kaifeng," 38. The reconstructed text is found in Sun, *Ruiying tu*, 4:19b–20a (2979, lower panel; 2980, upper panel). The use of a different edition, yet to be identified, remains a possibility in this case. This possibility appears all the more likely since *Taiping yulan* does not quote *Ruiying tu* at all under *qilin* 麒麟. See *Taiping yulan*, vol. 901, 889:9a–14a (28, bottom panel; 31, upper panel). *Han-i araha manju gisun*, 1708, 19:*gurgu-i hacin*:42a. Sun, *Ruiying tu*, 4:17b–18a (2978, lower panel; 2979, upper panel); cf. Yu ("*Yuzhi Qingwen jian*," 58), who imputes the Chinese characters *shui ying tu* 水英图 to the Manchu transcription.

In *Bencao gangmu*, it says:
"*gaha hengke* is called *chibaozi* [赤雹子], as well as *wanggua* [王瓜] and *ye tiangua* [野甜瓜]. Its root cures diseases such as thirst [disease, i.e., diabetes] and clotting of the blood. Its seeds cure diseases such as the coughing of blood and dysentery."

ben tsoo g'ang mu bithede, gaha hengke be, cy bao dz, geli wang guwa, ye tiyan guwa sembi. erei fulehe, kangkara, senggi melmenehe jergi nimeku be dasambi. use, senggi kaksire, ilhi hefeliyenere jergi nimeku be dasambi sehebi.[46]

Bencao gangmu does not present the alternative names of the plant in the order given here. The headword in *Bencao gangmu* is *wanggua*. It also gives other synonyms that are not listed here. Similarly, the description of the medical application of the gourd is more elaborate in *Bencao gangmu*.

In the entry for bitter melon we read,

In *Bencao gangmu* it says:
"*gaha oton* is called *laigua* [癩瓜], as well as *lai putao* [癩葡萄], *kugua* [苦瓜], and *jin lizhi* [錦荔枝]. It cures illnesses such as fevers and stomach burn. Its seeds nourish one's *qi*."

ben tsoo g'ang mu bithede, gaha oton be, lai guwa, geli lai pu too, ku guwa, gin li jy sembi. wenjere haksara jergi nimeku be dasambi. erei use, sukdun niyecebumbi sehebi.[47]

Bencao gangmu presents the synonyms for "bitter melon" in an order different from that in the quote in the *Mirror*. In Li Shizhen's book, the headword is *kugua*, which is glossed as *jin lizhi* and *lai putao*.[48]

---

46. *Han-i araha manju gisun*, 1708, 19:*orho-i hacin*:5b–6a. See also Corff et al., *Auf kaiserlichen Befehl*, vol. 2, 870 (lemma 4008.3).

47. *Han-i araha manju gisun*, 1708, 19:*orho-i hacin*:6a. See also Corff et al. in *Auf kaiserlichen Befehl* (vol. 2, 870 [lemma 4008.4]), which identifies it as *Momordica charantia* 'Bitter melon', the same gloss it gives for *lugiya hengke*.

48. Li Shizhen, *Bencao gangmu*, vol. 773, 28:21a–b (586).

The first thing to note regarding these two quotes from *Bencao gangmu* is that although they follow in succession in the *Mirror*, the two words are not listed close to each other in *Bencao gangmu*. The small red snake gourd is in the chapter "Herbs" (*cao* 草), whereas the bitter melon is listed under "Vegetables" (*cai* 菜).[49] The words' proximity here is thus not due to their arrangement in *Bencao gangmu*. On the contrary, the listing of bitter melon under "Herbs"—where Li Shizhen did not place it, as the compilers of the *Mirror* knew—was an independent editorial decision.

Moreover, it is possible that information from *Bencao gangmu* entered other entries as well, without an explicit citation of Li Shizhen's book. The entry for *monio* '(small) monkey (or macaque)' says that this short-tailed, yellowish creature "lacks a spleen in its belly; it digests what it eats by walking around." For this reason, perhaps, the small monkey was "a fickle troublemaker by nature" (*boco sohokon, uncehen foholon, hefeli dolo delihun akū, jeke jaka be yabuhai singgebumbi, banitai fiyen akū nungneku*).[50] Li Shizhen had said of the macaque (*mihou* 獼猴) that it "it has no spleen in its stomach, and digests by walking around" 腹無脾，以行消食.[51] Perhaps the editors of the *Mirror* got the information regarding the monkey's spleen from here, or from a related source.

Moreover, the *Mirror* contained an entry on *yuwan* (< Ch. *yuan* 猿) 'gibbon', whose habitat by the Qing period was restricted to the very non-Manchu southern parts of China. Details of the definition—the gibbon's "very long lifespan" (*jalgan umesi golmin*)—clearly come from Chinese sources, but this idea was so widespread that it is even harder in this case to pin it down to a specific text as the *Mirror*'s source.[52] Finally, although the point is obvious, *Bencao gangmu* postdates *Taiping yulan* by several centuries, so the inclusion of this material here has nothing to do with the compilers' use of that encyclopedia.

49. For a translation of the structure of *Bencao gangmu*, see Nappi, *Monkey and the Inkpot*, app. B.
50. *Han-i araha manju gisun*, 1708, 19:*gurgu-i hacin*:46b.
51. Li Shizhen, *Bencao gangmu*, vol. 774, 51a:36b (516, upper panel).
52. *Han-i araha manju gisun*, 1708, 19:*gurgu-i hacin*:46b; Gulik, *Gibbon in China*, 38, 92. Also on the gibbon in China, see Almonte, "Perception of Exotic Features."

The varied use of Chinese classical and more recent scholarly texts in the *Mirror* evidences a considerable degree of erudition derived from the Chinese *leishu* tradition, complemented by the earlier Manchu translations of Chinese classics. The *Mirror*'s compilers made occasional use of the pharmacopoeia tradition, but did not lavish attention on the medicinal use of plants. As we will see in chapter 3, the organizational schemes, editorial strategies, and epistemic premises on display in the *Mirror* were very much in line with earlier *leishu*, even as the tradition of *leishu* itself was going through considerable changes at the Kangxi court precisely during the years when the *Mirror* was being compiled.

Chinese Translations in the *Mirror*

In addition to the heavy use of Chinese scholarship, as discussed in the previous section, the botanical and zoological chapters in the Manchu *Mirror* include another kind of Chinese material. We mentioned that scholars have recently drawn attention to the difficulty of talking about plants and animals, whose habitat lay outside China proper, within the confines of the Chinese language in the early Qing period. Shen Qiliang, notably, listed some words relating to Manchurian nature in Manchu but was unable to provide them with a Chinese translation. The *Mirror* presents almost the inverse of that situation, or as close as one can get to it in a book written in Manchu.

The fact is that Kangxi's *Mirror*, its Manchu monolingualism notwithstanding, includes many Chinese words. They are, like the rest of the dictionary, written in the Manchu script and are thus not obvious when first browsing the book. Yet they are there, and in great number, with important consequences for how we should understand the dictionary. They have particular significance for the understanding of the flora and fauna chapters, as we will see.

When we talk about Chinese translations, we are not concerned with words that at one point or another were borrowed from Chinese into Manchu and then, as Manchu words, lemmatized in the dictionary. Some Chinese loans in Manchu were very old; some also entered the language from Mongolian. A word such as Chinese *boshi* 博士

'broadly learned scholar' (< Middle Chinese *pakṣhṛ*) might even have yielded two Manchu words: *baksi* 'scholar', via Turkic and Mongolian, and *faksi* 'craftsman', from Chinese via some other route.[53] We are not considering such words to be Chinese in the sense used here, and such words were not necessarily recognized as loanwords by Manchu speakers.

Nor are we concerned here with recent Chinese loans that are lemmatized in the dictionary, even if those words entered into the written Manchu lexicon only because they were included in the *Mirror*. A word such as *to gi* 'ostrich' falls into that category. As evidenced by its later history (explored more fully in chapter 7), this word was clearly perceived to be Chinese. Yet in Kangxi's *Mirror*, it is treated as any other Manchu word, as in the entry quoted earlier in this chapter.

The *Mirror* includes a third category quite unlike these two categories of ultimately Chinese words just discussed. These words are marked within the text of the dictionary itself. They always constitute a translation of the lemma, effectively rendering parts of the dictionary bilingual. Some of these entries are comparable to the entries with Mongolian translations listed in chapter 1. Such is the case when a Chinese word—apparently in use in Manchu—is listed as a synonym of a native word and supplied with yet another Chinese gloss. For example,

> *huhucu* 'Adenophera (bellflower)'. The vegetable called *mingzhe cai* [茗䔧菜] is called *huhucu*. [A] Chinese name is *shashen* [沙参].
> *ming je tsai se[m]e sogi be, huhucu sembi. nikan gebu ša šen sembi.*[54]

Some entries appear to have been crafted on the basis of the Chinese translations included in the definition. Consider *uncehen golmin buhū* 'long-tailed deer':

---

53. Reconstruction from Pulleyblank, *Lexicon of Reconstructed Pronunciation*. The glyph represented here by ṛ is an approximation of the one used in the lexicon. Schmidt, "Chinesische Elemente," 1932, 602; Doerfer, *Mongolo-Tungusica*, 243–44; Rozycki, *Mongol Elements in Manchu*, 23–24, 73 (casts doubt on *boshi* as origin of *faksi*).

54. *Han-i araha manju gisun*, 1708, 19:*sogi bogo-i hacin*:12b–13a.

Bigger than a deer. The tail is long. The Chinese call it *mi* [麋]. If the entire tail is boiled, then one can put the hairs together to make a fly whisk.

buhū ci amba, uncehen golmin, nikan mi sembi. uncehen be gulhun fuyefi, sika suwaliyame arfukū araci ombi.[55]

The paraphrastic name "long-tailed deer" contrasts with the many names that the inhabitants of Manchuria evidently had for different kinds of deer. Here, no such name is used, but instead a paraphrastic construction that might represent a *hapax legomenon* first introduced in this book. The cultural practice of making fly whisks, associated with the *mi* deer, named in Chinese, provides a rationale for the entry.

The Chinese translations in the *Mirror* stand out in that they are more numerous than translations into Mongolian. Yet they stand out in another regard as well. They are much more numerous in the chapters on flora and fauna than anywhere else in the dictionary. In fact, we count a full fifty-six entries describing species of plants and animals that include Chinese translations. The breakdown is uneven and quite telling.

In table 2.1, we list the number of entries with translations in relation to the total number of entries in the section in question, in increasing order of magnitude. This breakdown shows that the sections with the relatively higher percentage with translations are "Flowers," "Sea fish," and "Bugs."

The picture is similar if we compare the number of translations within a certain section with the total number of translations overall (see table 2.2). Yet, when considered this way, the dominance of the "Flowers" section disappears, and "Sea fish" and "Bugs" appear as the sections where Chinese translations are particularly numerous. In addition, there are also quite a few from "Flowers," "Herbs," "Vegetables and dishes," "Beasts," and so on.

These numbers suggest, first, that the Manchu and Chinese vocabularies for talking about marine creatures, insects, and flowers were already relatively integrated. At least lexical equivalences existed in these areas. It stands to reason that the existence of a Chinese-language

---

55. *Han-i araha manju gisun*, 1708, 19:*gurgu-i hacin*:44a.

Table 2.1 Chinese translations relative to total entries of section in 1708 *Mirror*

Section	Total entries	Number with translations	Percentage with translations
Small birds	50	1	2
Large birds	114	3	3
Trees	73	3	4
Beasts	75	4	5
Vegetables and dishes	90	5	6
Fruits	36	2	6
Herbs	76	6	8
Rice and grain	26	2	8
River fish	51	4	8
Bugs	75	8	11
Sea fish	44	13	30
Flowers	14	5	34

literature on these subject matters, reviewed in the beginning of this chapter, contributed to this integration. The recognition of this fact, however, leaves open the question why the compilers decided to add Chinese translations here.

We think that they did so probably because of their lack of confidence in their ability to provide a satisfactory description of the plants and animals in question. One reason we believe this is that six entries have only a Chinese translation—nothing else in the entry allows the reader to identify what the headword refers to.

The six botanical and zoological entries that consist of only a Chinese translation contrast markedly with the few translated entries we find elsewhere in the dictionary. For example, the chapters on medical conditions include two Chinese translations, but in both cases the entries also provide a definition in Manchu.[56] Similarly, *ku* 'soot' is both defined

---

56. *Han-i araha manju gisun*, 1708, 10:*nimere nidure hacin*:33b, s.v. *idarambi* 'to gasp for breath, to feel pain while breathing', translated as *ca ki*, i.e., *chaqi* 岔氣; 11:*yoo šugi-i hacin*:2b s.v. *šurtuku yoo* 'anal fistula', translated as *leo cuwang*, i.e., *loucang* 漏疮.

Table 2.2 Chinese translations within section relative to overall total in 1708 *Mirror*

Section	Number with translations	Percentage with translations
Small birds	1	2
Fruits	2	4
Rice and grain	2	4
Large birds	3	5
Trees	3	5
River fish	4	7
Beasts	4	7
Vegetables and dishes	5	9
Flowers	5	9
Herbs	6	11
Bugs	8	14
Sea fish	13	23
Total	56	100

and given a Chinese translation, as is *bokida* 'tassel', a few food items, and so on.[57] The translation is merely supplementary in all of these cases. In none of them does it constitute the entirety of the entry.

Yet eight flora and fauna entries have only a Chinese translation, in one occasion supplemented with a quote from the Chinese classics. Two of them are found among the "Herbs," the remaining seven among "Sea fish." First are the herbs:

[1] *kailari orho* 'oriental motherwort'. The herb that the Chinese call *yimu cao* [益母草] is called *kailari*.
kailari orho, nikasa i mu tsoo sere orho be, kailari orho sembi.[58]

57. *Han-i araha manju gisun*, 1708, 14:*tuwa šanggiyan-i hacin*:38a-b, 15:*mahala boro-i hacin*:15b, 18:*efen-i hacin*:19b, s.v. *tuhe efen* 'wheat pancake with a swirl in the dough', translated as *siowan bing*, i.e., *xuanbing* 旋餅; 19b–20a, s.v. *tampin efen* 'small rice pastry', translated as *jing el g'ao*, i.e., *zeng'er gao* 甑兒糕.
58. *Han-i araha manju gisun*, 19:*orho-i hacin*:5a; Corff et al., *Auf kaiserlichen Befehl*, vol. 2, 869 (item 4005.4).

[2] *onggoro orho* 'day lily' [lit. forgetting herb; cf. Ch. *xuancao* 諼草 idem]. The Chinese name is *hehuan* [合歡]. In the "Bo xi" [伯兮] chapter "Wei feng" [衛風] of the *Poetry Classic* it says: "How to get the forgetting herb [i.e., *xuancao*] and plant it in the garden."[59]

onggoro orho, nikan-i gebu ho hūwan sembi. ši ging ni wei fung ni be hi fiyelen de, adarame onggoro orho be bahafi amargi hūwa de tebure sehebi.[60]

Second are the seven sea fish:

[1] *kurce*. Lives in the sea, called *baigao yu* [白膏魚 'white-fat fish'] in Chinese.

kurce, mederi de banjimbi, nikan be g'ao ioi sembi.[61]

[2] *sotki* 'seabass [?]'. Lives in the sea, called *haiji yu* [海鯽魚] in Chinese.

sotki, mederi de banjimbi, nikan hai ji ioi sembi.[62]

[3] *giyaltu* 'A long and thin fish (*Trichiurus Haumela*)'. Lives in the sea, called *bai daiyu* [白帶魚] in Chinese.

giyaltu, mederi de banjimbi, nikan be dai ioi sembi.[63]

---

59. For the Chinese text and an English translation, see Legge, *Chinese Classics*, vol. 4, 105–6.

60. *Han-i araha manju gisun*, 1708, 19:*orho-i hacin*:6a-b. In addition, the following entry, *hūsiba orho* 'ivy (?)', consists of a Chinese translation (*pa šan hû* < *pashan hu* 爬山虎 'mountain-climbing tiger' [?]) and a Manchu synonym (*hedereku orho*, lit. rake herb). We have not counted this entry because it does not consist of only a translation.

61. *Han-i araha manju gisun*, 1708, 20:*mederi nimaha-i hacin*:36b–37a; Corff et al., *Auf kaiserlichen Befehl*, vol. 2, 980 (item 4487.1).

62. *Han-i araha manju gisun*, 1708, 20:*mederi nimaha-i hacin*:37a; Corff et al., *Auf kaiserlichen Befehl*, vol. 2, 980 (item 4487.2).

63. *Han-i araha manju gisun*, 1708, 20:*mederi nimaha-i hacin*:37a; Corff et al., *Auf kaiserlichen Befehl*, vol. 2, 980 (item 4487.3).

[4] *uyu* [?]. Lives in the sea, called *haiyan yu* [海鰻魚] in Chinese.
uyu, mederi de banjimbi, nikan hai yan ioi sembi.[64]

[5] *koojiha* [?]. The fish called *latun* [?] is called *koojiha*.
koojiha, la tun sere nimaha be, koojiha sembi.[65]

[6] *tama* 'sole'. Lives in the sea, called *xiedi yu* [鞋底魚] in Chinese.
tama, mederi de banjimbi. nikan hiyei di ioi sembi.[66]

[7] *ica* [?]. It comes up from the sea. The Chinese call it *miantiao yu* [麵條魚 'noodle fish'].
ica, mederi ci wesimbi. nikan miyan tiyoo ioi sembi.[67]

Although most of these entries specify that the fish in question either lives in the sea or comes up from the sea, they do not provide any new information. The entries are all placed in the section "Sea fish," so the reader will ipso facto know that the creatures live in the sea. The only new information in the entry is the Chinese translation (see fig. 2.1).

What should we make of the Chinese translations in the chapters on flora and fauna? That they are so much more numerous here suggests that these chapters, by virtue of their subject matter, are different from the rest of the dictionary, and not in the way one might at first assume. The Manchus evidently had closer contacts with Inner Asia and Manchuria than the Chinese did, and indeed this dictionary of the Manchu language reflects the natural world north of China. However, that does not mean that the compilers of the *Mirror* shared that knowledge. As we will see in part 2, northern Manchuria in particular

---

64. *Han-i araha manju gisun*, 1708, 20:*mederi nimaha-i hacin*:37a; Corff et al., *Auf kaiserlichen Befehl*, vol. 2, 980 (item 4488.2).
65. *Han-i araha manju gisun*, 1708, 20:*mederi nimaha-i hacin*:37b; Corff et al., *Auf kaiserlichen Befehl*, vol. 2, 981 (item 4491.3).
66. *Han-i araha manju gisun*, 1708, 20:*mederi nimaha-i hacin*:38a; Corff et al., *Auf kaiserlichen Befehl*, vol. 2, 981 (item 4494.1).
67. *Han-i araha manju gisun*, 1708, 20:*mederi nimaha-i hacin*:38b; Corff et al., *Auf kaiserlichen Befehl*, vol. 2, 981 (item 4494.3).

FIG. 2.1. A page with saltwater fish from the Kangxi *Mirror* (Beijing: Wuying Dian, 1708, 20: *mederi nimaha-i hacin*:37b). The definition of *koojiha* consists of only a Chinese translation (rendered in Manchu). Digital collections of the Staatsbibliothek zu Berlin – PK. http://resolver.staatsbibliothek-berlin.de/SBB0001F33E00000000.

was alien to all court intellectuals, bannermen and Chinese alike, and for some plants and animals, the Chinese-language gazetteer and travel records provided the most detailed information for a reader in the imperial capital. Although it is true that the *Mirror* describes Inner Asian flora and terrestrial macrofauna that the early Chinese lexicographer of Manchu Shen Qiliang was unable to translate into Chinese, the position of Inner Asian botanical and zoological knowledge in the *Mirror* is quite complicated. It is not simply that the *Mirror* did what earlier lexicographers could not do. On the contrary, as we saw earlier in the chapter, some of the Inner Asian information featured in the *Mirror* was clearly alien to the compilers, just as it had been to Chinese scholars such as Shen Qiliang.

The addition of Chinese translations—in some cases as a remedy for an entry that otherwise would have no definition whatsoever—similarly suggests that the compilers in Beijing were not especially familiar with the plants and animals they were trying to describe. Probably, in the cases when they gave only the Chinese translation, that translation was all they knew about that entry. The most plausible explanation is that parts of these chapters were compiled on the basis of Manchu-Chinese or Chinese-Manchu word lists. If the court indeed commissioned linguistic fieldwork among Manchus, which the prefatory material to the dictionary suggests, then such word lists might have been submitted to the compilers as the result of interviews with informants. Alternatively, the compilers were working from bilingual materials submitted from Manchuria as part of the project to compile a gazetteer for the region, as we discuss in chapter 4. Knowledge of saltwater fish, for example, might have been relatively specialized and not shared by Manchu and Chinese scholars at the court in Beijing. Presented with a bilingual word list of fish, the compilers might have had no other option but to reproduce the contents of the list while stripping it of Chinese characters. Unlike the Chinese literati sojourners and travelers in Manchuria, the *Mirror*'s compilers in Beijing would not, in that scenario, have had firsthand experience of the terrain.

This last consideration sheds further light on the note that introduced the section "Sea fish" discussed in chapter 1. We remarked that the note suggests an awareness that names for fish on the imperial periphery, in Chinese, and perhaps even in foreign languages greatly outnumbered those in Manchu. In light of the many Chinese translations and bare-bones definitions found in this section, it seems plausible that, when writing it, the compilers initially worked with a larger corpus of Chinese literature (probably including Jesuit texts) that they were ultimately incapable of completely rendering into Manchu, or bring into Manchu by associating the Chinese words with non-Manchu—but still Manchurian—vocabulary. In such a scenario, the note at the beginning of the section would be their way to excuse themselves for not entering all of their source material into the dictionary. The section on sea creatures, it follows, shows us that the scholarship that went into the *Mirror* was not so much Manchu as imperial. Both Chinese texts and vernacular Chinese discourse played an important role in the project.

In chapter 3, we further illustrate the salience of plant and animal knowledge among court scholarship in the 1700s by examining a group of Chinese texts. This new body of encyclopedic scholarship, we suggest, not only pushed against previous *leishu* traditions to impose new standards and rationale, but also later served as important references for the further modification of the Manchu lexicon in the Qianlong era.

CHAPTER THREE

Guang Qunfang pu *and the Reform of Literati Botany*

In the fifth month of 1708 (KX 47), another text concerning the study of plants was completed at the Kangxi court, its imperial preface dated only a month earlier than that of the *Mirror of the Manchu Language*.[1] Whereas the original work was titled *Qunfang pu* 群芳譜 (Catalog of myriad flowers), written by the late Ming scholar-official Wang Xiangjin 王象晉 (1561–1653, *jinshi* 1604), the Qing revision, *Guang* ("expanded, enlarged") *Qunfang pu*, replaced Wang's studio name *Er ru ting* 二如亭 (Pavilion of Double Resemblance) with the Kangxi emperor's more elegant studio name *Peiwen zhai* 佩文齋 (Studio of Literary Admiration).[2]

Serendipity alone cannot fully account for the contemporaneous completion of *Guang Qunfang pu* and the Manchu *Mirror* in 1708. We

---

1. For the dating of *Guang Qunfang pu*'s completion, see the "Imperial Preface" (*Yuzhi xu* 御製序) (KX 47/5/10), in Wang, *Yuding Peiwen*, vol. 845, *juan shou*:2b (208, upper panel). Yet this event did not appear in the *Veritable Records*, according to which the emperor returned to the Forbidden City from the Changchuan yuan 暢春園 (Flourishing spring garden) on that day and immediately left for a tour "beyond the border" (*saiwai* 塞外) the day after. This prompts the question of whether the preface had been prepared by courtiers and did not rise to the same level of importance as the Manchu *Mirror*. See *Shengzu Ren huangdi shilu*, vol. 3, 233:2a (327, upper panel). Entry dated KX 47/5/10–11.

2. Wang Xiangjin, *Er ru ting Qunfang pu*; Wang Hao, *Yuding Peiwen*.

propose that the two texts can indeed be read and analyzed together. The analytical move is justified first and foremost by the shared immediate context of the Kangxi court during the early 1700s: the newly operational publishing enterprise at the Hall of Martial Valor (Wuying Dian 武英殿),[3] a shared reference library at the Hanlin Academy, and an increasingly articulate and distinct cultural policy reiterated by the emperor himself and his acolytes. The two compilations also overlapped in their effort to name and describe plants from the newly integrated territory of northern Manchuria and, later, in the Mongolian lands. *Guang Qunfang pu* demonstrates the emperor's commitment to expanding—and dramatically revising, when necessary—the existing horizons of literati knowledge with imperial prerogatives, including in the subject matter concerning plants. In part 2, we will examine how certain plants growing in Manchuria and Inner Asia found their way into *Guang Qunfang pu* and other Chinese-language texts just as well as into the Manchu *Mirror*.

Indeed, the coincidence of *Guang Qunfang pu*'s completion along with the Manchu *Mirror* in 1708 invites us to rethink the question of botanical knowledge across linguistic divides. Both were, to be sure, among the numerous scholarly projects undertaken at the Kangxi court during the 1700s. Designated teams of Chinese and Manchu officials delivered one compendium after another, featuring prominent encyclopedic *leishu* such as the *Yuanjian leihan* 淵鑑類函 (Categorized boxes of the Profound Mirror Studio, completed in 1701 [KX 40]) as well as innovative lexicographical works such as the *Kangxi zidian* 康熙字典 (Character standard of the Kangxi Reign, completed in 1716 [KX 55]). The imperial studio name *Peiwen* adorned both the official title of *Guang Qunfang pu* and also another imperially commissioned compendium of rhymes, namely, *Peiwen yunfu* 佩文韻府 (Storehouse of Rhymes from the [Studio] of Literary Admiration), which was compiled in the period from 1704 to 1711. Given the emperor's polymathic interests, the late Kangxi court also dedicated effort to establish its authority by commissioning texts on agronomy, astronomy, mathematics, and music, often leaning heavily on the technical expertise of

---

3. Hucker, *Dictionary of Official Titles*, 574 (item 7840). Cf. Brunnert and Hagelstrom, *Present Day Political Organization*, 22 (item 94).

Jesuit missionaries at court.[4] In this context, *Guang Qunfang pu* is often mentioned in passing as but one specialized court project in the 1700s and has not received too much scholarly attention in its own right.

For our purposes here, however, it is worth noting how *Guang Qunfang pu* was the first and only example of late Kangxi court scholarship dedicated exclusively to the subject of plants, or what might retrospectively fall into the category of natural-historical erudition. The court-appointed editors not only appropriated and expanded Wang Xiangjin's *Qunfang pu* to showcase imperial grandeur, but in so doing also revised existing norms of how individual plants (and some animals, as we shall see) were discussed in an encyclopedic framework. In chapter 2, we showed that the Manchu *Mirror* drew from a deep-seated tradition of Chinese encyclopedic writing with utility at the center. Yet many of the conventions that characterized that tradition, and that were maintained in the Manchu *Mirror*, were also undergoing significant revision at the same time through works such as *Guang Qunfang pu*. The year 1708 thus marked a time when Manchu lexicography and Chinese encyclopedism both evolved under tremendous political pressure at the Qing court. The study of plants appears as a unique lens to examine this epistemic conjuncture.

Of particular interest here are the ways in which plants assumed prime attention as objects of study in *Guang Qunfang pu*, which shifted away from a cosmically resonant exploration of categories (*lei* 類 as in *leishu*) toward an exhaustive lexicographical investigation of individual names for things that started to assume a *historical* character. Although the editors of *Guang Qunfang pu* did not consult—and could not have consulted—the Manchu *Mirror*, the book set a new standard for textual and experiential investigation that would have a lasting impact on both Chinese-language encyclopedism and Manchu lexicography, at least in the arena of plant and animal knowledge.

This chapter first examines how the court-commissioned *Guang Qunfang pu* departed in many significant ways from Wang Xiangjin's original *Qunfang pu*, focusing, in turn, on the authorship, organizational schema, and finally the epistemic premises in the discussion of

---

4. See discussion on part 1 in the introduction.

each entry that stands for a single type of plant life. We then contextualize the emergence of lexicographical encyclopedism at the Kangxi court in the early 1700s by comparing *Guang Qunfang pu* with the discussion of plants in earlier court-commissioned *leishu* such as *Yuanjian leihan* and other encyclopedic projects such as the first draft of *Gujin tushu jicheng* 古今圖書集成, focusing on its early phase of compilation by the Chinese scholar Chen Menglei 陳夢雷 (1650–1741) under the patronage of a Qing prince. This chapter ends with Kangxi's reflection on the purpose of encyclopedic learning by the 1710s, which sought to realign the command over words with that over things.

Expanding *Qunfang pu*: Authorship and Authority

Guang Qunfang pu is unique among the numerous court-commissioned scholarly texts in the 1700s in that it directly follows the lead of a recently deceased author and his scholarly work in the late Ming. Indeed, one may see this text as an unmistakable imperial favor toward Wang Xiangjin's grandson, Wang Shizhen 王士禛 (1634–1711, *jinshi* 1658), who served the Qing in a long-storied official career and who had recently retired at the time when the book was published. In what would be his last collection of scholarly miscellany, the septuagenarian poet and celebrated author duly copied the imperial preface in full.[5] For Wang and his kinsmen, the completion of *Guang Qunfang pu* served as an ultimate vindication of their family honor under the new dynasty. In other words, the continued popularity and rehabilitation of *Qunfang pu* became a symbol of lineage status sustained over the traumatic losses of several family members during the Ming-Qing transition.[6] As we shall see later in this chapter, Wang Shizhen and

---

5. Wang Shizhen, *Fen gan yu hua*, 1–2 (chap. 1).
6. For Wang Shizhen's reminiscence of his illustrious lineage and his grandfather Wang Xiangjin in his nineties, see Wang Shizhen, *Chibei outan*, vol. 1, 108–9, 113–15, 140, 220, 220–26. For Wang Yuyin's 王與胤 (Xiangjin's son, Shizhen's uncle, who assisted Wang Xiangjin in editing the manuscript of *Qunfang pu*) martyrdom in 1644, see Wang Shizhen, *Chibei outan*, vol. 1, 115.

other Chinese officials strived to honor their ancestors by taking the leadership over several court-sponsored encyclopedic projects. But their efforts also helped consolidate Kangxi's image as a sagacious ruler who patronized classical scholarship and reputable scholar-official families.

The estate of the once illustrious Wang clan was also the place where Wang Xiangjin first composed *Qunfang pu* during a period of forced retirement from office.[7] Estranged from late Wanli officialdom, Wang Xiangjin devoted his time to growing food from the "more than a hundred *mu* of modest farmland" 薄田百畝 and tending to "a small garden outside the city gate" 郭門外有園一區, which according to him yielded enough food to feed his growing family of eight people. At the center of the garden, Wang had a pavilion built to entertain guests, where they sat and appreciated rows of various vegetables, verdant trees, and shrubs including pine, jujube, apricot, bamboo groves, as well as many more kinds of herbaceous flowers.[8] Wang Xiangjin came up with the idea of *Qunfang pu* based on his intimate knowledge of plants and his interest in studying references to any plants in books, especially history and literature. Working steadily through his library and gardens, Wang's manuscript eventually grew to encompass several hundred entries within a decade (1615–1625).

In 1628—and at the beginning of the Chongzhen reign (1628–1644)—Wang was reinstated back in office following the demise of his factional foes. He then held powerful official positions in the lower Yangzi region, where he befriended local luminaries and made it known that he had a great treatise on horticultural matters ready to greet more readers. The first print edition of *Qunfang pu* appeared around 1634 (in 31 *juan*), produced by the prestigious publishing enterprise operated by Mao Fengbao 毛鳳苞 (who later changed his given name to Jin 晉). The elegantly produced text (see fig. 3.1) is adorned by flowery prefaces written by a motley lineup of cultural celebrities and high officials including Chen Jiru 陳繼儒 (1558–1639);

---

7. According to Wang Xiangjin's postscript (*bayu* 跋語) dated to 1621 (Tianqi 1), his refusal to obsequiously please those in power resulted in his prolonged exclusion from office. See Wang Xiangjin, *Er ru ting Qunfang pu*, *bayu*:1a–b (852, lower panel).

8. Wang Xiangjin, *Er ru ting Qunfang pu*, "author's preface":1b (35, lower panel).

FIG. 3.1. Late Ming edition of *Qunfang pu* (Changshu: Mao Jin, 1634–1644, *Tianpu* 1:1a), opening of *juan* 1 showing editorial input and discourses on Heaven (*tian*). Courtesy of the East Asian Library and the Gest Collection, Princeton University Library.

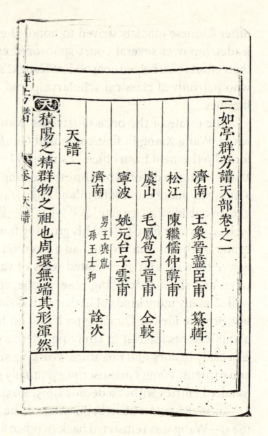

Shen Yongmao 申用懋 (sometimes written as *mao* 楙) (1560–1638), the son of former Prime Minister Shen Shixing 申時行 (1535–1614); Zhu Guosheng 朱國盛 (*jinshi* 1610); Fang Yuegong 方岳貢 (*jinshi* 1622); the Restoration-Society (Fushe 復社) leader Zhang Pu 張溥 (1602–1641); retired scholar and polymath Xia Shufang 夏樹芳 (c.1551–c.1635); and a minor official Xu Rixi 徐日曦 (*jinshi* 1622).[9] Reprints of *Qunfang pu* during the second half of the seventeenth century, however, avoided

9. There is some confusion over the precise dating of *Qunfang pu*'s early print history. The most complete late Ming copy we have seen is the one in Princeton University's Rare Book collection with all the prefaces included (call no. TC283/718). The Library of Congress holds another late Ming edition but one with fewer prefaces (LOC control no. 2012402063).

Qing censorship by omitting some prefaces written by those who became Ming loyalists.[10]

Wang Xiangjin unequivocally oriented the entire enterprise of gardening and horticultural writing around his thwarted political ambitions. Reversing the famous dictum in the *Analects*, where Confucius deferred to "old farmers" and "old gardeners" for their experience, Wang Xiangjin argued that he became "exactly like" (*ru*) experienced farmers and gardeners, and named his pavilion *Er ru ting*, meaning "Pavilion of Double Resemblance." In his words,

> To partake in the transformative and generative processes [of Heaven and Earth and stand, by virtue of this capacity, as a third term between them in the cosmic continuum], so that the old are at peace the young cherished.[11] These tasks one can carry out in one's present station in life; why leave them up to someone else?[12] Behold the harvest at my farm! Out there in the suburbs and in here at this garden, the eyes see nothing but verdant growth. To keep the eight members of my family warm and fed; this too is to partake on a small scale in [the transformative and generative processes] and in bringing peace [to the old] and cherishing [the young].

10. In this chapter, we refer to the *Qunfang pu*'s text using the much more commonly available edition reproduced in the *Siku cunmu congshu bubian* series, also available via the *Scripta Sinica* (*Hanji quanwen*) database. Yet we became aware that this so-called late Ming edition is in fact early Qing, given its inclusion of the taboo character *xuan* 玄 at the beginning of multiple volumes, and that the "great-grandsons" (*xuansun* 玄孙) of Wang Xiangjin were credited as having participated in the revision of the text. For another slightly different early Qing edition with the same taboo characters, see a copy held at Princeton University Library (call no. C283/863). For more detailed description of these editions, see Meiguo Pulinsidun daxue Dongya tushuguan, *Pulinsidun daxue tushuguan cang*, vol. 2, 543, 546. Later on, *Qunfang pu* continued to be included in Wang Shizhen's literary anthologies.

11. Regarding "generative processes": Wang is paraphrasing the *Doctrine of the Mean*. Translation from Plaks, "*Ta Hsüeh*" and "*Chung-Yung*," 44. Translation of "so that the old" from Lau, *The Analects*, 44–45.

12. Wang is again paraphrasing the *Doctrine of the Mean*. See Plaks, "*Ta Hsüeh*" and "*Chung-Yung*," 32.

叅贊化育，老安少懷。素位可行，豈異人任。子試觀吾農圃功成，出而郊坼，入而苑囿，滿目菁芿。八口溫飽，是亦一小叅贊、小安懷也。[13]

This analogy between the home and the world hinged on the dominant political culture of Ming scholar-officials, namely, that their claim to political participation was built not primarily on birth or performance in office, but on a kind of complex cultural supremacy demonstrated through learned self-cultivation (and by inference the cultivation of myriad things). Yet this asserted analogy between horticulture and politics was not meant to be interpreted literally. For all his professed familiarity with weeding, tilling, watering, and planting, Wang still relied on farmhands with "shaved heads and long beards" (*pingtou changxu* 平頭長鬚) to transplant flowers in his garden every spring and autumn.[14] The emphasis was on the symbolic—and cosmically meaningful—actions of cultivation and nurturing life, but not necessarily the valorization of the hands-on labor itself.[15]

Judging from the prefaces, *Qunfang pu* found a highly receptive audience among fellow literati and scholar-officials. Chen Jiru, aesthete and prolific author in his own right, highlighted the exalted nature of this book in that it was "not just intended only for the farm and garden," but in fact "encompassing the great men's affairs over Heaven, Earth, and Humankind" 托名農圃而大人三才之能事畢具矣.[16] Noting the difference of Wang Xiangjin's large and encyclopedic work from previous *pu* that were shorter and focusing on witticism or exoticism, Chen Jiru advocated for understanding the term *pu* with an exalted sense of family genealogies (*jiapu* 家譜)—with the detailed portraits of plant "lives" being analogous to biographies in an extended lineage.[17]

---

13. Wang Xiangjin, *Er ru ting Qunfang pu*, *bayu*: 2a–b (853, upper panel).

14. Wang Xiangjin, *Er ru ting Qunfang pu*, *juan shou*, in the section "Fragrant traces" (*fangzong* 芳踪): 23a–b (50, lower panel).

15. This serious but always philosophically resonant devotion to technical subjects is seen widely in numerous literati writings during the early seventeenth century, such as the discussion of Song Yingxing in Schäfer, *Crafting of the 10,000 Things*.

16. Wang Xiangjin, *Er ru ting Qunfang pu*, 2a, of Chen Jiru's preface (31, upper panel).

17. Lists, sometimes with lyrical and witty annotations, of flowers, exotic fruits, and domestic pets started in earnest back in the twelfth and thirteenth centuries

Pushing the analogy further, Chen evoked the familiar Confucian trope of the microcosmic relationship between the family and the polity (*guo* 國), arguing that family genealogies were "exactly like" (*you* 猶) dynastic histories (*guoshi* 國史). Here again, this cosmic analogy between the home garden and the imperial realm is established on the author's erudite and intense meditation on their mutual resonance.[18] We return to the valorization of the genre of *pu* later.

It is this cosmically situated authorship—and the sense of authority associated with it—that Qing academicians had to tackle and tame in their expansive revision effort less than a century later. According to Wang Shizhen, the temporary office (*guan* 館) for *Guang Qunfang pu* started operating in 1705 (KX 44), headed by four Hanlin academicians along with dozens of scribes and proofreaders. They completed the work swiftly in a little less than two years, and disbanded afterward.[19] Even though no singular authorial voice prevailed over *Guang Qunfang pu*, much initiative was likely to have come from the leading academician on the editorial team, Wang Hao 汪灝 (b. 1651), who received his *jinshi* degree in 1703 (KX 42) not through the usual channels of examinations, but directly via a special imperial decree.[20] Wang, who had only been a humble scholar in Huizhou Prefecture hoping to be placed in a magistrate position, allegedly owed his meteoric rise to an opportunistic presentation of a lyrical rhapsody (*fu* 賦) to the emperor during the latter's Southern Tour in 1702 (KX 41). Accompanying the emperor back to Beijing, Wang Hao performed so well in literary compositions at court that the emperor placed him in charge of many court

---

and has remained a popular genre, loosely known as *pulu*, throughout the Ming dynasty. For a history of the *pulu* genre and its keen attention to material culture, see Siebert, *Pulu*.

18. Wang Xiangjin, *Er ru ting Qunfang pu*, 2a, 4a, of Chen Jiru's preface (31, upper panel; 32, upper panel). For more discussion of the centrality of self-image in late Ming political culture, see Zhang Ying, *Confucian Image Politics*. For the role of print culture in facilitating the imagination of the domestic sphere resonating with the wider world, see He, *Home and the World*.

19. For *Guang Qunfang pu*'s starting (KX 44/6/12) and finishing (KX 46/2) dates, see Wang Shizhen, *Fen gan yu hua*, 1 (chap. 1).

20. *Shengzu Ren huangdi shilu*, vol. 3, 147 (211:18b, top panel). Entry dated KX 42/3/29.

literary projects in the 1700s.[21] It speaks to Wang Hao's stature that the three other academicians whose names appear after him on *Guang Qunfang pu*'s editorial team were all more senior in age and experience than he (Zhang Yishao 張逸少 [fl. 1687–1716], Wang Long 汪淥 [d. 1742], and Huang Longmei 黃龍眉, all of whom obtained *jinshi* degrees in 1694).[22] Aided by a team of fifteen "personnel in charge of proofreading and transcribing" (*jiaolu renyuan* 校錄人員) with various lower degrees, it was probably Wang Hao who was primarily responsible for proposing the overall editorial strategies based on which they proceeded to revise and expand *Qunfang pu*.

In contrast to Wang Xiangjin's *Qunfang pu*, which owes its publication to his close relationship with the Jiangnan literati who also promoted it, *Guang Qunfang pu* was produced by professional artisans at court but apparently paid for by voluntary donation. For unknown reasons, Liu Hao 劉灝 (*jinshi* in 1688 [KX 27], who bore no relationship to Wang Hao despite their identical given names) had been banished from the Hanlin Academy in 1700 (KX 39) and later appointed to be an itinerant censor in Henan. Probably in an attempt to regain imperial favor, Liu "voluntarily" donated funds and personally supervised the printing of *Guang Qunfang pu*, a process that probably began earlier in 1708 and concluded shortly after Kangxi wrote the imperial preface (July 18, 1708 [KX 47/6/1], preface dated June 27, 1708 [KX 47/5/10]).[23] In

---

21. See Zhao Hong'en, *Jiangnan tongzhi*, vol. 511, 167:28a (81, upper panel). For more details about Wang's good fortune, including a house, a sable fur coat, and other luxury items he received as imperial favor, see Fang and He, *Xiuning xian zhi*, vol. 3, 12:74a–b (1091–92).

22. Zhang Yishao was only able to keep his appointment at the academy as a special favor to his father, the high official Zhang Yushu 張玉書 (1642–1711). See Zhao Erxun, *Qingshi gao*, vol. 12, 3165.

23. This observation comes from a well-informed late Qing anecdotal source: Zhu, *Anle kangping shi suibi*, 166–67. For a more nuanced analysis of the mechanism of officials' publishing of imperial texts in the Kangxi era, see Weng, *Qingdai Neifu keshu yanjiu*, vol. 2, 69–74. Weng thinks that in most cases, the hefty cost was more likely to have been shared between the court and the official, and that this phenomenon persisted even after the Hall of Martial Valor (Wuying Dian) took over the lion's share of court text output. For Liu Hao's biography, which euphemized his role in *Guang Qunfang pu* and later the *Kangxi zidian* as one of "editor" (*zuanxiu* 纂修) even though his name is not found in the list of formal editors, see Tu, *Jingyang xian zhi*, 7:28b–29a (338–39).

sum, the cultural nexus between Wang Xiangjin and his literati publishers in the late Ming was replaced by the enterprising schemes of various scholar-officials to earn and maintain imperial favor.

Expanding *Qunfang pu*: Organizational Scheme

One of the most praiseworthy features of Wang Xiangjin's *Qunfang pu* in the eyes of late Ming literati readers was its wide scope that embraced the full diversity of plant life. According to Chen Jiru, *Qunfang pu* transcended artificial boundaries set forth in previous agricultural treatises (*nongshu* 農書), which tended to only feature staple crops relevant to agriculture and sericulture, or the pharmacopoeia (*bencao* 本草), which emphasized the pharmacological properties and in addition included substances of mineral and animal origins. By contrast, *Qunfang pu* not only discussed cereals and mulberry trees, but also covered in great detail hemp, cotton, tea, and miscellaneous vegetables, flowers, and grasses.

The 1708 imperial preface for *Guang Qunfang pu* endorsed the expansive view of botanical life. It begins thus:[24]

> Ever since the Divine Husbandman tasted herbs and distinguished various grains, the people started to know how to plant, grow, and cure [with medicinal herbs]. Ever since the ancient king Yiqi ordered Xihe to create a calendar so as to grant the seasonal rhythm, the people started to till the land and harvest the crops without error, and then a hundred kinds of artisanal works prospered and flourished. O how magnificent! They enabled the inception of things and the accomplishment of human affairs; they opened the gate to usher in the new beginnings of the future—such achievements of sagely emperors are comparable to Heaven and Earth.

---

24. The possibility that the leading editors participated in drafting this preface cannot be ruled out, but we still refer to it as the emperor's preface.

粵自神農氏嘗草辨穀，民始知樹藝醫藥。伊耆氏命羲和推步定歷以授時。民始知耕穫之不愆，而百工績熙。偉哉！開物成務，啟牖來茲。聖帝之功，與天地並矣。²⁵

Even though Kangxi linked *Guang Qunfang pu* to previous traditions of pharmacopoeia (through his reference to the Divine Husbandman), agricultural treatises, and calendrical sciences, his preface sought to attribute the "inception of things and the accomplishment of human affairs" to teachings originated from ancient sage kings, not the exalted principles of a cosmos subject to philosophical speculations. To elevate the supreme stature of sage kings thus came at the expense of latter-day scholars, who merely observed the processes of cosmic procreation but could never truly grasp the essential truths embedded in them.²⁶ Quite explicit is Kangxi's positioning of himself as on the same footing as the ancient sage kings, with the implicit assumption that the investigation of the "myriad variances and vicissitudes" 極萬變消長之情 in plant life for profitable human ends belongs to the proper task of the imperial authorities. In other words, one of the principle aims of *Guang Qunfang pu* is to transform plant knowledge from a meditative appreciation of cosmic principles into a means of governance with intrinsically normative intentions. Although Wang Xiangjin was able to "locate and compile the strengths of previous works" 蒐輯衆長，義類可取 and thus set up a worthy precedent in the literary tradition, it would, according to Kangxi, need to be adapted for a very different political program and brought to perfection according to the new standards.²⁷

Unsurprisingly, the Qing editors found much fault in Wang Xiangjin's way of organizing content throughout *Qunfang pu*, which they chastised in their revisions as representative of the self-aggrandizing tendency of literati learning. First, they found it unacceptable that Wang should begin with two long fascicles discussing matters pertaining to

---

25. Wang Hao, *Yuding Peiwen*, vol. 845, 1a (207, lower panel).
26. Cf. the late Ming vision for the "inception of things" as discussed in Schäfer, *Crafting of 10,000*.
27. "Imperial preface" (*Yu zhi xu* 御製序), in Wang Hao, *Yuding Peiwen, juanshou* 1b (207, lower panel).

the Heavens (*tian* 天). Seen from the cosmic worldview characteristic of literati learning, it was only natural for Wang to contemplate on the making of weather and seasonality that ultimately conditioned all cycles of plant life on earth. *Guang Qunfang pu*, by contrast, drastically reduced the discussion of macrocosmic elements, combining the two fascicles on Heaven (*tian*) and annual cycles (*sui* 歲) into one shorter treatise on "Heavenly rhythms" (*tianshi* 天時). In other words, the Qing editors argued that to understand plants one does not need to discuss at length the behavior of the sun, the moon, and the stars; the only relevant knowledge was to be found in a discussion of how climate and seasonality conditioned the growth and decay of living things. Especially suspect was Wang Xiangjin's intense interest in the ominous interpretation of weather events as auspicious or inauspicious (*zaixiang* 災祥), which the Qing editors deemed irrelevant and deleted.[28]

Second, the Qing editors ridiculed Wang for ending his book on "myriad flowers" with a somewhat peculiar "Catalog of cranes and fish" (*He yu pu* 鶴魚譜). Although these animals, largely appreciated for their elegant beauty, no doubt furnished an integral part of Wang's lifestyle in retirement, they were obviously not plants and so did not belong in *Guang Qunfang pu*.[29]

Third, the original *Qunfang pu* included an odd section titled "Fragrant traces of past luminaries" (*wangzhe fangzong* 往哲芳蹤), in which Wang "exhaustively collected" (*shou er luo zhi* 收而羅之) short biographies of worthy gentlemen, ranging from ancient hermits to contemporary luminaries including Chen Jiru. The profiles ended with two poems that Wang composed. Although Wang and his late Ming readers saw no problem in allegorically placing persons (mostly educated men) alongside flowers, cranes, and goldfish—after all, all creatures in their view embodied the "essence between Heaven and Earth" (*liang jian jingying* 兩間菁英)—the nexus of cultural power that granted legitimacy to literati political agency in the late Ming had become deeply inappropriate in the eyes of Qing officials a hundred years

---

28. Wang Hao, *Yuding Peiwen, fanli*:1a–b (211, lower panel).
29. Wang Hao, *Yuding Peiwen, fanli*:3a (212, lower panel).

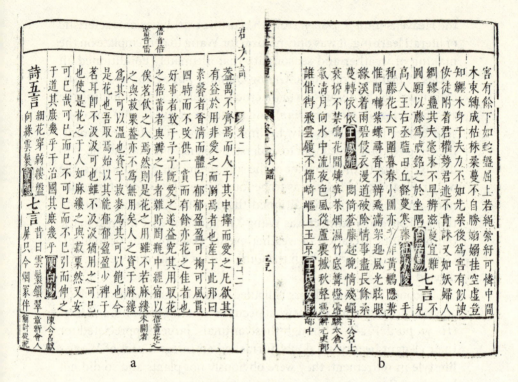

FIG. 3.2. Pages from *Qunfang pu* (Changshu: Mao Jin, 1634-1644), showing top panel glosses and bottom panel notes on biographical information of Ming authors (a, *Huapu* 2:42b), including a poem by a "maiden of the Wang family" (*Wangshi nülang*) (b, *Mupu* 2:55a). Courtesy of the East Asian Library and the Gest Collection, Princeton University Library.

later (see fig. 3.2).[30] This section was also eliminated from *Guang Qunfang pu*.

In sum, the Qing court scholars made decisive organizational changes to *Qunfang pu*, which moved the overall character of the text

30. Wang Xiangjin, *Er ru ting Qunfang pu*, 39, upper panel ("Short preface" [*xiaoxu* 小序] to the section "Fragrant traces"). Most luminaries included here were known for their literary fame and elegant lifestyle, including Tu Long 屠隆 (1542-1605) and Chen Jiru, who was referred to by his sobriquet Chen Migong 陳麋公, "Lord Moose Chen." Throughout his work, Wang paid close attention to make sure he included the full name, sobriquets, home town, and official position of the "illustrious gentlemen of our dynasty" (*guochao minggong* 國朝名公). See Wang Hao, *Yuding Peiwen, yili*:2b (37, lower panel).

away from literati general learning to something more specific: a text that claimed to be a comprehensive study of plants should be exclusively about plants, and nothing else. The unruly associative character of the original book was thereby tamed and contained, with the attention of readers redirected away from the charismatic author and toward plants themselves.

To be sure, utility to the socioeconomic life of humanity remains the primary principle by which plant life was classified in both *Qunfang pu* and *Guang Qunfang pu*. Wang Xiangjin conceived utility through the eyes of literati scholars such as himself: first, plants for food and drink—"Grains," "Vegetables," "Fruits," and "Tea"; plants for making clothing—bamboo, mulberry, hemp, kudzu, cotton; plants with medicinal uses; and finally, plants (and other creatures) to be appreciated for their beauty—"Trees," "Flowers," "Grasses," and "Cranes and fish." *Guang Qunfang pu*, however, begins with the most routinely cultivated crops for agricultural production, namely, "Grains" followed by fibrous crops.[31] Next, the expanded sections "Vegetables," "Tea," "Fruits," "Trees," "Flowers," and "Grasses" were intended to "aid the nourishment of life" (*zihou sheng* 資厚生) and "broaden the exploitation of their uses" (*pu liyong* 溥利用). Medicinal plants, now narrowed down and revised according to the "specialized books" (*zhuanshu* 專書) of *bencao* pharmacopeia, were moved to the end to showcase the court's "valuing of the people's lives" (*zhong ren ming* 重民命). As Kangxi noted in the preface, the goal of the book was decisively not to "boast of one's erudition and lavish the audience with embellished prose" 固不惟矜淹洽侈藻麗也.[32] Overall, the hierarchy of utility shifted from one imagined from the perspective of a cultured literatus toward the livelihood of the common people (*min* 民) (see table 3.1).

Having reset the agenda of plant learning as imperial science, Kangxi—and the Hanlin academicians—demonstrated the superiority

---

31. Wang Hao, *Yuding Peiwen*, vol. 845, imperial preface, 2a (208, upper panel). For more on the political iconography of tilling and weaving, see Hammers, *Imperial Patronage*.

32. Wang Hao, *Yuding Peiwen*, vol. 845, imperial preface, 2b (208, upper panel). For the importance of literary "embellished prose" in Wang Xiangjin's *Qunfang pu*, see Huang, "Yuan, pu, wenxue yu Huaxu guo."

Table 3.1 Corresponding sections and entry counts in *Qunfang pu* (QFP) and *Guang Qunfang pu* (GQFP)

QFP	Entries	GQFP	Entries
Heaven 天譜	n/a	Heavenly Seasons 天時記	n/a
Year 歲譜	n/a		
Grains 穀譜	16	Grains 穀	27, including beans
		Mulberry and Hemp (including cotton) 桑麻	7
Vegetables 蔬譜	36	Vegetables (and edible fungi) 蔬	113 (75 new entries in *juan* 17)
		Tea 茶	1
		Flowers 花	187 (116 new entries in *juan* 53)
Fruits 果譜	43	Fruits 果	91 new entries in *juan* 67
		Trees and Vines 木藤	77 new entries in *juan* 80; 90 new entries in *juan* 81
Tea 茶譜	1		
Bamboo 竹譜	1	Bamboo 竹	same
Mulberry, Hemp, and Kudzu 桑麻葛譜	5		
Cotton 棉譜	1	*merged into* Mulberry and Hemp	
Medicinal Plants 藥譜	53		
Trees 木譜	24		
Flowers 花譜	47		
Grasses 卉譜	42	Grasses 卉	180 (138 new entries in *juan* 92)
		Medicinal Plants 藥	427
Cranes and Fish 鶴魚譜		*deleted*	

of *Guang Qunfang pu* over its predecessor via the explicitly marked "added" (*zeng* 增) entries, as well as "added" sources quoted in existing entries. Likely due to the short duration of time they spent on the project (less than two years), the Qing editors did not radically rearrange entries within each section, but merely grouped their added entries toward the end of each section. As table 3.1 also shows, the newly added entries were concentrated in a few spots (*juan* 17 for added vegetables, *juan* 53 for added flowers, *juan* 67 for fruits, *juan* 80–81 for trees, *juan* 92 for herbaceous plants).

Where, then, did these added entries come from? In chapter 5, we discuss the Manchurian and Inner Asian plants that were rarely mentioned in earlier literary sources if at all. For now, we focus on the fact that the majority of added entries concerned plants mentioned in the accumulated literary tradition in classical Chinese, ranging from mythical trees that appeared in ancient texts such as *Shanhai jing* and *Zhuangzi* 莊子 to middle-period collections of jottings (*biji* 筆記), while including late Ming treatises on edible plants.

Aside from classics, dynastic histories, and literary anthologies, *Guang Qunfang pu* also surveyed a number of local gazetteers, some of which were rather recently compiled, and culled them for novelties. For example, *Guang Qunfang pu* cited *Huangshan zhi* 黃山志 (Gazetteer of the yellow mountain), which was compiled and expanded in the mid-seventeenth century, as well as updated gazetteers of Yunnan province and its prefectures.[33] The editors furthermore combed through *Da Ming yitong zhi* 大明一統志 (Unified gazetteer of the Great Ming) and added items of interest where Wang Xiangjin had neglected them. For instance, an edible sprout grown in Jiangxi was associated with a famous monk in *Da Ming yitong zhi*:

> Arhat's weed. Grows on West Mountain, Nanchang prefecture, Jiangxi. Its leaves resemble bean seedlings. It is thus named because the Exalted

---

33. Wang Hao, *Yuding Peiwen*, vol. 846, 53:22b (529, upper panel [Yellow Mountain flora]); vol. 847, 92:35a–b (477, lower panel [drug lore from Xinjin and Shixing counties]).

Monk with an Efficacious Vision brought it over from the West Mountain.

羅漢菜。出江西南昌府西山。葉如豆苗。因靈觀尊者自西山持至故名。[34]

To this brief record, however, *Guang Qunfang pu* combines a different lore from the neighboring Huguang (comprising the Qing provinces of Hubei and Hunan), likewise recorded in *Da Ming yitong zhi*:[35]

It is also found on Double-Horn Mountain, Qizhou subprefecture, Huguang. Old lore says that a wonderful monk planted it there. If one eats it with meat, then its good flavor will be lost.

湖廣蘄州二角山亦有之。舊傳有異僧所種，若雜葷物即無味。[36]

Guang Qunfang pu editors clearly intended to survey local records from various provinces and consolidate scattered records in which plants with the same name appeared, an approach that did not figure in Wang Xiangjin's original work.

It is worth noting that most encyclopedias in the tradition of "classified writings"—including *Yuanjian leihan*, completed at court in the early 1700s—rarely made much use of gazetteers, which were considered preparatory material for dynastic historiography. The local and preliminary nature of gazetteers perhaps made them appear to have little value in terms of offering good allusions and aiding literary composition, relative to the classics, official histories, and acclaimed literary works. By drawing extensively and purposefully from recently compiled gazetteers, especially from frontier regions (as we will see in chapter 5), *Guang Qunfang pu* set up a court-sanctioned precedent that was beginning to alter the character of encyclopedism as not so much a timeless encapsulation of all knowledge past and present, but as a

---

34. Li Xian, *Ming yitong zhi*, vol. 473, 49:11b (6, lower panel).

35. Hunan had been split off and made into its own province in 1676. See Guy, *Qing Governors*, 52.

36. Li Xian, *Ming yitong zhi*, vol. 473, 61:35b (276, lower panel); Wang Hao, *Yuding Peiwen*, vol. 845, 17:48b (591, upper panel).

historiographical project that must draw from the latest and immediately relevant sources, and one that would inevitably evolve in time.

In sum, the organizational vision presented in *Guang Qunfang pu* dovetails with the Qing court's political goal of resetting scholarly standards and reforming literati learning. Not only were sections rearranged and greatly expanded, the discussion of the same plant was also meticulously revised on the level of each entry, as we shall see.

Expanding *Qunfang pu*: Structure of Entries

In Wang Xiangjin's *Qunfang pu*, the design of each entry reflects the cultural view of plants as cosmic creatures closely entwined with human utility. His description of a plant begins with a short summary of names and appearances and moves on to the plant's growth cycle and auspicious or inauspicious qualities. The most extensive discussion, in Wang's words, details the cultural methods of planting, grafting, cultivation, irrigation, pruning, harvest, and so on. The following passage, for instance, gives vivid instructions on how to grow eggplants. This account, like many similar observations in Wang's book, was unique and not borrowed from previous agricultural treatises:

> Directly sow seeds in the ground during the second month. It requires fertile and nutritious soil and frequent watering. After the seedling grows into four or five leaves, transplant it along with the soil [around the root] and place them about one foot (*chi*) from each other. The root must be compactly situated in the soil, otherwise wind will enter into the crevices and cause hardship to the plant. The area around eggplants must not have loose dirt, lest rain splash up muddy residue on the leaves, which will then wither and not flourish. Transplant in sunny weather. Hoeing and good maintenance of the soil are indispensable.
> 二月下子，須肥熟地，常澆灌之。俟四五葉，帶土移栽，相離尺許。根宜築實，虛則風入難活。區土不宜有浮土，恐雨濺泥污，葉則萎而不茂。宜天晴栽，鋤治培壅，功不可缺。[37]

---

37. Wang Xiangjin, *Er ru ting Qunfang pu*, vol. 80, "Catalog of vegetables" (*shu pu* 蔬譜), 2:38b (343, upper panel).

Elsewhere in the treatise, Wang also demonstrated his familiarity with economically significant crops unique to the Shandong region. Clover (*muxu* 苜蓿), for instance, received a detailed description that has not appeared in previous works such as the *bencao* pharmacopoeia:

> Its height is one foot or so. Its slender stem grows into forked branches, its leaves resemble those of peas but smaller. Three leaves grow from one spot. Purple flowers bloom from the top tip.
> 高尺餘，細莖分叉而生，葉似豌豆頗小，每三葉攢生一處，稍間開紫花。[38]

Interestingly, Wang Xiangjin also remarked in the same paragraph that this plant was not a familiar sight for southerners.

Drawing from his experience, Wang Xiangjin models the cosmic process of ceaseless transformation by documenting not only the ways in which humans facilitated the growth of a certain plant, but also the "clinical uses" (*liaozhi* 療治) of the plant in question.[39] As we can see from the left column of table 3.2, Wang goes through the culinary and medicinal uses of a plant with extensive quotations followed by literary references. Overall, the figure of the Confucian-gardener stands central in the discussion of each plant. It is this deeply personal understanding and appreciation of the cosmic process that become the unifying thread in each entry.

Replacing this literati-centered cultural view of plants, *Guang Qunfang pu* opted for a different way of demonstrating erudition that could be said to have become more lexicographical. After all, the first classical reference invoked in Kangxi's 1708 preface is the ancient dictionary *Erya*, which sought to name the myriad things in the world with unique characters. Although Kangxi also praised the "clarity and completeness" (*mingbei* 明備) of the *bencao* pharmacopeias that

---

38. For Wang's original description, see Wang Hao, *Yuding Peiwen*, vol. 845, 14:1a–b (510, lower panel).

39. See the discussion of "pharmaceutical objecthood" in Bian, *Know Your Remedies*, 1–18.

had become the predominant genre of natural erudition since the Tang dynasty, he did not make *Guang Qunfang pu* into a new version of *bencao*.[40] To the contrary, *Guang Qunfang pu* made it clear that not all plants have medicinal uses and thus discussions of possible "clinical uses" for plants that were not classified as "Medicines" (*yao* 藥) should be removed. The Qing editors explicitly stated that they worried about the danger of readers "blindly following the recipes and mistreating people" (*ni fang yi wu* 泥方貽悮).[41] Here, it is clear that the Qing editors no longer saw all creatures as potential agents of cosmic transformation. Either a plant could be legitimately used as medicine or it could (and should) not.[42]

Overall, the Qing editors by and large did not alter or override Wang Xiangjin's comments on horticultural practices or his quotations from previous sources. Yet it is notable how they meticulously and systematically altered the order in which each component appears within an entry. As we can see in table 3.2 (right column), entries in *Guang Qunfang pu* started with a short section "Names and appearances" (*mingzhuang* 名狀), yet the key verb in this phrase was changed from "to catalog" (*pu* 譜) to "to explicate" (*shi* 釋). The term "the explication of names" alludes to an ancient lexicographical work titled *Shi ming* 釋名, which focuses on the philological and etymological analysis of common names of things.[43] True to this mission, each entry in *Guang Qunfang pu* began with a reference to *Shuowen jiezi*, another ancient lexicographical work, that glossed the name of a plant, following it with various names gleaned from earlier dictionaries and other sources. The emphasis thus shifted away from the mode of literati writing in pursuit of literary elegance and cosmic coherence, and invoked

---

40. Wang Hao, *Yuding Peiwen*, vol. 845, imperial preface, 1b (207, lower panel). Kangxi had ordered the reprint of a Ming pharmacopeia as recently as in 1701, but notably did not mention it in this imperial preface. See Bian, *Know Your Remedies*, 115–17.

41. Wang Hao, *Yuding Peiwen*, vol. 845, *fanli*:2b–3a (212, upper and lower panels).

42. Cf. discussion of "pharmaceutical objecthood" in Bian, *Know Your Remedies*, 1.

43. Miller, "*Shih ming*."

Table 3.2 Structure of entries in *Qunfang pu* (QFP) and *Guang Qunfang pu* (GQFP)

QFP	GQFP
Name and appearance 譜物名物形	Explication of name and appearance 釋其名狀
Growth cycle, divination, auspicious and inauspicious signs 譜占候, 譜休徵咎徵	"Comprehensive investigation (*huikao*)" based on collected facts (*shishi*) 次徵據事實, 統標曰彙考
Planting, grafting, cultivation, irrigation, pruning, and harvest 譜種植, 譜接插, 譜壅培灌溉, 整頓收採	
Culinary and medicinal preparation 譜製用, 譜療治	
Historical references and literary embellishments 譜典故, 譜麗藻	"Collection of Embellishments" including verse and prose on the subject 詩文題詠統標曰集藻
	Side Notes (*bielu*) including culinary (not medicinal) preparation, and methods of planting and grafting, and so on 製用移植諸法, 統標曰別錄
Appendix 譜附錄 (similar species)	Appendix 附錄 (similar species)
Pronunciations in upper register, glosses in lower register 一音釋標之上層, 訓詁列之下格	Pronunciations and glosses 音釋, 訓詁 are moved into main text as double-lined commentary

instead the tradition of *Erya* and the inquiry into the names of flora through a historical investigation of sources.[44]

Consistent with this vision, the Qing editors also adopted a more scholarly format to present glosses of pronunciation and etymology not as top-of-the-page notes but instead as double-lined commentaries directly following the character or phrase in question. As a result, the visual layout of the text in *Guang Qunfang pu* assumes a much more dense and scholastic appearance than the late Ming editions of Wang Xiangjin's text (see fig. 3.3).

44. The lexicographical character of the *Guang Qunfang pu* has been noted by Emil Bretschneider, who proposed translating the title as "Enlarged Thesaurus of Botany." See Bretschneider, *Botanicon*, 70–71.

佩文齋廣羣芳譜卷第五十三

花譜

萬年花

增 萬年花草本小朵如盞一莖百朵其色粉紅而有紅絲雖經久乾枯及沾泥汙
顏色鮮新不變

御賜名曰萬年花

金蓮

增 金蓮花出山西五臺山塞外尤多花色金黃七瓣兩層花心亦黃色碎蘂平正
有尖小長狹黃瓣環繞其心一莖數朵若蓮而小六月盛開一望徧地金色爛然
至秋花乾而不落結子如粟米而黑其葉綠色瘦尖而長五尖或七尖

彙考

〔遼史營衛志道宗每歲先幸黑山拜聖宗興宗陵賞金蓮乃幸子河避暑

〔洛陽花木記〕金蓮花出嵩山頂 〔周伯琦上都紀行詩註上都草多異花有名
金蓮花者似荷而黃 〔五臺山志山有旱金蓮如眞金挺生綠地相傳是文殊聖
蹟

御製金蓮花賦 俾嘉名於華頂結異質於清涼冠方貢之三品賦正色於中央寨芙蓉
而在陸麗菌笤於崇岡顧柳池之非偶登蘋澗之可方煥彪炳而戒文散梅檀而結

集藻

賦

Within the framework of lexicographical explication, the Qing editors displayed a zeal for research in a new, sprawling section titled "Collective investigations" (*huikao* 彙考) found in each entry. Designated as the totality of "facts as evidence" (*zhengju shishi* 徵據事實), the "collective investigations" showcased a more systematic approach to bibliographical rigor than the late Ming original. The "added" (*zeng*) quotes interspersed with Wang's "original" (*yuan*) quotes are first grouped by bibliographical category (classics followed by histories) and then within each category in chronological order. More interesting is the distinction imposed between sources representing "facts and deeds" (*shishi* 事實), which command a somewhat higher epistemic value, and "literary embellishments" (*jizao* 集藻). A final section, which did not exist in Wang's original work, is called "Side notes" (*bielu* 別錄) and relegated toward the end of each entry, where the Qing editors grouped what they considered to be the most frivolous and least relevant sources.

In his discussion of "emergence of evidential discourses" in Qing philological studies, Benjamin Elman referred to the rise of rigorous empirical research first within late Ming literati circles, and then demonstrated its further elaboration in the hands of later "Han-Learning" scholars such as Qian Daxin 錢大昕 (1728–1804) and Jiang Fan 江藩 (1761–1831). The key thread that these scholars followed in their scholarship was the "search for the truth in actual facts" (*shishi jiushi* 實事求是).[45] Yet subsequent interpretations of Elman's study have often omitted the role of Qing court scholarship in valorizing and invigorating precisely this research ethos, especially during the intervening decades between 1700 and 1760 (Ori Sela, notably, stresses the importance of government sponsorship for scholarly projects particularly in the second half of the century).[46]

Here we see an example in *Guang Qunfang pu* in which the court editors highlighted not so much the epistemic premise that a certain truth could be attained from facts (*shishi*), but instead the painstaking effort to establish them and use them as evidence (*zheng*). Although Wang Xiangjin did try to be as comprehensive as possible in his search

---

45. Elman, *Philosophy to Philology*, chap. 3.
46. Sela, *China's Philological Turn*, 10.

for all relevant quotes that mentioned, for instance, the word for Chinese chive (*jiu* 韭), the Qing editors explicitly sought to outshine Wang's ability by locating seventeen more references than the ten in Wang's original work.[47] When Wang neglected to present a quotation verbatim but simply paraphrased, the Qing editors would correct him by restoring the original quote with a complete citation. In sum, the Qing imperial library provided a crucial material setting in which *Guang Qunfang pu* and other works helped establish the "best practice" of evidential research.

Without digressing too far afield, suffice it to say here that the rise of "evidential learning" as a comprehensive ethos of research not only had multiple origins beyond philological investigation of the Confucian classics alone, but also involved the investigation of historical facts and mundane aspects of life, including plants. Whereas in Wang Xiangjin's work, Chen Jiru used the notion of history and historiography to praise the worthy members of society and establish norms of behavior, by the early eighteenth century, historical scholarship took on a more robust outlook of textual investigation beyond praising and denigrating figures past and present. In this sense, the *Guang Qunfang pu* adoption of "compiled investigations" (*huikao*) to summarize trustworthy facts represented a court-sanctioned method of historical research that was channeled through the investigation of plants.

In sum, the experience of reading each entry in *Guang Qunfang pu* led not to the appreciation of cosmic resonances that permeated all creatures as Wang's original work intended, but to the pedagogical experience of time travel through different scholarly genres, as well as the appreciation of sources as "facts" apart from literary renditions. If, in Wang Xiangjin's work, plants as the object of inquiry became enmeshed in the cultured experience of the literati self, the Qing *Guang Qunfangpu* instead rendered plants as lexicographical entities that were knowable via a judicious examination of sources that were worthy (or not so worthy) of being included as evidence. The historical dimension of plants resided not in their analogical resemblance with human characters and their embedded habitat within a cosmically resonant world,

---

47. Wang Hao, *Yuding Peiwen*, vol. 845, 13:17a–21a (498, lower panel; 500, lower panel).

but in the historicity of the accumulated textual record of plants that could be rigorously studied. That record included—as we have seen—local sources such as gazetteers.

Plants and Animals in Late Kangxi Encyclopedism: *Yuanjian leihan* and *Gujin tushu jicheng*

We have thus far suggested that the Qing court's revision of Wang Xiangjin's *Qunfang pu* resulted in a larger epistemic shift away from categorical resonance and toward a lexicographical mode of investigating names, which was in turn buttressed by a critical examination of ancient and more recent sources. Exactly how and when this shift took place, as well as the potential impact beyond *Guang Qunfang pu*, is the subject of our discussion in the rest of this chapter.

The most immediate precedent of encyclopedic texts originated from the Kangxi court in the early 1700s was *Yuanjian leihan*. Like *Guang Qunfang pu*, *Yuanjian leihan* was also inspired by a late Ming author's work.[48] An older generation of learned Chinese scholar-officials, led by Wang Shizhen and Zhang Ying 張英 (1637–1708), presented *Yuanjian leihan* to the Kangxi emperor in 1701 (KX 40). The sections on foodstuff, medicine, various kinds of plants, and animals are located toward the end of the book.[49] According to Dai Jianguo's 戴建國 count, *Yuanjian leihan* added a few hundred more "categories" (*lei*) and "entries" (*mu* 目) to the Ming original, most of which were concentrated in the sections "Frontier affairs" (*biansai* 邊塞) and "Official posts" (*sheguan* 設官), but more surprisingly in the final sections on flowers, grasses, trees but to a lesser extent animals.[50] In sum, in this immediate precedent to *Guang Qunfang pu*, we can already detect a marked interest in

---

48. The original work is *Tang lei han* 唐類函 (Categorized boxes of Tang sources) compiled by Yu Anqi 俞安期, a commoner who lived through the Wanli era (1572–1620).

49. See the table of contents (*mulu* 目錄) in Zhang Ying and Wang, *Yuding yuanjian leihan*, vol. 1, *mulu* 4:10b–30b.

50. See Dai, Yuanjian leihan *yanjiu*, 173–75, 188–93, 200–205. Additional entries include 235 to the "Frontier affairs" section, mostly ethnonyms in official histories, and ninety-two to "Official posts." Altogether, *Yuanjian leihan* added two hundred

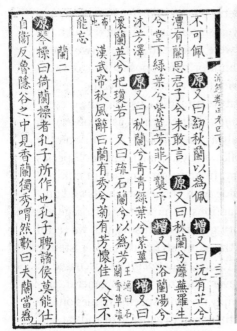 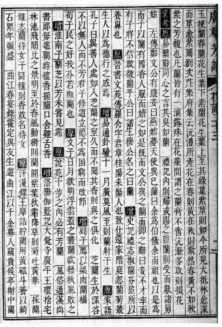

FIG. 3.4. Layout of "original" and "added" in *Yuanjian leihan* (Beijing: Nei fu, 1710, 408:21b). Courtesy of the East Asian Library and the Gest Collection, Princeton University Library.

FIG. 3.5. Layout of "original" and "added" in *Guang Qunfang pu* (Shanghai: Jinzhang tushuju, Republican period, 44:2b). Lithographic edition. Courtesy of New York Public Library.

expanding the existing encyclopedic tradition to offer a more detailed and comprehensive coverage of plants and animals.

On a closer look, *Yuanjian leihan* also anticipated *Guang Qunfang pu* in its systematic cull of references from historical and literary texts, providing a convenient overview of sources not unlike keyword searches of today. New entries and sources not seen in the Ming original were clearly marked with the character "added" (*zeng* 增), a design feature that as we have seen would be adopted in *Guang Qunfang pu* as well (see figs. 3.4 and 3.5).

---

various kinds of plants distributed in the sections "Trees," "Flowers," and "Grasses," but only twenty-four new entries in "Birds" and nineteen in "Insects."

Despite the many similarities between *Yuanjian leihan* and *Guang Qunfang pu*, completed only eight years apart at the Kangxi court, the ways in which plants and animals were discussed in these two works differed substantially. First, they stand for two opposite ways of organizing information, general categories (*lei*) versus the specific type, or even "species" (*zhong* 種), of the creature under discussion. The presentation of plants and animals in *Yuanjian leihan*, which as we observed, follows earlier *leishu* conventions, always begins with an entry on the generic category (e.g., vegetables), followed by a small number of species belonging to the preceding category. In addition, important cultural concepts are also included side by side with entries on plants and animals themselves; for example, the actions of harvest and storage are discussed under the general category grains. In this regard, we recall that the Manchu *Mirror* followed the same structure and epistemic premises foregrounding the categories (*lei*, Ma. *hacin*), as discussed in chapter 1. *Guang Qunfang pu*, however, does not discuss general concepts such as flowers or grass but simply presents specific names one by one. This tendency to enumerate, rather than to generalize and categorize, is an important feature seen in the genre of catalogs (*pu*).[51]

Second, despite its substantive attention devoted to plants, *Yuanjian leihan* prioritized a thorough review of classical and literary sources over what it perceived to be technical works. For example, although a vast literature on *materia medica* including over a thousand kinds of botanical drugs was fully available for use, *Yuanjian leihan* only discussed eighteen kinds of medicinal plants whose names appeared prominently in the Confucian classics. Similarly, *Yuanjian leihan* only included three specific kinds of vegetables (curly mallow [*kui* 葵], shepherd's purse [*ji* 薺], and allium [*cong* 蔥]), notwithstanding a much longer list readily available in agricultural treatises. *Guang Qunfang pu*, on the other hand, paid much more attention to such technical treatises, and as we have shown above, valued "factual" sources more than *belles lettres*.

A third important difference between *Yuanjian leihan* and *Guang Qunfang pu* lies in the latter's explicit emphasis on novelty, which in turn cultivated an appreciation for the "unusual/strange" (*yi* 異) that

51. Siebert, *Pulu*.

also harbored a strong sense of marvel or wonder. On reading *Guang Qunfang pu,* Wang Shizhen, one of the two leading editors for *Yuanjian leihan,* was impressed by its discussion of many flowers he considered "unusual or strange" and copied all 116 names in his last miscellany.[52] The search for the *yi* also resonated with another early 1700s encyclopedic project taking place adjacent to the Kangxi court, namely, Chen Menglei's early draft of *Gujin tushu jicheng,* as we show next.

One example of this emphasis on novelty and curiosity can be seen from the example of "Beans" (*dou* 豆, alternatively "Pulses"). *Yuanjian leihan* covered the subject with hundreds of quotations from ancient and modern sources, carefully arranged in four sections. Here we can already detect the hierarchy of factual, anecdotal, and literary sources that waxed lyrical over the humble crop. Taken together, however, *Yuanjian leihan* provides a treasure trove of references for someone intent on composing a learned investigation of the subject of beans. *Guang Qunfang pu,* by contrast, selected from this massive discourse only a few records that introduced rare names of beans. The "Great efficacious bean" (*Daling dou* 大靈豆), for instance, was discovered from a rare Song *biji* as a fantastical famine food, one kernel of which could allegedly keep one satiated for forty-nine days without additional intake of any sustenance.[53] As a result, the reader's attention shifted away from the august and elegant textual references to drift instead toward the wonderful existence of specific things out there in the world. By juxtaposing the ageless trees in *Zhuangzi* to frontier plants known only by their vernacular names, the genre of the catalog (*pu*) transformed the categorical learning in earlier *leishu* that focused on cosmic resonance into a show room of marvels retrieved from the past and present, as we explore in greater detail in chapter 5.

Shortly after or at the same time when *Yuanjian leihan* was completed at the Kangxi court, Kangxi's third son, Prince Cheng (1677–1732), whose personal name was Yinchi 胤祉 (in Chinese) or In c'y (in

---

52. Wang Shizhen, *Fen gan yu hua,* 94 (chap. 4).
53. For the general discussion of beans, see Zhang Ying and Wang Shizhen, *Yuding yuanjian leihan,* vol. 6, 395:9a–17b. For the two kinds of beans added to *Guang Qunfang pu,* see Wang Hao, *Yuding Peiwen,* vol. 845, 10:22a (446, upper panel).

Manchu), had a similar idea.[54] In 1701 (KX 40), Yinchi, who already established a reputation as an avid patron of geometry and mathematics, recruited the Fujianese scholar Chen Menglei upon the latter's recent return from exile and offered to have Chen compile a new encyclopedia. In Chen's account, he was overjoyed to have a new patron, after a long ordeal that he saw as unjustly imposed on him by his foes accusing him of treason. Recalling his discussion with Yinchi, Chen recalled the prince as saying,

> Books such as the three comprehensive references and the *Extended Meaning* [of the Great Learning] (*[Daxue] yanyi* 大學衍義, 1229) contain a lot of details on the institutions of government, but do not extend to minuscule things such as insects, fish, grasses, and trees.[55] Books such as the *Assorted Compilations* of the Tang and *Imperial Digest [from the Era of Rousing the State through] Great Tranquility*, by contrast, could only help with vocabulary and literary embellishment, but not grasp the grandeur of Heavenly virtue and the Kingly Way. How good it would be to have one book that consistently discussed the great and the small, high and low, and ancient and modern times, while sorting out the content by categories and sections and tying together everything [into an orderly system] with ropes and threads. Compiling a book in this manner would help aggrandize the cultural governance of our Sacred Dynasty.

---

54. Current dictionaries give the pronunciation *zhǐ* for the second character in the prince's name. However, as the Manchu spelling suggests, the character must have been pronounced as *chǐ* in the early eighteenth century. Consider that Yinchi had a brother, Yinzhi 胤祉 (1672–1734, where zhi is likewise *zhǐ*), whose Manchu name was In jy. Their names can hardly have been homophonous, and were indeed not, if we judge by the Manchu spelling. Furthermore, the pronunciations given in *Yuding Kangxi zidian*, vol. 230, 21:32a (403, lower panel) suggest that the normative reading in this period had the initial *ch-* rather than *zh-*. Note that Yinchi's name was changed on the Yongzheng emperor's succession to the throne. He became Yunchi in Chinese and Yūn c'y in Manchu.

55. The three comprehensive references or *santong* refer to the *Tongdian* 通典 (Comprehensive institutions, ca. 800 CE), *Tongzhi* 通志 (Comprehensive treatises, 1161), and *Wenxian tongkao* 文獻通考 (Comprehensive examination of textual and discursive sources, 1307).

三通衍義等書，詳於政典，未及蟲魚草木之微。類函、御覽諸家，但資詞藻，未及天德王道之大。必大小一貫，上下古今，類列部分，有綱有紀。勒成一書，庶足大光聖朝文治。[56]

Chen Menglei agreed to take up Yinchi's task "with a combination of ecstasy and fear" (*xi ju jiao bing* 喜懼交並).[57] Working with the prince's library in addition to his own collection of more than fifteen thousand fascicles of books, Chen immediately started to work. Yinchi also covered additional expenses and hired assistants to transcribe and make copies (*shanxie* 繕寫). Although no evidence indicates that Chen Menglei had access to *Yuanjian leihan*—which was only printed in 1710–1711, after Chen's work concluded—Chen took a similar approach to improve *leishu* by intense and exhaustive textual searching, claiming that "not a single character escaped my attention" (*zhi zi bu yi* 隻字不遺) in the thirteen Confucian classics and twenty-one histories. By 1706 (KX 45), the ten-thousand-fascicle draft was completed; Chen referred to it as simply a "comprehensive compilation" (*huibian* 彙編) in his letter to Yinchi.[58]

We know what happens next: Yinchi presented the book to his father, the emperor, who granted it a new title, *Gujin tushu jicheng*. After Kangxi's death in 1722, Yinchi lost to his fourth brother, Yinzhen, over a brutal struggle over imperial succession, and Chen Menglei was again exiled to Manchuria. The victorious Yongzheng emperor took over Chen's manuscript and had his own court scholars revise and publish *Gujin tushu jicheng* with copper movable type in 1728. This book, standing at ten thousand fascicles, remained the largest encyclopedic *leishu* in the entire Chinese tradition and is still a vital reference work for Sinology today. In chapter 5, we ponder the extent of revision done during the Yongzheng reign and its connections to *Guang Qunfang pu*. For our purposes here, it is important that already in 1706, Chen Menglei described his draft as including "six collections

---

56. Chen Menglei, *Songhe shanfang wenji*, 1416:2:4b–5a (38, upper and lower panels).
57. Chen Menglei, *Songhe shanfang wenji*, 1416:2:4b–5a (38, upper and lower panels).
58. Chen Menglei's work was carried out from the tenth month of KX 40 (1700) to the fourth month of KX 45 (1705). See Chen Menglei, *Songhe shanfang wenji*, vol. 1416, 2:5b (38, lower panel).

(*huibian*), thirty-two sections (*zhi*), and more than six thousand headings (*bu*)," which essentially remained unchanged in the Yongzheng final edition.[59]

Gujin tushu jicheng's peculiar approach to flora and fauna was thus already set in the early 1700s. True to Yinchi's vision, the encyclopedia made a point of collecting not only the grand affairs of government, but also the minute creatures that merited naming and describing. Plants (*caomu* 草木 'grass and trees') and animals (*niaoshou* 鳥獸 'birds and beasts') appear in Chen's encyclopedia as two major "sections" (*dian* 典) that in turn belong to the "Collection on the erudition of things" (*bowu huibian* 博物彙編), one of six such "collections" that made up the encyclopedia. Rather than grouping his sources under general "categories" (*lei*), Chen Menglei further divided the section on plants into seven hundred "headings" (*bu*), and the section on animals in 317 "headings." Recalling that *Gujin tushu jicheng* has altogether thirty-two "sections" and more than six thousand "headings," individual plants and animals took up an outsized share of individual headings in the encyclopedia.[60] Because of the specificity with which Chen Menglei organized his material, *Gujin tushu jicheng* started to assume a lexicographical character in the description of plants and animals, which in fact resemble *Guang Qunfang pu* much more than they do earlier *leishu* such as *Taiping yulan* and even *Yuanjian leihan*.[61]

A closer comparison between *Gujin tushu jicheng* and *Yuanjian leihan* in other aspects of their approach to plant and animal knowledge confirms this move toward lexicographical encyclopedism. First, whereas *Yuanjian leihan* still folded the human-centered *cultural* aspects (e.g., storage, cultivation) into the discussion of particular plants, *Gujin tushu jicheng* clearly separated the section "Arts and

---

59. Chen Menglei, *Songhe shanfang wenji*, vol. 1416, 2:5b (38, lower panel). *Zhi* was later changed to *dian* 典.

60. Other minutely divided sections include "Clan and family names" (more than 2,700 headings), "Foreign countries" (542), "Mountains and rivers" (401), and "Local jurisdictions" (223).

61. Similar observations to *Gujin tushu jicheng*'s "gigantic" coverage and meticulous attention to particular names have also been made in Bretschneider (*Botanicon Sinicum*, 71–72) and Needham and Lu (*Biology and Biological Technology*, 6:206–7).

techniques" (*yishu* 藝術) from those on flora and fauna.[62] Second, *Gujin tushu jicheng* went far beyond *Yuanjian leihan*'s coverage of sources to include various technical texts (e.g., *materia medica*, agricultural treatises) and more importantly local gazetteers, including recently compiled titles, as well as contemporary travelogues (for more, see chapter 5). The "Collection on the erudition of things" even included illustrations and descriptions taken from Verbiest's *Kunyu tushuo* (which, as we argue in chapter 1, might already have influenced the Manchu *Mirror*), including creatures like the toucan, rhinoceros, chameleon, salamander, beaver, hyena, and giraffe.[63] In its wide-ranging use of sources, *Gujin tushu jicheng* emphasized the validity and veracity of evidence, particularly those gleaned from recent and contemporary works.

The proximity in time and space for all four encyclopedic projects—*Yuanjian leihan*, *Gujin tushu jicheng*, *Guang Qunfang pu*, and the monolingual Manchu *Mirror*—thus points to an emergent consensus at the Kangxi court and its immediate surroundings, which sought to resituate plant and animal knowledge away from the traditional framework of *leishu* and toward a new kind of lexicographical erudition that exulted in the particularity of names and therefore things. This episode of court scholarship in the 1700s has resulted in a decisive push away from literati culture of the seventeenth century, featuring instead new ideas about rigorous research, evidential learning (*kaozheng*), and a renewed reverence toward imperial authority in the realm of scholarship.

Conclusion

By the time the Kangxi emperor finally bestowed a preface on the 450-fascicle *Yuanjian leihan* in 1710 (KX 49), he expressed clear impatience with conventional *leishu*'s exclusive focus on "literary composition"

---

62. Interestingly, the four sections included under the "Erudition of things" (*bowu huibian*) were "Arts," "The divine and extraordinary" (*shenyi* 神異, which Giles erroneously translates as "religion"), and "Plants" and "Animals."

63. Walravens, "Konrad Gessner," 91, including the notes; Zou, *Zai jian yishou*, 193–201.

(*wenju* 文句), which he did not regard as sufficient to count as "scholarship" (*xueshu* 學術):

> I have once said that although the ancient authors divided their work on governance from their literary compositions, the latter was used to articulate moral principles (*li*), whereas the former could manifest the close-at-hand principles in far-away places. How can it be called "scholarship" (*xueshu*) when one exclusively focuses on mere literary texts? Even so, literature is indeed where the principles find their place in a concentrated manner. It is not advisable to simply abolish them.
> 嘗謂古人政事文章雖出於二，然文章以言理，政事則理之發邇而見遠者也。豈僅以其區區文句之間而可以自命為學術乎？自六朝乃有類書，而尤盛於唐。此豈非求之文句之間者哉？雖然，理之所寓，於斯萃焉，弗可廢也。[64]

In rhetoric at least, the emperor construed the *leishu* tradition that culminated in Tang times as originally only intended for literary composition, and praised the new encyclopedic texts compiled under his watch as newly pertinent for "scholarship" (*xueshu*). This comment in 1710 thus cemented his effort during the previous two decades to "dominate learning from above," as Willard Peterson argued.[65] The court now felt confident enough to not only discipline literati learning, but also set out new standards that overrode long-standing traditions. The subject of encyclopedism, as well as the description of plants and animals embedded in previous *leishu*, became a particularly active domain for new imperial interventions that would have far-reaching effects later on in the eighteenth century.

In this chapter, we have used *Guang Qunfang pu* to illustrate the complex impetus for changes in scholarship during the last decades of the Kangxi reign. On the one hand, as noted in chapter 2, earlier examples of encyclopedic texts in Chinese provided an essential framework for the unprecedented fashioning of Manchu learning in this period; on the other hand, what it meant to compile an encyclopedic

64. Zhang Ying and Wang Shizhen, *Yuding yuanjian leihan*, vol. 1, imperial preface, 1a–b.
65. Peterson, "Dominating Learning."

book was likewise going through profound changes in other projects that scholars at the Qing court were quickly churning out in the 1700s.

We can by now safely conclude that despite its relatively modest title and less grandiose scope (plants only), *Guang Qunfang pu* in fact did much more than simply expand Wang Xiangjin's work on plants; it also sought to qualitatively redefine what counted as legitimate plant knowledge, and, by inference, what it meant to conduct an "investigation of things" (*gewu*). We have situated *Guang Qunfang pu* in the context of the mid- to late Kangxi years (1700–1710) and showed how it broke new ground, which rendered it different from both its Ming original text and also other court encyclopedias' discussion of plants. Specifically, we highlighted the ways in which lexicographical erudition drew attention to particular plant names and histories in *Guang Qunfang pu* at the expense of categorical coherence and literati taste. The erudition pursued in *Guang Qunfang pu* thereby became lexicographical in nature.

In his study of *Kangxi zidian*, Nathan Vedal remarked on the expressed need to "actively assert the authority of the court and delegitimize other elite and commonly used reference works." Being an important precedent that predated *Kangxi zidian* by only a couple of years, *Guang Qunfang pu* also demonstrated this urge to revise and combine previously well-recognized genres and create new reference works.[66] From the perspective of the throne, the officials in charge of compilation and publishing were all instruments—dispensable and interchangeable, if need be—that served to articulate and propagate the imperial agenda. Wang Hao's extraordinary career as editor in chief for a number of important Kangxi-era reference works ended abruptly in 1712 (KX 51), only a few years after the completion of *Guang Qunfang pu*. Implicated in the literary inquisition against Dai Mingshi, Wang Hao narrowly escaped death sentence by strangulation only at the mercy of the emperor, who had Wang's entire household demoted to servitude under the Eight Banners.[67]

---

66. Vedal, "Preferring Omission," 5.

67. The sentencing was considered lenient by the emperor on account of Wang's long service in "inner court compilation projects" (*neiting zuan xiu* 內廷纂修). See *Shengzu Ren huangdi shilu*, vol. 3, 250:3a (473, lower panel; entry dated KX 51/4/*renxu*).

The political urgency for visible, pompous action and the specific pattern of literary activities at court had rendered *Guang Qunfang pu* an appropriate and convenient project to take on. Yet, on closer scrutiny, we see how the lofty agenda articulated by the imperial preface in fact hinged on a hodgepodge of editorial decisions performed by actors who often had to improvise on very short notice. On the one hand, the short duration of the compilation process in the palace rendered it a library project restricted to the palace offices. That means the Qing editors did not have access to actual plants—unlike Wang Xiangjin, who did tend to his garden year after year. On the other hand, aside from their considerable effort to collect "textual plants" (Martina Siebert's term), the editors did try to document new species of plants even when no previous textual references could be located, including entries on Inner Asian and Manchurian plants.[68] To better understand how *Guang Qunfang pu* and other court encyclopedic texts such as *Gujin tushu jicheng* accomplished this curious blend between ancient and modern, the classical and the vernacular, and above all bookish and experiential evidence, we need to turn our attention away from the court in Beijing to see how this epistemic negotiation had already started on the margins of the Qing empire, and would ultimately inform the later permutations of Manchu lexicography during the latter half of the eighteenth century.

68. See Siebert, "Animals as Text."

PART II

Plurilingual Nomenclature in Manchuria

CHAPTER FOUR

Naming Manchurian Things in Early-Qing Chinese Writings

On the third day [of KX 27/8, 1688/8/28] we set out early across even mountains for sixty-six *li*. We camped at Zhi-xi, where mountains rise up all around. Lush grass filled the plain. Gazing out in the distance, our eyes were struck by verdant green; an uncommon sight beyond the border. From here to our outposts (Ch. *galu* < Ma. *karun*), the land belongs to the Sunud (*Seniu*).[1] The enfeoffed prince of the Sunud represents one of the forty-eight [sic] banners of the vassal states of our dynasty. In their land, the desert gradually diminishes as the grass grows thicker. It gets closer and closer to the central plain, thus the vapors of the earth gradually thicken. It is not that barren land of the Khalkha. Alas! Water and grass are things that the world easily generates; and yet, at its furthest boundaries it is lacking. If an armchair scholar discussing the investigation of things, who had not been himself to that

---

1. This interpretation follows Fang Jianchang in "Yi-liu-ba-ba nian Faguo Yesuhui shi" (47). Gerbillon appears to describe the same location. Gerbillon recorded the visit to this location as happening on August 20, eight days earlier than Zhang. Gerbillon wrote (Halde, *Description géographique, historique, chronologique*, vol. 4, 155–59) that they walked for eighty *li*, and that the camp with the stunning view over a vast plain was surrounded by mountains on all sides except to the northeast. Gerbillon described August 28 as a day of rest, on which the high officials in the party left for a hunt with the local prince (*taiki* < early Mandarin *taiki* 台吉, ultimately from Mo. *tayiji* 'prince'), who might have been the prince mentioned by Zhang. On this day, an official of the Lifan Yuan told Gerbillon and others about the forty-nine banners of the Mongols, which echoes with Zhang's recollections from that day.

land, said that the land of the Khalkha has exhausted all its water and grass, who would have believed it?

初三日，早行踰平山六十六里。[2] 駐至喜，環疊皆山，茂草盈野，遠望青翠豁目；此出塞所罕見者也。[3] 自噶祿至此為色紐地。色紐爵王為本朝屬國四十八旗之一也。其地沙磧漸少，而草生茂密。與中原漸近，故地氣漸厚。非夫喀爾喀不毛之地也。嗚呼！水草天下易生之物也，而絕塞且無之。苟非身歷其地而閉戶談格物，謂喀爾喀地絕水草，其誰信之？[4]

—Zhang Penghe in his Mongolian travelogue from 1688.

In 1682, the Kangxi emperor embarked on his second tour of Manchuria a decade after his first trip there in 1671. Among his entourage was the Hanlin academician Gao Shiqi 高士奇 (1645–1703), a southerner and learned Chinese official who duly recorded the day-to-day proceedings of the imperial tour from Beijing to Mukden—and back. In her vivid account of Kangxi's eastern tour, in part reconstructed from Gao's writings, Ruth Rogaski emphasized the dual function of the imperial visit as both pilgrimage and scientific expedition: To know Manchuria, from the Qing perspective, was also to invest meanings in the Manchu ancentral homeland "where the dragon arose."[5]

---

2. As far as we know, no critical edition of Zhang's text exists, and the different editions show variations. We have used the 1974 edition (Zhang Penghe, *Fengshi Eluosi riji*, 1974, 9:44b–45a [5816, bottom panel; 5817, top panel]), but we note variants we have seen. Moreover, the punctuation provided in the 1946 edition (Zhang Penghe, *Fengshi Eluosi riji*, 1946, 32) has been helpful. The 1985 edition has *zao* 蚤 (Zhang Penghe, *Fengshi Eluosi riji*, 1985, 97:30b [104, bottom panel]).

3. The text between Xi-zhi and this location is missing in the 1995 reprint edition (He, *Shuofang beisheng*, vol. 742, 42:29a [15, top panel]).

4. The 1995 reprint edition has *jing* 境 following *di* 地 in this sentence (He, *Shuofang beisheng*, vol. 742, 42:29a [15, top panel]).

5. Rogaski, *Knowing Manchuria*, 2022, 61–104; for Gao Shiqi's account, see 65–67, 93–96.

Toward the end of Gao Shiqi's diary is an appendix (*fulu* 附錄) that touches on less exalted subject matter than dragon veins and display of imperial martial valor. Composed simply of thirty Manchu words transcribed phonetically using Chinese characters and each described with a Chinese gloss, Gao Shiqi intended to record such "names and facts" (*mingshi* 名實) to illustrate the "difference" of the "indigenous people's everyday foodways, livelihood, and production" 土人日用飲食生殖之殊 as he observed them during his stay of ten days or so among the residents of Ula along the Sunggari River.[6] Gao Shiqi's list has not attracted much scholarly attention, aside from an early article by Giovanni Stary (1946–2022), who was mainly interested in the linguistic features that illustrate how early Qing Ula residents might have spoken and used specific Manchu words.[7] For our purpose here, however, it is of interest to see how Gao Shiqi grappled with the names of plants and animals, across the linguistic divide, with a clear ethnographic interest.

Like the Manchu *Mirror*, Gao Shiqi's word list focused on utility, featuring first homely items for the household such as eating and cooking utensils, furniture, boats, buckets, carts, candlesticks, chimneys, and so on. Starting from number seventeen on the list, Gao described a dozen or so plants and animals including grasses, fruits, water creatures, vegetables, and various foodstuffs made of grains and oil. Many of these words would later on appear in the Manchu *Mirror* but with slightly different descriptions. Taking Gao Shiqi's word list seriously would compel us to reckon with the important mediating role of language, and in particular lexicography, in discourses about Manchuria and Inner Asia's flora and fauna decades before the Kangxi *Mirror* was completed in 1708.

As chapter 2 pointed out, parts of the chapters on plants and animals in Kangxi's *Mirror* relied on vernacular Chinese sources of some kind or other. Numerous entries consist of imprecise and at times

---

6. Gao Shiqi, *Hu cong dongxun rilu*, appendix (*fulu*), 1a–4a. On Gao, see Fang Chao-ying, "Kao Shi-Ch'i."

7. Giovanni Stary studied Gao's list to reveal idiosyncrasies in pronunciation between Manchu speakers of the early Qing versus later times, noting also the occasional inclusion of Mongolian words. See Stary, "Manchu Word List."

virtually nonexistent definitions, whose sole utility stems from the presence of a Mandarin-Chinese translation. These translations are much more numerous in the chapters on flora and fauna than elsewhere in the dictionary, confirming, in a sense, Schlesinger's insight—discussed in chapter 1—that it was the natural world in particular that vexed Manchu lexicographers. Yet, unlike the case of the unfinished entries in Shen Qiliang's bilingual dictionary that Schlesinger discussed, here it is not the lack but instead the abundance of Chinese translations that suggests a difficulty on the part of the compilers.

The numerous Chinese translations, even for words referring to plants and fish that were found in Manchuria, suggest that the Inner Asian natural world did not enter into the *Mirror* only through the kind of interaction mandated by imperial decrees such as those discussed in chapter 1, which brought northern Tungusic or Nivkh words into the dictionary. On the contrary, much information on the natural world of the empire's new northern periphery reached Beijing and the court in Chinese linguistic form. We show in this chapter that, by the early 1700s, a considerable body of Chinese writings, mostly travelogues by court officials and literati exiles, offered rich descriptions of notable species in Manchuria, including their names in Manchu or vernacular Chinese. These Chinese works revealed a rising curiosity toward the material livelihoods of the Manchu and Inner Asian lands and their peoples, and anticipated efforts to fix and sanctify knowledge about Manchurian flora and fauna in the following decades. In some instances, this Chinese literature peppered with Manchu expressions provided information on natural products that were not even covered in the *Mirror*. Later in the eighteenth century, the court would make efforts to unite these disparate but partially entangled Manchu and Chinese discourses through various editorial projects that we discuss in subsequent chapters.

We begin this chapter by looking at how different early Qing Chinese sources grappled with the question of naming unknown plants and animals found in Manchuria or on the Inner Asian frontier, examining in turn the compilation of a gazetteer for the region, travelogues across the Khalkha Mongols' homeland, and last miscellaneous prose writings by Chinese scholars who by choice or by fiat found themselves in Ningguta and other outposts of Manchuria. We cross-examine all these sources to reveal how certain species emerged as foci of attention in this early stage with sometimes incongruent descriptions.

A Chinese Gazetteer for Manchuria

The production of Chinese-language texts on northern Manchuria followed the integration of the region. Decades before the *Mirror* was published, an official gazetteer describing various aspects of Manchuria had already been published in Chinese, and several Chinese individuals had described the region in their private writings. In this section, we discuss *Shengjing tongzhi* 盛京通志 (Comprehensive gazetteer of Mukden) as an early attempt to describe Manchurian nature before turning to the travelogues and notes.

The writing of local gazetteers long predated the Qing. As one authority on the topic has written, they "were a uniquely native genre, thoroughly Chinese."[8] However, the first "unified gazetteer" (*da yitong zhi* 大一統志 or *yitong zhi*) for the entirety of the imperial realm, undertaken under court auspices, was compiled under Mongol rule in the late thirteenth century, and some Mongols and Central Asians participated in that project.[9] The Chinese Ming regime that succeeded the Mongols continued to commission gazetteers.[10] Indeed, the tradition of the central government to commission gazetteers on the scale of the empire was the lasting contribution the Mongols made to the genre, as opposed to its linguistic and scholarly conventions.

---

8. Brook, "Native Identity," 243.

9. The translation of *yitong zhi* is a bit tricky. The Qing court understood it to mean *uherileme ejehe bithe* 'Book of united records'. See Ihing, *Qingwen buhui*, 2:11a. Willard Peterson translated *yitong* as "integrated domain," meaning centrally appointed administration "without delegation or toleration of inherited control over militarily autonomous regions," and traced the idea back to ancient history toward the end of the Warring States era. See Peterson, "Introduction."

10. Dennis, "Early Printing in China," 128–33; Dennis, *Writing, Publishing, and Reading*, 35–48.

Because the Qing empire already in the late seventeenth century greatly exceeded the Ming realm in size and degree of natural and—arguably—cultural diversity, the established practices of gazetteer writing could not easily yield a "unified gazetteer" for the Manchus' dominion. The Qing emperors, however, like their predecessors on the throne in Beijing, wanted such a book. *Shengjing tongzhi*, the gazetteer for Manchuria, was part of that project to extend the standardized presentation of localized knowledge to the non-Chinese territories of the empire.

Kangxi initiated a project of producing a *Da Qing yitong zhi* 大清一統志 (Unified gazetteer of the Great Qing) in 1672.[11] At the time "the project commenced, Chinese-language court scholarship on Inner Asia was effectively outsourced to a cohort of monoglot Han editorial officials" who had "difficulty securing information sufficiently detailed to meet the needs of their project."[12] It was initially not an easy task to produce and present knowledge of Manchuria in the Chinese format of the gazetteer. Yet it was in the interest of the newly established Qing court to put its homeland onto the cultural map of imperial territories, and the *Shengjing tongzhi* served its purpose to revamp the Ming record of this land as a barbarian frontier (see fig. 4.1).[13]

Information accrued. The institutions established in Manchuria produced documents in both Chinese and Manchu that were sent to Beijing. The discourse on natural products that occurs in these documents to some extent parallels that of the Chinese-language gazetteer. For example, words like *yengge* 'wild cherry' occur in Manchu documents sent to the Mukden branch of the Imperial Household Department.[14] We discuss the description of the wild cherry in Chinese

---

11. Xu Qianxue started compiling the *Yitong zhi* in the 1660s drawing from the notes by Gu Yanwu, his uncle. See Peterson, "Advancement of Learning," 542. Yongzheng later ordered provincial gazetteers to be updated in the late 1720s and at regular intervals thereafter.

12. Mosca, "Literati Rewriting of China," 97.

13. Elliott, "Limits of Tartary."

14. Other words are *siyang sui li* (*xiangshui li* 香水梨) *sulhe* 'fragrant water pear' (in 1667) and *hiyangsui* (also < *xiangshui*) *šulhe* 'fragrant water pear' (in 1741). This "fragrant water pear" is listed in the gazetteer as *xiangshui li* (under *li* 'pear'). Lateral communications from the Imperial Household Department and from its Department

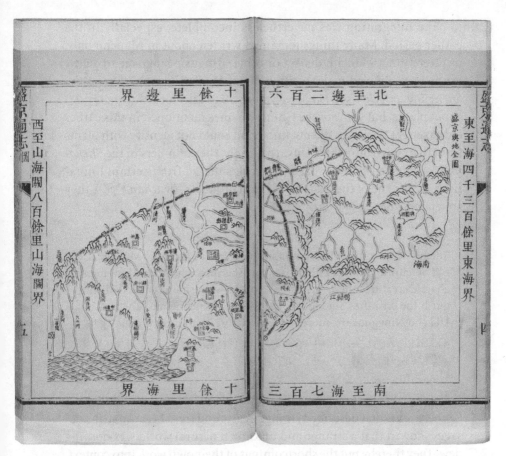

FIG. 4.1. 1684 *Shengjing tongzhi,* map (4b-5a). Digitized local gazetteers, Harvard-Yenching Library. https://id.lib.harvard.edu/alma/990080625400203941/catalog.

sources like the gazetteer in detail in chapter 5. The occurrence of such words in sources in both languages shows the integration of the Manchu and Chinese vocabulary for talking about the natural world of the northeast.

of Ceremonial (Zhangyi Si 掌儀司) to captains in Mukden, KX 6/7/16 (1667/9/3), QL 5/1/9 (1740/2/6), and QL 6/9/3 (1741/10/12), Zhao and Liaoning sheng dang'an guan, *Heitu dang: Kangxi,* vol. 1422; Zhao and Liaoning sheng dang'an guan, *Heitu dang: Qianlong,* vol. 2, 3–4, 270; *Shengjing tongzhi,* vol. 8, 21:10b, 11b.

The integration was nevertheless incomplete, especially in the earlier period. Many things remained written about in Manchu only, or uttered in a Manchu dialect or other Tungusic language to a government agent and never written down at all. The compilers of the *Shengjing tongzhi* used *Da Ming yitong zhi* to ask "locals" (*turen* 土人) about plants, but some of the Chinese names mentioned in this earlier old gazetteer the locals did not know and could not identify with plants from their environment.[15] Unsurprisingly, the compilers of the *Shengjing tongzhi* recognized that their work suffered from certain limitations. At the end of the chapter "Natural products" (*wuchan* 物產), they appended a note:

> The hidden depths of the mountains and seas of [greater] Mukden are full of natural products. There is not a fisherman who knows how many aquatic creatures there are. Not one of the woodsmen is able to distinguish the creatures among the multitude of living beings. The lands of Ula and Ningguta are, moreover, especially rich in rare [products].
> 盛京山海奧區，品物充牣，水族之夥，漁者莫識其名。芸生之繁，樵者莫辨其族。而烏喇、寧古塔之地，珍異尤多。

Following this passage, the editors quoted ancient and middle-period Chinese texts that described strange beasts and trees because such passages showed that a strange and unknown natural world was nothing new. They thereby put the shortcomings of their own work into context. Commenting on the excerpts, the editors continued,

> These are all things that have never appeared before our modern eyes and ears. We are not able now to discuss them in detail but only record the resources that can be used or made into tools, or picked out for prescriptions and medications. We leave to future encyclopedic scholars (*bowu zhe*) to record their local-language [names] (*fangyan*) and separate their different kinds.

15. *Shengjing tongzhi*, vol. 8, 21:8b.

皆非今人耳目所及見者，今不能詳舉，第紀其器用所資，方藥所取者，著其方言，區其種類，以俟博物者詳考云。[16]

The editors of *Shengjing tongzhi*, lacking the Chinese-language documentary resources available to scholars working on China proper, in this way acknowledged their inability to record the natural products of Manchuria in sufficient detail. They deflected responsibility on the native fishermen and woodsmen, who allegedly did not have perfect knowledge either. Yet their acknowledgment that the "local language" (*fangyan*) remained beyond their grasp suggests that they knew that many names and knowledge did exist in the communities of the empire's northeast, but the genre of the gazetteer and the way in which it was compiled prevented this knowledge from reaching the printed page.

Nevertheless, the editors did include a certain amount of flora and fauna that they named using words from the local languages but still associated with older Chinese literature. The *tuolu* 駝鹿 'moose', literally "camel deer," was one such word. The editors apparently got this word from Lu Dian's 陸佃 (1042–1102) *Piya* 埤雅 (Supplement to Approaching Perfection), whom they either paraphrased or quoted from using an edition (or a gazetteer or encyclopedia) unknown to us. Lu had written that

> Furthermore, among the barbarians of the north there are *mi* deer and camel deer. They are extremely big and deep green [or blue] in color. The caudal patch is yellow and without spots. These too are kinds of deer. Their antlers are big, striped, and hard and smooth like jade. Their pilose antlers can be used as well.
>
> 又北方戎狄中有麋鹿、駝鹿，極大而色蒼，尻黃而无斑，亦鹿之类，角大而有文，堅瑩如玉，其茸亦可用。[17]

Lu provided a single description for both "camel deer" and *mi* deer, effectively equating the two. The name for the latter, *mi* deer, was a

16. *Shengjing tongzhi*, vol. 8, 21:28a–b.
17. Lu Dian, *Piya*, 37.

word still current in the Chinese language of the late seventeenth century, as we see later on. The editors of the gazetteer did not mention the name *mi* deer in this context, and the description of the camel deer differed somewhat from the authoritative version of Lu's text. The editors quoted Lu as saying that it is "shaped like a camel, greenish yellow and without spots. The antlers are hard and smooth like jade, and marbled" 形似駝，色蒼黃而無斑。角堅瑩如玉，有文理. They further glossed the lemma as *kan-da-han* 堪打漢, that is, *kandahan* 'moose', and said that in their day it was found north of the Ussuri in Ningguta.[18] *Kandahan* was later included in the Manchu *Mirror* with a definition more detailed than the one in the gazetteer in some respects (description of a skin flap under the chin, shape of the antlers) but less so in others (no description of the habitat).[19]

The publication of *Shengjing tongzhi* suggests that by the time the Kangxi *Mirror* was completed in the 1700s, Manchu names of flora and fauna, especially certain species exclusively found in Inner Asia, were already not an entirely unfamiliar subject. As we will see in the next chapter, *Shengjing tongzhi* would become an important reference for the making of a much more ambitious and expansive program of imperial *gewu* that Kangxi championed himself during the last decade of his reign. At the same time, however, this cutting-edge knowledge of less-known plants and animals in Manchuria already extended to Chinese scholars beyond the immediate court circle. We now turn to travelogues from an imperially mandated trip through the Khalkha lands of Outer Mongolia on the eve of their submission to the Qing.[20]

18. *Shengjing tongzhi*, vol. 8, 21:20b.

19. Schlesinger's translator Guan Kang (*Diguo zhi qiu*, 19) translates *kandahan* in the context of the *Mirror* as *mi* deer.

20. During the decades that followed the first edition of *Shengjing tongzhi*, the Qing court explored and mapped the region of Manchuria further. In 1690, the military governor of the Amur was ordered to survey the region and submit a report to Beijing to act as source material for the planned *Da yitong zhi*. Local products and tribute items were among the topics that the Grand Secretariat in the capital wanted to know more about. The survey work was delegated to banner colonels and Solon officers throughout the region. Reports and records were reportedly kept both in Manchu and Chinese. In 1695, the military deputy lieutenant-governor of the area sent a report to the Board of War, which routed such communications. The report

A Travelogue from the Land of the Khalkha

A journey through the Khalkha lands of Outer Mongolia in 1688 happened without Kangxi's presence but on his order. The Qing court was seeking a peace settlement with Russia at a time when the Manchus' attention was taken up by the Dzungars. In the late spring of 1688, a thousand riders departed Beijing for the Russian border with Mongolia. They never reached it because the Dzunghars attacked the Khalkha that summer, and the party was ordered to turn back to China after a little less than two months on the road.[21] The journey was largely through arid and inhospitable terrain. Three members of the party wrote travelogues: the official Zhang Penghe 張鵬翮 (1649–1725) and the poet Qian Liangze 錢良擇 (b. 1645), who were both Chinese (from Sichuan and Jiangsu, respectively), and Jean-François Gerbillon (Zhang Cheng 張誠, 1654–1707), a French Jesuit (who wrote his travelogue in French). The party split up into two groups ("banners" [*qi* 旗]), with Zhang and Qian in one and Gerbillon in the other. The three did not have an opportunity to compare notes, and their accounts are not always easy to reconcile.[22] Qian's diary provides the most detail in terms of natural-historical observations.

At the time of the expedition, the Khalkha had not yet submitted to the Qing. Qian described their land as foreign territory. It is worth entertaining whether the description of the inhospitable Khalkha environment to some extent reflected the area's political status. At least, the geopolitical boundary made outer Mongolian nature more inaccessible:

---

included some information on the animals, birds, and fish that the local people lived by. A second report submitted in 1709, however, was entirely focused on geography, with no information on natural products. Perhaps the authorities at this stage still needed primarily to gain a knowledge of the lay of the land. Detailed information regarding flora and fauna might not have been a priority. See Matsuura, *Shinchō no Amūru*, chaps. 1–2; Kicengge, *Daichin Gurun to sono jidai*, 217–33; Kicengge, "Manwen Wula deng chu difang tu kao," 194.

21. Bossiere, *Jean-François Gerbillon*, 31–33.
22. Qian, *Chusai jilüe*, 8:41b (4120).

From the moment we entered Mongolia, the local language is different from in China. However, they have already submitted to our court and come to present tribute every year. Many among the clerks and officials of the Board of Government of the Outer Regions (Lifan Yuan) have learned to understand [their language], and thus the names of the places we passed through and of the local products can still be successfully translated into Manchu or transcribed into Chinese characters. Once we entered Khalkha territory, however, even though communications have been open for a long time, we were unable to completely ascertain the place names.

自入蒙古，其方言雖異中華，然既歸附本朝，每歲朝貢，其土語及字義，理藩院官吏多備曉之。故所歷地名與方物等，猶能以滿語及中國字音通譯之。暨入噶爾噶界，則雖通好日久，而地名不能全曉矣。[23]

Following from this state of affairs, the few descriptions of plants and animals for which we can find an echo in the later Manchu *Mirror* refer to the Inner Mongolian territories that were already part of the empire. On June 8 (KX 27/5/11), Qian encountered a bird, which serves as an example:

There is a bird called *elan*.[24] It is as big as a pigeon, but brown. It flies against the wind and descends once its energy is exhausted. Members of our retinue often found its fledglings in the grass.

有鳥名阿蘭。其大如鳩，而褐色，迎風飛翔，力竭乃下。從人多於草間探得其雛。[25]

The bird described here is probably the North China crested lark, if not perhaps the Eurasian skylark. A bird named *elan* 阿藍 (a different

---

23. Qian, *Chusai jilüe*, 8:7b (4052).
24. The character transcribed as *e* here is nowadays usually (but not always) read *a*. It ought to have been read *e* in the mid-Qing period, however. The sound *e*, i.e. [ʏ], is close to the final of *wo*, which is a syllable likewise seen in orthographic variants of the name for this bird, as we explain below. The pronunciation *a* is more difficult to reconcile with that alternative orthography. In *Yuzhi zengding Qingwen jian* (vol. 233, 30:36b), the characters *alan* 阿蘭 are indeed transcribed as *o lan* in the Manchu script, which corresponds to our transcription *elan*.
25. Qian, *Chusai jilüe*, 8:9a (4055).

orthography) is listed in *Shengjing tongzhi*. There is no overlap with the description here, but the two are not irreconcilable. The gazetteer says that *elan* is a generic name for several kinds of small sparrows, or small birds (*xiao que* 小雀) and listed several names for different varieties. Large ones, for example, were called *fengtou* 鳳頭 'Phoenix heads', which referred to the crest on birds such as the crested lark. The gazetteer then said that locals raised the birds for their amusement, but did not describe the birds' appearance or behavior further.[26]

The bird is known from early Manchu dictionaries as well, but without any description. Shen Qiliang translated the Manchu word *sama* [sic] *cecike* 'small shaman bird' as *wolan que* 窩藍雀 '*wolan* sparrow' (and under *cecike* 'small bird': *saman cecike* 'small shaman bird' as *wolan* 窩藍), another variant of the same vernacular Chinese name (*elan* ~ *wolan*).[27] The Manchu *Mirror*, by contrast, gave a partially circular definition of *saman cecike* (like the *wenderhen*, but with a crest on its head, varied song) and *wenderhen* (somewhat similar to the Phoenix-head *elan* but without a crest).[28] Li Yanji listed *wolan* 窩攔 as one of the names for crested *wenderhen* and *wolan que* as the translation of *saman cecike*, in both cases using this different Chinese character for *lan*.[29] Qian's description was more detailed than, but seemingly also disconnected from, both the Chinese discourse of the gazetteer and the Manchu dictionaries.

This discussion shows that in the case of the lark, encountered not far from Chinese territory, Qian had a Chinese name to attach to the creature, if only a vernacular one with an uncertain orthography, cut off from the classical tradition. By contrast, on July 2 (KX 27/6/5), several days' march into Khalkha territory, Qian described a plant for which he had no name at all. We use this plant to show that the difficulty to attach labels to the foreign nature of Mongolia, present in Qian's text, was later alleviated by Manchu dictionaries. Qian wrote,

26. *Shengjing tongzhi*, vol. 8, 21:18a.
27. Hayata and Teramura, *Daishin zensho*, vol. 1, 97 (line 0631b2), 176 (1045b3).
28. In Qianlong's Manchu-Chinese *Mirror*, these were later translated as *fengtou elan* 鳳頭阿蘭 'Phoenix head *elan*' and *elan* 阿蘭, respectively (*Han-i araha manju gisun*, 1708, 19:*cecike-i hacin*:35b; *Yuzhi zengding*, vol. 233, 30:36b–37a [230, upper and lower panels]).
29. Li Yanji, *Qingwen huishu*, 4:19b (86, upper panel), 12:30a (224, upper panel).

Close to the road there are trees as tall as twenty feet or more and as short as three or four feet rising up like a forest in the desolate wilderness. Their branches are of the type of the goji and their leaves of the type of the tooth-pick pine. I do not know the name of it.

道傍有樹，高者二丈餘，短者三、四尺，林立荒郊。枝類枸杞，葉類剔牙松。不識其為何名。[30]

The Manchu *Mirror* later gave a description of a plant that was probably the same tree Qian had seen. The habitat and height of the tree are the same in both descriptions, and the spidery branches of goji shrubs resemble those of the bushes identified with the name used in the corresponding comparison in the *Mirror*. Importantly, the *Mirror*'s editors had a name for the plant:

jak moo 'jay (boundary?) tree'.[31] Grows in wasteland. Seen from afar, it looks like dead wood. The leaves are like [those of] *budurhūna* (< Mo. *buduryan-a*, some kind of bush growing on arid land).[32] Tall ones reach close to twenty feet in height. When very green, it will burn if fire is set to it, like coal.

jak moo, gobi de banjimbi. goro tuwara de bucehe moo-i adali, abdaha budurhūna de adali, den ningge juwe jang hamimbi. umesi niowanggiyan de tuwa dabuci uthai tayambi, yaha-i adali.[33]

---

30. Qian, *Chusai jilüe*, 8:34a (4105).
31. Rozycki (*Mongol Elements in Manchu*, 119) follows Lessing and colleagues (*Mongolian-English Dictionary*, 1022) in saying that Mo. *jay* refers to a bush. Kowalewski (*Dictionnaire mongol-russe-français*, vol. 3, 2296), however, lists *jay modon* as a translation of Ma. *jak moo* and says that it is a tree. Tsintsius (*Sravnitel'nyĭ slovar'*, vol. 1, 242) does not list Ma. *jak moo* under *jay* (< Mo.) in the sense of "boundary," but it would fit with Qian's description of the plants lining the side of the road.
32. Rozycki, *Mongol Elements in Manchu*, 36. Different species are given in Kowalewski (*Dictionnaire mongol-russe-français*, vol. 22, 1184) and Lessing and colleagues (*Mongolian-English Dictionary*, 131).
33. *Han-i araha manju gisun*, 1708, 19:*moo-i hacin*:13b.

Li Yanji did not translate the name of this tree, but Juntu did, as *cuisheng mu* 催生木 'hasted-growth wood'.[34] Kangxi discussed it in Chinese using a transcription of the Manchu name.[35] Thus this plant from Outer Mongolia that Qian Liangze was unable to name entered the Chinese lexicon in the eighteenth century through court writings and bilingual lexicography.

The example of Qian Liangze's description of the desert plant that the Manchu *Mirror* called *jak moo* further corroborates what we posited at the beginning of chapter 1, namely, that Kangxi's Manchu *Mirror* should not be seen as a pristine depository of Inner Asian natural knowledge inaccessible to Chinese readers. This impression gets even stronger when we look at examples from the emerging discourse about Manchurian nature among the Chinese cultural elite, as represented by the exile and travelogue literature that emerged or entered circulation around the same time as the completion of the Kangxi *Mirror* in 1708, or even earlier. Yet this literature in itself contained dissonances, as the following comparisons with the *Mirror* demonstrate.

Manchurian Travelogues and Miscellany by Exiled Literati and Officials

The earliest Chinese literati exiles arrived in Ningguta, a newly settled outpost, in the late 1650s. They immediately started to document the place after the established conventions of local gazetteers, which commonly included a list of local products. The scholar-official Fang Gongqian 方拱乾 (1596–1666, *jinshi* 1628), who was punished along with his son for currying favor in the provincial civil service examinations, lived for three years in Ningguta and composed a short "treatise" (*zhi* 志) of his place of exile after returning home. In it, Fang

---

34. Li Yanji, *Qingwen huishu*, 9:12a (167, lower panel); Juntu, *Yi xue san guan Qingwen jian*, 44a (139, lower panel).

35. The Kangxi emperor refers to this tree as *chake* 查克 in his miscellaneous notes (see *Shengzu ren huangdi yuzhi wenji*, vol. 1299, ser. 4, 27:8b [209, lower panel]). It was later given a different Chinese name in Qianlong's Manchu-Chinese *Mirror* (*Yuzhi zengding*, vol. 233, 29:33b [194, lower panel]). The new Chinese name for *jak moo* was *zhuomu* 灼木.

described Ningguta as a desolate place one had "no reason to stay [in] or return [to]" 無住理亦無還理, and where "not a single grain of rice had been seen since the settlement" 開闢來未見稻米一顆. Because neither Manchu nor Chinese settlers had lived there for long, the customs there also varied greatly, and Fang did not record Manchu terms except for certain place names and official titles.[36]

Further literary inquisitions in the 1660s sent more Chinese officials and literati to Ningguta. Among them was the former Ming Minister of War Zhang Jinyan 張縉彥 (1600–1672), who went with a sizable book collection in tow and built a social scene from the growing group of literati exiles there. The verses and prose they composed in Ningguta presented their correspondents in the capital and back home with glimpses of the treacherous, marshy roads up north, where the mountains and rivers abounded with wild and threatening creatures such as bears, fish, and eagles.[37] Among Zhang's close acquaintances was the celebrated poet Wu Zhaoqian 吳兆騫 (1631–1684), whose verses inspired Ruth Rogaski's moving account of the "landscapes of exile" at the onset of a long tradition of Manchurian nature writing that would bear important repercussions in the twentieth century.[38]

For the most part, exiled literati chose to document the Chinese vernacular names of local species. They expressed no further curiosity in knowing the Manchu name for haw (*shanzha* 山楂), for instance, or those for pine nut and hazelnut. They also highlighted the intimate connections to court ritual in Beijing, paying heightened attention to the so-called tribute (*gong* 貢) items from the region. For instance, Fang Gongqian in the 1660s already took note of a kind of crawfish (*hasima*) that was made into a special kind of paté, which was collected by the authorities to dispatch to Beijing for the imperial house's

36. Fang returned south in 1662, where he wrote this account (Fang Gongqian, *Ningguta zhi*, vol. 1, 448, upper and lower panels). For the long editorial history and different standards of inclusion used in the collectanea where we find Fang's text, see Bian, "Re-Collecting the Glorious Age."
37. Zhang Jinyan, *Ningguta shan shui ji*.
38. Rogaski, *Knowing Manchuria*, 19–60.

ancestral worship rituals.³⁹ By the late 1690s and early 1700s, however, the constructed image of Manchuria as an uncivilized and barren land had given way to a slightly different tenor.

Part of this shift stems from a generational change between, on the one hand, the older literati who had fallen under suspicion of sedition by the conquest elite and, on the other hand, the older generation's children and younger officials who had come of age during the Kangxi reign and displayed no outward hostility toward the Manchus. Notable examples of this later generation include Yang Bin 楊賓 (1650–1720), who visited his exiled father in Ningguta from 1689 to 1691 (KX 28–30) and composed his popular travelogue about crossing the "Willow Palisade" (*liubian* 柳邊).⁴⁰ Fang Shiji 方式濟 (*jinshi* 1709), whose official career was abruptly forestalled because of his father's implication in the case against Dai Mingshi 戴名世 (1653–1713)—a vocal scholar-official who invoked in his poetry Southern Ming reign names that had been deemed treasonous offense—was exiled to the military garrison at Bukui 卜魁 (Qiqihar).⁴¹ While there, Fang composed a detailed account of Manchuria with the intention of providing more accurate and up-to-date information than the 1684 *Shengjing tongzhi*.⁴² Last but not least, Wu Zhaoqian's son Wu Zhenchen 吳振臣 (n.d.) composed a loving memoir of Ningguta, where he was born, raised, and lived until adulthood. In contrast to the cold, barren landscape in the elder literati's eyes, Wu Zhenchen's account teemed with the fragrance of wild roses, abundant hazelnuts and pine nuts, and the ecstasy of fishing by the clear stream.⁴³ All three texts mentioned here circulated widely in

---

39. See Fang Gongqian, *Ningguta zhi*, vol. 1, 449, upper panel; Stary, "Manchu Word List," 583. Cf. the entry for *hasima* in the Kangxi *Mirror* (*Han-i araha manju gisun*, 1708, 20:*birai nimaha-i hacin*:35a).

40. Yang, *Liubian jilüe*; see also Edmonds, "Willow Palisade."

41. On Dai Mingshi, see Durand, *Lettrés et pouvoirs*.

42. See also Fang Shiji, *Longsha jilüe*, 1418–19. The text was also included in *Siku quanshu*. Fang Shiji's son Fang Guancheng 方觀承 (1696–1768) survived a fatherless childhood and eventually rose to high official position later in his life and became known for his water management and agronomy.

43. Wu Zhenchen, *Ningguta jilüe*, vol. 2, 1716–23. The description of flora and fauna is on page 1718. The author's preface is dated 1721 (KX 60).

manuscripts and appeared in print in various literary collectanea by the early nineteenth century.

This positive portrayal of a rigorous, resourceful wilderness presented a marked contrast to the bitterness and lingering grievance in Wu Zhaoqian's earlier verse. It also paid more attention to particular plants and animals, together with their various names. We now turn to examine three case studies that shed light on local flora in Manchuria that became known more widely through these sources.

On Mushrooms, Grasses, and Berries: A Cross-Examination of Early Qing Sources

In some instances, Chinese accounts recorded species with Manchu names in such a vague way that it raises questions regarding their sources. Yang Bin, for instance, referred in his travelogue to two fruits named *fa-fo-ha* 法佛哈 and *mi-sun-wu-shi-ha* 米孫烏什哈, describing them as "sweet and sour, edible ... not found in the central land [of China]" 味甜酸可食 ... 皆中土所無者.[44] The most likely entry in the Manchu *Mirror* that matches this description is a fruit named *fafaha*, which "grows on a tree, with red color, flat seeds, and sour taste" (*tubihe-i gebu, moo de banjimbi, boco fulgiyan, faha halfiyan, amtan jušuhun*).[45] Furthermore, *mi-sun-wu-shi-ha* appeared in the *Mirror* as *misu hūsiha*, a red, sour fruit that grows in bunches, like grapes, in the forest and served as an incense fragrance (*hiyan-i wa bi*).[46] It might have referred to magnolia wine, or, less plausibly, rowan (which, however, is neither sweet nor fragrant).[47] In sum, it is quite likely that Yang Bin never really saw or tasted these fruits, but instead copied the names

---

44. Yang, *Liubian jilüe*, 3:13a.
45. *Han-i araha manju gisun*, 1708, 18:*tubihe-i hacin*:49a.
46. *Shengjing tongzhi*, 29:12a.
47. *Han-i araha manju gisun*, 1708, 18:*tubihe-i hacin*:49a; Hu, *Xin Man-Han*, 539, s.v. *misun huusiha*; Fèvre and Métailié, *Dictionnaire Ricci des plantes*, 469, s.v. 五味子 wǔ wèi zǐ. This assumes that the identification of *misun huusiha* with Chinese *wuweizi* is correct. The word has this Chinese translation in Mingdo, *Yin Han Qingwen jian*, 19:224b. However, *Qinding Shengjing tongzhi* lists *wuweizi* separately from *misu hūsiha* (vol. 2, 27:18b, 1326).

NAMING MANCHURIAN THINGS   141

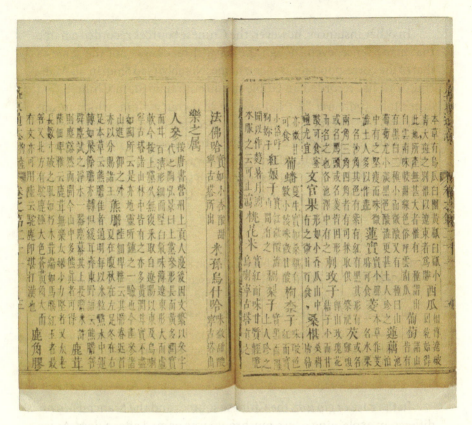

FIG. 4.2. 1684 *Shengjing tongzhi*, local product list, showing *fafaha* and *misu usiha* (21:11b-12a). Digitized local gazetteers, Harvard-Yenching Library. https://id.lib.harvard.edu/alma/990080625400203941/catalog.

and descriptions of these Manchu-named fruits from a source none other than *Shengjing tongzhi*, discussed earlier in this chapter. The gazetteer similarly recorded *fa-fo-ha* and *mi-sun-wu-shi-ha* as two sour fruits resembling each other that grew in Ningguta (see fig. 4.2).[48] Here the Chinese sources indeed appeared unable to interpret the name beyond recording the sounds of Manchu words. By contrast, the Manchu lexicographer Mingdo 明鐸 in 1735 translated *fafaha* as *hong ying* 紅櫻 'red cherry'.[49]

48. *Shengjing tongzhi*, vol. 8, 29:12a.
49. Mingdo, *Yin Han Qingwen jian*, 19:224b.

In other instances, however, the Chinese sources recorded a name that made sense, even though the sense it made was not the same as that of the corresponding Manchu name. Notably, Gao Shiqi in his Manchu word list recorded an edible mushroom named, in Chinese, "*mi deer's tail*" (*mizi wei* 麋子尾) and, in Manchu, *gio ura*, which meant "rear end of Siberian roe deer." Gao speculated that the name of the mushroom was due to its unique smell, reminiscent of that of the wild animal.[50] In the Manchu *Mirror*, *gio ura* was listed in the section "Vegetables and dishes" (*sogi booha*) along with six other edible mushrooms and fungi.[51] The *Mirror* provided a very different explanation of the name *gio ura*:

> Grows from hardwood. The color is whitish and [the texture] tough. On the surface, it has a kind of white hair resembling those grown on the crease between a roe deer's hind legs. Edible when cooked. Also edible when preserved with fermented soybean paste.
>
> *mangga moo de banjimbi, šahūkan bime silemin, oilorgi de gio-i ulan-i ba-i šanggiyan funiyehe-i adali banjihabi. urebufi jembi. misun de gidafi inu jembi.*[52]

In the case of *gio ura*, the Chinese sources and the Manchu *Mirror* differed in explaining why a mushroom was named after a deer. As we

---

50. Gao, *Hu cong dongxun rilu*, appendix (*fulu*):3b; Stary, "Manchu Word List," 584.

51. They are *megu* (< Ch. *mogu* 蘑菇) 'mushroom' (a generic term: "there are very many kinds" [*hacin umesi ambula*]), *sisi megu* 'hazelnut mushroom' (found among hazelnut trees, deep yellow in color, and smallest of all the mushrooms), *coko megu* 'chicken mushroom' (grows in dirt and on decaying wood, white on top and with blackish folds [cf. Imanishi, Tamura, and Satō, *Gotai Shinbunkan yakukai*, vol. 1, 805 (item 14218): "black interior" 内部は黒い]), *hailan megu* 'elm mushroom' (growing on elms, deep yellow color), *kūwara megu* 'encircling mushroom' (whitish and pea green), *sanca* 'wood ear mushroom' (grows on trees after rain, only those growing on sandalwood trees are eaten). See *Han-i araha manju gisun*, 1708, 18:*sogi booha-i hacin*:9b–10a.

52. *Han-i araha manju gisun*, 1708, 18:*sogi booha-i hacin*:10a. For a Japanese translation, see Imanishi, Tamura, and Satō, *Gotai Shinbunkan yakukai*, vol. 1, 805 (item 14221).

discuss in later chapters, *gio* was usually translated into Chinese as *pao* 麅, not *mi* deer, which as mentioned had an uncertain identity, its notable tail notwithstanding. *Shengjing tongzhi*, for one, listed *mi lu* '*mi* deer' among the animals and defined it using quotes from older Chinese literature. The gazetteer's editors likewise wrote, however, that "the locals distinguish them according to size, sex, and the number of antlers, but *mi* is not distinguished from *lu*" 土人以大小、牝牡及角之多寡分名，而麋與鹿無辨者, which reads as an acknowledgment that Chinese terminology was at odds with the nomenclature used in the languages of the region.[53] Yet suffice it to say here that Gao probably used *mi* in reference to some kind of musk deer, which were characterized by their strong smell. He inferred that the smell was why the mushroom was named after it. The Manchu *Mirror*, by contrast, listed a Manchu name that associated the mushroom with the Siberian roe deer, not the musk deer. The Siberian roe deer has a white caudal patch, as described in the *Mirror*.

In the preceding case, one of the two sources was probably mistaken, and we are inclined to think that the *Mirror* provided the more accurate explanation of the name. In other cases, however, Chinese exile literature and travelogues provided information on plants and animals that were neither included in the Manchu *Mirror*, nor necessarily highlighted in later court compilations regardless of language. In those cases, they were the only sources available for knowledge on Inner Asian nature.

One such name was "Ula grass" (*Wula cao* 烏喇草), which Gao Shiqi rendered in his Manchu word list as *hūwaitame gūlha foyo* 'tied-up boot grass'. As exiles and their visitors trekked along the marshy roads to Ningguta, the long, sturdy grass was an essential item to wrap around one's boots and keep the feet warm (to form a puttee of sorts).[54] The Kangxi *Mirror*, however, introduced four kinds of *foyo* but did not elaborate on this indispensable way of using them.[55] Paradoxically, the

---

53. *Shengjing tongzhi*, vol. 8, 21:20a.

54. Gao Shiqi, *Hu cong dongxun rilu*, appendix (*fulu*):2b; Stary, "Manchu Word List," 582.

55. *Han-i araha manju gisun*, 1708, 19:*orho-i hacin*:2b. The term *foyo*, translated as *wula cao* 烏拉草 'Ula grass', was added in the 1772 *Expanded and Emended Manchu*

Chinese travelogues, which predated the Manchu *Mirror*, nevertheless went beyond it in describing a plant that was essential to everyday use for Ningguta residents, but perhaps less familiar to the Manchu elite who had long gotten used to lives in Mukden or Beijing, where the editors of the *Mirror* also lived and worked.

One last interesting species that was mentioned in the Kangxi *Mirror*, but more elaborately described in early Chinese writings, is the so-called *ilhamuke*, lit. flower water. The *Mirror* listed it among the fruits and gave it the following description:

> Name of a fruit. From the root, it grows two or three branches with leaves, and then a small stem sprouts out and bears fruit. The taste, while sour, is quite pleasant.
>
> *tubihe-i gebu, fulehe ci juwe ilan gargan abdaha banjifi, ajige cikten tucifi tubihe banjimbi. amtan ambula jancuhūn bime jušuhuken.*[56]

Gao Shiqi rendered the Manchu name phonetically as *yi-er-ga-mu-ke* 一兒噶木克 (*ilga* was an alternative pronunciation of *ilha* 'flower'). He reasoned that the name probably derived from the pretty appearance of the ripe fruits, which he likened to "lightly red and fresh" (*qian hong er xian* 淺紅而鮮) flower petals, noting how local people "rushed to eat them" (*ren zheng shi zhi* 人爭食之) when they were in season.[57] Wu Zhenchen, for his part, compared the *yi-er-ha-mu-ke* 衣而哈目克 fruit to a "small *yangmei* (bayberry), but pitless" 似小楊梅而無核. Wu described the plant as "herbaceous, with reddish vines, growing among other grasses" 草本紅藤，生雜草中, which makes us think that he was most likely referring to raspberries.[58] Neither Gao nor Wu offered any hint as to whether a vernacular Chinese name was in use at the time.

Mirror, however, likely due to the influence of Chinese travel writings about this item. See *Yuzhi zengding Qingwen jian*, vol. 233, 29:9b (182, bottom panel).

56. *Han-i araha manju gisun*, 1708, 18:*tubihe-i hacin*:49a–b.

57. Gao Shiqi, *Hu cong dongxun rilu*, appendix (*fulu*):3a; Stary, "Manchu Word List," 583.

58. Wu Zhenchen, *Ningguta jilüe*, vol. 2, p. 1718, middle panel.

The 1684 *Shengjing tongzhi*, however, attempted to provide a more poetic Chinese name to the fruit. It called it "peach blossom water" (*taohua shui* 桃花水), but the character *shui* was unfortunately misprinted as *mi* 米 'rice', which would indeed have been more expected as the head of a noun phrase referring to berries.[59] Later on, Yang Bin, who as we have shown read *Shengjing tongzhi* and used it as his source, continued using the name "peach flower water" but added a lively and detailed account of how this fruit figured in the life of Ningguta locals:

Peach flower water. Herbaceous. Looks like a bayberry (*yangmei*) but without a kernel. Its color is red and tastes sweet; texture light and fragile. Goes bad after one touches it. It grows all over the ground in the fifth and sixth months. The local residents pick out the most densely grown area, where they put up tents or sheds, and pool money to buy wine. Men and women then each form their own groups and rush to pick and eat [the berries], moving to another spot on the following day, until the fruits have been completely consumed.

桃花水。草本，狀若楊梅而無核。色紅味甘，質輕脆。過手即敗矣。五六月間遍地皆是。居人擇最多處設帳房或棚子釀分載酒，男女各為群爭採食之。明日又移他處，食盡乃已。[60]

By the 1710s, according to Fang Shiji, this fruit had become sufficiently well known beyond Ningguta for fresh berries from Aigun to be preserved with honey and dispatched to Beijing as a tribute item.[61]

---

59. *Shengjing tongzhi*, vol. 8, 21:11b. Note that the Chinese name *tao hua shui* (lit. peach blossom water) was matched to *ilhamuke* for the first time here, and not without a certain awkwardness, because the Chinese term already existed in classical literature, where it referred to spring flooding of the Yellow River that results from melting snow in the third lunar month. See Chen Yuanjing, "Suishi guangji," 1:6b–7a (22–23). Accordingly, Shen Qiliang used *taohua shui* to translate a different Manchu term, *nimašan muke* 'melting snow water' that had nothing to do with this herbaceous berry. See Hayata and Teramura, *Daishin zensho*, vol. 1, 47 (0337a4).

60. Yang, *Liubian jilüe*, 3:13a.

61. Fang Shiji, *Longsha jilüe*, 2:1418, middle panel. Fang used the plain Chinese term *huashui* 花水 'flower water' for this product, which avoided confusion with "peach blossom water."

Our purpose in going through all the nuanced differences in these early Chinese accounts of *ilhamuke* and other local species in the preceding paragraphs was to shed a different light on the Manchu *Mirror* as not necessarily constituting the most authentic account of Manchurian nature from the time around 1700. On the contrary, when the *Mirror* was being planned and compiled at court, Chinese authors had already seized on their own experience in the region, melding it with important precedents in classical and more recent literature, including the historiography of local products in gazetteers. Throughout the eighteenth century, then, readers from near and far could savor the exoticized difference of the region that was rendered more and more familiar.

The entangled nomenclature of plants and animals, as well as the ethnographic narratives of how local people used them, reveal the complex linguistic reality of everyday life in early Qing Manchuria, where Manchu, vernacular Chinese, and other languages intermingled and evolved together. Unsurprisingly, the ambiguities and discrepancies between the Chinese sources and the Manchu *Mirror* later generated more discussion at the Qing court, at that time with the purpose of imposing greater order and consistency across the languages involved. In chapter 5, we examine how court-commissioned encyclopedias in the 1710s and 1720s brought an imperial aura to the documentation of frontier species. The humble and fragile fruit *ilhamuke*, meanwhile, returns to the center of our attention under a different guise in chapter 8.

CHAPTER FIVE

Manchurian and Inner Asian Plants in Guang Qunfang pu *and* Gujin tushu jicheng

> Blue eyebrows: The flowers are bluish in color. Swinging under the oblique sunlight, they look striking and lovely.
> 翠蛾眉。花翠色，隨日側映，鮮妍有致。[1]

Among all the "unusual flowers" (*yihua*) that intrigued Wang Shizhen in *Guang Qunfang pu,* what is so special about this dainty blue flower? In chapter 3, we noted the meticulous way in which *Guang Qunfang pu* cited previous sources to add to Wang Xiangjin's already erudite work. This entry for blue eyebrows is odd because it did not have any explicit citations, and yet the entire sentence was in fact copied verbatim from the 1684 *Shengjing tongzhi*'s list of local products.[2] Elsewhere, *Guang Qunfang pu* cited *Shengjing tongzhi* for five vegetables, five fruits, and eight grasses native to Manchuria, including *fafaha* and *misu hūsiha,* the two fruits bearing Manchu names that we discussed in chapter 4.

1. Wang Hao, *Yuding Peiwen,* vol. 846, 53:38a (537, upper panel).
2. *Shengjing tongzhi,* 21:10a.

We also established in chapter 3 that *Guang Qunfang pu*, an imperially commissioned text, reformed literati learning of plants in late Ming times by decentering the figure of the literati author and his literary flourishes from Wang Xiangjin's original treatise. Elevating the genre of "catalogs" (*pu*) to unprecedented coverage and depth of research also allowed the Qing editors to depart from previous conventions of *leishu*, to focus instead on the individual histories of each plant rather than its cosmic resonance with other entities. This approach, which we call "lexicographic encyclopedism," can be seen not only in *Guang Qunfang pu* but also other late Kangxi-reign encyclopedic texts such as *Gujin tushu jicheng*.

But as the silent citation of *Shengjing tongzhi* about a bluish flower indicates, there is more than meets the eye when we examine the epistemic interventions made in *Guang Qunfang pu* and other late Kangxi encyclopedic texts concerning flora and fauna in the Manchu homeland and Mongolian territories beyond the Great Wall (*saiwai* 塞外). In this chapter, we interrogate the processes by which the latter term, commonly used at the time to indicate the Qing domain beyond the historically Chinese regions known as "the interior" (Ch. *neidi* 內地; Ma. *dorgi ba*), assumed heightened political significance as an exceptional epistemic space toward the very end of the Kangxi reign and continuing beyond the emperor's death in 1722. First, we highlight the centrality of experience by court scholars such as Wang Hao (the chief editor of *Guang Qunfang pu*, as we recall) and others as they accompanied the emperor for tours beyond the Great Wall in the 1700s and 1710s. In contrast to flora and fauna recorded in the earlier body of travel writings by Chinese literati exiles and court officials described in chapter 4, these later trips witnessed Kangxi's conscious effort to reshape plant and animal knowledge in front of his Chinese courtiers. Beyond *Guang Qunfang pu*, this new body of knowledge also found its way into the final form of *Gujin tushu jicheng*. Using plants and animals as our point of entry, we shed new light on the unresolved controversy surrounding the authorship and revision process of this last—and largest—imperially commissioned Chinese encyclopedia.

In the second half of this long chapter, we trace further how this intense effort to stabilize frontier flora and fauna's presence in court scholarship was further edified in Kangxi's posthumously published opus *Kangxi ji xia gewu bian* 康熙幾暇格物編 (Collection of the investigation of things during times of leisure from the myriad affairs [of state] in the Kangxi period).³ This new development took place around the time when the emperor oversaw the construction of the Rehe summer resort and its grounds beyond the Great Wall, which became the space in which some of those frontier plants and animals could be found and admired in a setting imbued with imperial grandeur.

Imperial Inspections of Flora and Fauna beyond the Great Wall

In chapter 3, we examined all the explicit display of bookish erudition in *Guang Qunfang pu*. This makes it all the more surprising that quite a few plants in this book, such as the "Blue eyebrows," actually stand out for their bare-bones description that did not include any textual references. Consider this "yellow flower" (*huang hua* 黃花), for instance, that was abruptly introduced toward the end of the revised "Catalog of flowers" (*huapu* 花譜):

> Its name is unknown. The flower has four conjoined petals, resembling a golden cup. The color is goose yellow. Grows to the north of the Han-Temur-mountain range. Only one blossom grows on one stem, and its solitary beauty swings elegantly. Most grasses wither after frost, and yet this flower shines even more brightly.
> 不知其名。一朵四瓣，合抱如金盞，其色鵝黃。出汗帖木兒嶺以北，每一莖只一朵，翩然獨秀，眾草經霜凋萎，是花經霜，顏色愈覺鮮明。⁴

With no explicit acknowledgment of sources, the single clue in this entry that helps us locate this nameless yellow flower is the place

---

3. On this book, see Jami, "Investigating Things under Heaven," especially 175n9 for the title.

4. Wang Hao, *Yuding Peiwen*, vol. 846, 53:27a (531, lower panel).

"Han-Temur mountain range" (*Han Tiemu'er ling* 汗帖木兒嶺).[5] Where is it located and how did this flower enter this court-commissioned encyclopedia of plants?

In chapter 3, we described the meteoric rise of Wang Hao, who obtained his *jinshi* degree by a special edict in 1703 (KX 42) and went on to work on multiple court projects including *Guang Qunfang pu*. Immediately after receiving the prestigious degree and official title as a Hanlin bachelor, Wang Hao accompanied the Kangxi emperor for an imperial tour and seasonal hunt to "avoid the heat" (*bishu* 避暑) beyond the Great Wall. Together with six other renowned scholar-officials, Wang joined the imperial entourage for a long trip for over a hundred days, during which the "Southerners" (*nanren* 南人) rode on horseback, feasted on deer meat and other northern delicacies, and conversed with the emperor and the heir apparent on a daily basis. We get to know much of their experience in detail because Wang Hao recorded the trip in a short travelogue titled *Suiluan jien* 隨鑾紀恩 (An account of grace during my trip accompanying His Majesty's carriages).[6] Also among the group of Hanlin scholars on the move was the renowned poet Zha Shenxing 查慎行 (1650–1727, *jinshi* 1703), who similarly obtained his degree owing to Kangxi's special favor.[7] In his miscellany titled *Ren hai ji* 人海記 (My account in the sea of humankind), Zha recorded all sorts of anecdotes gathered during his long sojourn in the imperial capital, including what he saw during the imperial hunting trips. Like Wang Hao, Zha also kept a diary during the 1703 tour that only survived in an early twentieth-century manuscript copy.[8] These sources shed light on the entangled relationship between the hunting trips and seemingly unrelated court texts such as *Guang Qunfang pu*.

The yellow flower just discussed appears in Wang Hao's diary on September 16, 1703 (KX 42/8/6), when the imperial entourage crossed

---

5. This is the Manchu transcription of the name as it appears on the the 1721 atlas of China, searchable at *Qing Maps* (https://qingmaps.org/), where it is recorded, however, as a "valley" (*holo*) rather than as a mountain range.

6. Wang Hao, *Sui luan ji en*, ser. 1 (*zhengbian* 正編), *zhi* 帙 1, 1a, 6b.

7. For Zha Shenxing's career at court, see Chen Jingzhang, *Zha Shenxing nianpu*, 25–28.

8. Zha, "Peilie biji."

the "Han Tiemu'er da-ba-han" 汗鐵木耳打把漢 (< Mo. *dabaya[n]* 'mountain pass, mountain range', via Ma. *dabagan* idem) and entered "Mongol territory" (*Menggu jie* 蒙古界).⁹ Nine days later, after an extravagant hunt in which more than 1,300 Khorchin Mongol warriors participated, the emperor celebrated the Mid-Autumn Festival not far from the snow-covered Khingan Mountains, summoning the Hanlin scholars to his great tent for a feast.¹⁰ There, according to Wang's record, they were first presented with a wild boar that the emperor had just shot with an arrow, whose canine teeth resembled "steel hooks" (*ganggou* 鋼鉤). After a few rounds of fine pastries and fresh fruits brought in from the imperial gardens of Beijing, the emperor "passed down to show" (*banshi* 頒示) several stems of a kind of "yellow flower":

> Its stem and leaves resemble the common poppy (*yumeiren*), yet its five petals are closed like a golden cup. The emperor instructed us, saying: No flowers under Heaven could resist withering after frost, except for those blossoms beyond the border that are resistant to frost. The heavier the frost, the more vivid their colors become. They are all of this kind [of flowers].
>
> 又頒示黃花數枝，其莖及葉皆如虞美人，而五瓣環合若金盞。上諭：天下花卉，未有經霜不萎者，惟塞外各花能拒霜，霜愈重則色愈鮮，胥此類也。¹¹

It is quite clear that Wang Hao composed the "yellow flower" entry in *Guang Qunfang pu* based on this encounter, which was orchestrated by the emperor himself. Because the words describing the flower had no precedent, Wang had no choice but to give this account without a

---

9. Wang Hao, *Sui luan ji en*, ser. 1 (*zhengbian* 正編), *zhi* 帙 1, 5b, and relatedly Zha, "Peilie biji," 224. For the words *dabaya(n)* and *dabagan*, see Rozycki, *Mongol Elements in Manchu*, 52. Kangxi was pronouncing this word as a Manchu speaker. The transcription of intervocalic Manchu -*g*- as a voiceless velar fricative (Pinyin *h*-, IPA *x*-) might reflect Manchu pronunciation at the time. See Kam, "Romanization of the Early Manchu," 136. By contrast, Qian Liangze (*Chusai jilüe*, 8:7b [4052]) was transcribing the Mongolian word when he wrote *daba* 打八 (cf. modern Mongolian *dabaa*).

10. Zha, "Peilie biji," 227–29.

11. Wang Hao, *Sui luan ji en*, 7b; Zha, "Peilie biji," 229.

textual reference to earlier Chinese literature. It is also notable that in the later *Guang Qunfang pu*, the number of petals was changed from five to four, which is a more accurate description of common poppies. It is possible, as Wang Zhao 王釗 suggests in his study of Qing palace paintings, that court scholars later reexamined similar species in the imperial capital in their pictorial and textual rendering of frontier species.[12] This minute discrepancy between Wang Hao's diary and *Guang Qunfang pu* suggests, at least, that additional natural-historical observations were conducted following the conclusion of the imperial tour in 1703.

Historians have commented extensively on imperial tours beyond the Great Wall undertaken beginning in the 1680s. Natalie Köhle noted the interwoven threads between pilgrimage and patronage that situated a long-standing religious destination such as Mount Wutai as an eminent stop in Kangxi's western tours, which included five visits to the site (twice in 1683, and again in 1698, 1702, and 1710).[13] Situating the Manchu imperial hunt as part of a long-standing tradition of royal hunting across Eurasia, scholars have further detailed the complex logistical orchestration that made those trips possible, as well as the careful stewardship of the hunting grounds.[14] Our understanding of High Qing rulership has by now become firmly associated with the imagery of the "peripatetic sovereign," a "court on horseback." Although most elaborate examples come from Kangxi's southern tours and the Qianlong reign, it is nevertheless clear that Qianlong was consciously imitating his grandfather in conducting those tours and using them as occasions to strengthen Qing control over the Chinese cultural elite.[15]

---

12. Wang Zhao, "Guan hu dong zhi," 73, 78, 83. Wang uses Jiang Tingxi's scroll painting of a "yellow flower," which he identifies as "wild poppy" (*ye yingsu* 野罌粟) to substantiate the claim that Jiang did not paint the flower while he was touring in the frontier region, given that the leaves were misplaced on the scape of the flower rather than toward the base of the stem, a feature much more often observed in common poppies.

13. See Köhle, "Why Did the Kangxi Emperor," esp. 91–100.

14. Elliott and Chia, "Qing Hunt at Mulan"; Allsen, *Royal Hunt in Eurasian History*, 45–46.

15. Chang, *Court on Horseback*, 81–86; Elliott, *Emperor Qianlong*, chap. 5.

The 1703 hunting tour, then, became the crucial occasion that introduced Manchu and Mongolian flora into Chinese-language encyclopedic works such as *Guang Qunfang pu* a few years later. The diaries by Wang Hao and Zha Shenxing departed from previous travelogues written by scholar-officials such as Gao Shiqi, who followed Kangxi on tours in the 1680s to Manchuria and to Mount Wutai.[16] As we saw, Gao's list of Manchu words contains a substantial number of plants and animals, but he largely collected those ethnographic facts from the "local people" (*turen*). Although Kangxi also hunted during his 1683 tour to Mount Wutai, as Gao's diary records, the focus of Gao's account was always to carefully gloss the cultural landscape of the imperial itinerary by quoting previous historical and geographical accounts in classical Chinese.[17] By contrast, twenty years later, it is very clear that Kangxi used the court audience (conducted under a tent) to bring his entourage's attention to particular species of plants and animals unknown to the southerners. The scholars were not just there to gloss the cultural landscape in order to further exalt the imperial tour. They were there to bear witnesses to objects that they could not yet name but were tasked to render legible for a wider audience. The power dynamics was inverted on this tour. Rather than the learned Chinese scholar quizzing the local people for new facts, it is the Manchu emperor (and occasionally the heir apparent) who glosses and describes wonderful creatures to the awestruck scholars.

Having examined the significance of the 1703 imperial hunting tour and the inspection of frontier flora and fauna at the moving court, we turn to concrete examples of how *Guang Qunfang pu* described these new plants and their relationship to the Chinese courtiers' private writings. In each example, we also cross-examine the Chinese accounts with those of the 1708 Manchu *Mirror*, as well as the complex ways in which some Inner Asian plants acquired a stable presence in literary Chinese discourse. At the same time, *Guang Qunfang pu* generated cross-references between those Inner Asian plants and similar species previously documented in Chinese sources. The result is a

---

16. Gao was already part of the entourage that headed north to Uliyasutai in 1683, but because of illness he returned to Beijing. See Gao Shiqi, *Saibei xiaochao*.
17. For a related discussion, see Rogaski, *Knowing Manchuria*, chap. 1.

multilayered narrative that not only highlighted the distinct character of a habitat beyond the Great Wall, but also engendered possible links and identifications to more familiar plants in China proper.

Iconic Species: Golden Lotus and Luminescent Wood

Much as he praised the fresh appearance of the yellow flower in the face of harsh frost, Kangxi was fond of making moralizing allegories out of plants and animals encountered during his tours. Even a seemingly mundane plant, such as one that bears numerous small pink blossoms on one stalk, received a new name from the emperor. This "Ten-thousand-year flower" (*wannianhua* 萬年花) he named to reflect the fact that the flower "does not change its color even after drying for a long time or when touched by mud" 雖經久乾枯及沾泥汙，顏色鮮新不變.[18] The imperial tours provided ample opportunities for Kangxi to try out lyrical tropes long used in the Chinese tradition on new species found in Manchuria or the Mongol lands. The results, either short poems or more elaborate genres such as a rhapsody (*fu*), found their way into *Guang Qunfang pu* and the emperor's growing anthology of verse and prose (*shiwenji* 詩文集). The presence of imperially composed verse in *Guang Qunfang pu* thus represents a very conspicuous effort to insert new referents into the well-established Chinese literary tradition of "poems on things" (*yongwu* 詠物), in which mundane objects were lyricized and inscribed with sublime virtue.[19]

One such iconic plant that received extensive attention in *Guang Qunfang pu* was the "Golden lotus flower" (*jinlianhua* 金蓮花). Encountered during Kangxi's trip to Mount Wutai, the flower's golden color (cf. Ma. *aisin* 'gold' in Aisin Gioro, the name of the imperial clan) and niche habitat beyond the Great Wall made it an ideal subject. This flower is described as follows in the entry in *Guang Qunfang pu*:

---

18. Wang Hao, *Yuding Peiwen*, vol. 846, 53:1a (518, lower panel).
19. The translation of the term is from Xiao and Knechtges (*Wen xuan*, 1:12), where it occurs in the singular.

Golden lotus flower. Grows on Mount Wutai in Shanxi. Particularly abundant beyond the border. The color of the flower is golden-yellow, it has seven petals arranged in two layers. The center of the flower is likewise yellow, with densely but neatly arranged stamens surrounded by sharp-tipped, tiny, narrow, and yellow-colored petals. There are several flowers on one stalk. Each flower resembles a lotus but smaller. Blossoms bloom in the sixth month, and one can appreciate the spreading and shiny golden color all over the ground with one glance. In the fall, the flowers dry out but do not fall. The seeds are millet-sized and black. The leaves are green, narrow, point-tipped and long, with five or seven lobes.

金蓮花。出山西五臺山，塞外尤多。花色金黃，七瓣兩層，花心亦黃色，碎蕊平正，有尖小長狹黃瓣環繞其心。一莖數朵，若蓮而小。六月盛開，一望徧地，金色爛然。至秋花乾而不落，結子如粟米而黑。其葉綠色瘦尖而長，五尖或七尖。[20]

In addition to this meticulous description of the plant itself, *Guang Qunfang pu* referred to a record in *Liaoshi* 遼史 (History of the Liao) about how Khitan rulers several hundred years earlier had appreciated this flower during similar hunting trips, thus establishing a continuity between Qing rulership and previous conquest dynasties in Chinese historiography. The text also cited a gazetteer of Mount Wutai that associated this flower, which resembles "true gold emerging uprightly from the verdant land," with a "sacred miracle" (*shengji* 聖蹟) performed by the Mañjuśrī Bodhisattva, protector of the Mount Wutai temples.

All these references helped the reader comprehend the intricate allusions included in the twenty-seven-couplet rhapsody that was likewise included in *Guang Qunfang pu* following the quoted entry. Composed by the emperor himself, the rhapsody praised the flower's "rectifying color" (*zhengse* 正色), which constituted a metonymic reference to Qing rule.[21] Su Ya-fen 蘇雅芬 first traced appearances of the golden lotus flower across a diverse range of literary texts and paintings that included scrolls and albums.[22] Wang Zhao further highlighted the plant's central role in a new kind of natural-historical (*bowu* 博物)

---

20. Wang Hao, *Yuding Peiwen*, vol. 846, 53:1b (518, lower panel).
21. Wang Hao, *Yuding Peiwen*, vol. 846, 53:2a–3a (519, upper and lower panels).
22. Su, "Jiang Tingxi *Hua qunfang*," 64–65, 72–74.

learning fostered by the Kangxi emperor.[23] In multiple scroll paintings and albums produced by courtiers such as Jiang Tingxi 蔣廷錫 (1669–1732), the golden lotus flower featured prominently and later was widely transplanted in Qing palaces in and beyond Beijing.

It would be wrong, however, to interpret Kangxi's strong voice in glorifying this flower as an authentic contribution to Chinese encyclopedic learning that stemmed from the emperor's Manchu upbringing. Anecdotal evidence suggests that the "terrestrial golden lotus" (*han jin lian* 旱金蓮) as a unique product of Mount Wutai had become a commodity that "travels far" (*zhiyuan* 致遠). During the early years of his sojourn in Beijing, Zha Shenxing recorded the enjoyable occasion when he and his friends made tea by pouring boiling water over this dried flower, which unfurled its lovely golden petals inside a tea cup. It was only during his 1703 trip accompanying the emperor beyond the Great Wall that he saw the flower growing in the wilderness for the first time.[24] It was thus by no means an unfamiliar creature to the Chinese before the eighteenth century. Furthermore, we note that the flower was not recorded in the Manchu *Mirror* of 1708. Only after its iconic status was cemented in *Guang Qunfang pu* and other court texts in the 1710s and 1720s did the flower belatedly, during the Qianlong reign, acquire an invented Manchu name (*aisin šu ilha*).[25]

Another plant in *Guang Qunfang pu* that acquired the edifying praise of an imperial rhapsody is not a glittering flower but a rotten tree that glowed at night. The entry for the "night-glow wood" (*yeguang mu* 夜光木) reads as follows:

23. Wang Zhao, "Guan hu dong zhi," 83–86.
24. Zha Shenxing, *Renhai ji*, 24.
25. *Yuzhi zengding*, vol. 233, supplement, 3:35a–b (395, bottom panel): "Grows outside the border [i.e., in Mongolia]. The flower is somewhat similar to the lotus [*šu ilha*, i.e., Ch. *lianhua* 蓮花] but smaller. As it emerges from the stem it is a golden color, has seven petals in two layers. Several flowers can bloom from one stem" (*aisin šu ilha. jasei tulergi ba bade banjimbi, ilha šu ilha de adalikan bime ajige, jilha ci aname gemu haksan boco, nadan fiyentehe juwe ursu, emu cikten de ududu ihan ilambi*).

(Addition:) Night-glow wood, also called "shining wood." Beyond the border, trees die and wither after a long time. Their roots do not rot, but instead entangle themselves deep in the earth. With water seeping through them, the roots become luminous and transluscent inside and out, just like the legendary "pearl tree" that glows at night. Its rays can illuminate objects. In order to get it, one must find—during daylight—the damp, dim and blackened parts of decaying trees, and remove the bark. If light emanates from the trunk, then it is the right kind. When washed clean of mud and sand, the entire material becomes crystal clear. Only months later will it dry up and its luminescence diminish.

〔增〕夜光木一名亮木。塞外木經久而枯，其根不朽，蟠廻連理于土中。水漬木根，光輝透徹，中外一色，有類珠樹夜明，餘可以燭物。取者晝日從朽株中，驗其滲濕昏黑，往取去其皮而光發則是矣。泥沙淨盡，竟體晶熒，經月水乾，光乃漸減。[26]

The encounter with this tree likewise occurred during Kangxi's frequent tours beyond the Great Wall. Unlike the situation for the "golden lotus flower," *Guang Qunfang pu* had no earlier sources on which to draw in this case. Kangxi wrote an eighteen-couplet rhapsody to "praise the marvel" (*jia qi you yi* 嘉其有異) of this "ancient wood" (*gumu* 古木) burning bright at night, as he explained in a short prefatory note. As early as 1683, Kangxi brought back a sample of the luminous rotting wood from a tour beyond the Great Wall and made a point of showing it (along with some big-horned rams) to the scholar-official Gao Shiqi, who, because of illness, had not accompanied Kangxi on the journey. Twenty years later, the emperor "passed on and showed" (*chuanshi* 傳示) this luminescent wood to Wang Hao, Zha Shenxing, and other Hanlin scholars as they stopped for the night near the Gubei Pass of the Great Wall, beyond which Mongol herders lived. Likely to the satisfaction of the emperor, the learned scholars all expressed their awe at this marvel that "the 'principle of things' cannot explain" 物理之莫解者。[27]

26. Wang Hao, *Yuding Peiwen*, vol. 847, 81:27a–28a (255, lower panel; 256, upper panel).

27. Gao, *Saibei xiaochao*, 4a–b. Cf. Wang Hao, *Sui luan ji en*, 2a; Zha, *Renhai ji*, 111. Rogaski discussed the marvel of petrified wood in *Knowing Manchuria* (3).

Despite the iconic status the luminescent wood acquired in *Guang Qunfang pu*, the name and nature of this tree were by no means straightforward to establish. Despite our best efforts, we could find no Manchu names for it in Qing dictionaries; nor does any investigation seem to have been carried out by the authorities to ascertain what kind of tree it was before the rotting and decay commenced. Yet the court-generated discourse regarding this "marvelous" (*yi* 異) phenomenon traveled beyond the court. During his trip as imperial envoy to the Ryūkyū kingdom in 1719 and 1720, the Chinese official Xu Baoguang 徐葆光 (*jinshi* 1712) observed the local people cutting open freshly caught fish from the ocean. Examined at night, the fish flesh glimmered with "bright and translucent" (*mingtou* 明透) greenish, flame-like light, which immediately reminded Xu of the "night glow trees of Rehe" 如熱河夜光木, of which he had probably learned from court texts, such as *Guang Qunfang pu*, which emperor's frequent hunting trips had generated.[28] A new discourse on biofluorescence thus grew to encompass both flora and fauna based on observations derived from different corners of the Qing empire.

In chapter 6, we examine how the genre of the rhapsody gained further significance in the hands of the Qianlong emperor as a genre for discussing Manchurian nature. Here, we turn now to examining three more mundane species: a Mongolian nettle, a crabapple-like fruit, and a broad-leafed Manchurian tree in *Guang Qunfang pu* that, albeit without the celebration of imperially composed verse, similarly reveal significant entanglement between Chinese sources and those recorded in other languages. As true for other case studies in this book, readers may choose to read one or all of the subsections or skip along to follow the main narrative.

Inner Asian Plants in Guang Qunfang pu *(1): Nettles*

As mentioned, *Guang Qunfang pu* was completed in the same year (1708) as the Manchu *Mirror*. It appears that the former's Chinese editors had no formalized channel to communicate with the latter's editorial team

---

28. Xu, *Zhongshan chuanxin lu*, 251.

even though some of the same plants were featured in both works. Furthermore, because Wang Hao worked with his extensive notes taken in imperial audience during the trips beyond the Great Wall, there is in fact no basis for assuming that his account of Manchurian plants should be considered less authoritative than that of the Manchu *Mirror*. Instead, these entries, compiled in Chinese, often include more detailed information than the corresponding *Mirror* entries.

For example, the *gabtama*, a kind of nettle discussed in chapter 1 as bearing a Mongolian name in the Manchu *Mirror*, appears in *Guang Qunfang pu* in a very different light. This plant does not appear to have been particularly well known to Manchu lexicographers living and working in China proper, given that neither Shen Qiliang (in 1683) nor Sangge (in 1700) listed it.[29] Li Yanji (in 1724), translating the definition in the *Mirror*, did not provide a Chinese name.[30] Daigu's dictionary from 1722, however, listed *gabtama* and translated it as "scorpion grass" (*xiezi cao* 蝎子草).[31] *Guang Qunfang pu* described it this way:

> Scorpion grass. Grows abundantly beyond the border. Height four to five feet, intermingles with miscellaneous weeds. Its leaves are most poisonous. Should humans touch it by mistake, the skin will immediately become red and swollen as if bitten by a scorpion. Horses do not even dare to touch it. Only camels can eat it. It is similar in kind to nettles.
> 蝎子草。塞外多有之，高四五尺，叢生亂草間，其葉最毒。人悞觸之，立即紅瘇如蝎子所螫，故名。馬亦不敢近之，唯駝能食，與蕁相類。[32]

Although the irritating effects of the prickly leaves here bear clear resemblance to the entry on *gabtama* in the Manchu *Mirror*, the

29. Hayata and Teramura, *Daishin zensho*; Sangge, *Man-Han leishu*.
30. Li Yanji, *Qingwen huishu*, 2:44a (50, lower panel).
31. Daigu, *Manju gisun-i yongkiyame toktobuha bithe*, 5:56a. This translation postdates both the *Mirror* and *Guang Qunfang pu*, but predates Qianlong's expanded Manchu-Chinese *Mirror*. The latter book upheld the translation, but clearly did not invent it. The two Manchu and Chinese terms were thus by that time already matched in commercial lexicography, so there is no reason to believe that they were not referring to the same plant.
32. Wang Hao, *Yuding Peiwen*, vol. 847, 92:35b (477, lower panel). For the common name for *qianma*, see Fèvre and Métailié, *Dictionnaire Ricci des plantes*, 361.

differences are important in this description, which is here given under a vernacular Chinese name. The first fact provided by the *Mirror* emphasizes that the plant, when young and tender, can be "blanched and eaten" (*ajige de hethefi jembi*). By contrast, *Guang Qunfang pu* renders it as a formidable creature that can be consumed only by camels (not even horses). Moreover, the *Mirror*'s phrase "stinging the hand" (*gala nukambi*) did not sound as dangerous as *Guang Qunfang pu*'s claim that its leaves were "most poisonous." At first glance, it seems that *gabtama* as an innocuous—if annoying—plant in Manchu became more menacing when presented in Chinese.

Looking at the literati writings associated with the 1703 tour, however, reveals more ambiguity and a greater multiplicity of meanings generated by the encounter between Chinese court scholars and this prickly grass. The same day that the imperial entourage crossed Han-Temur mountain range, Kangxi sent word to Wang Hao and others to alert them to the presence of scorpion grass in the area. In his diary, Wang recorded the emperor "ordering the guards to look for this grass and show us" 命侍衛覓此草以示 in case they were unaware that the plant was as "poisonous as a bee sting" (*du ru fengshi* 毒如蜂螫). Zha Shenxing confirmed the geographical distribution of the grass as extending beyond the mountain range, repeating the same analogy ("more poisonous than a bee sting"). However, Zha also pointed out the seasonality of the grass's growth cycle: "When tender, it can be eaten as food by horses; after the first frost, however, it becomes prickly, stinging and untouchable" 嫩時可供馬秣，經霜則辛螫不可觸.[33]

In these descriptions, we thus see that what the Chinese scholars were hearing about this grass in 1703 was not that different from what was said in the Manchu *Mirror*. So where exactly did they get the more extreme notion of the grass being "most poisonous"?

Let us return to the last comment made by *Guang Qunfang pu*, where it was said that scorpion grass is "analogous" (*xianglei* 相類) to the *qian* grass, which was known from Chinese records. What, then, is *qian* 蕁? The name appears at another entry in *Guang Qunfang pu* within the "Catalog of medicine" (*yao pu* 藥譜), where the editors inserted a new entry for *qianma* 蕁麻 'nettle' by quoting the corresponding content in

33. Wang Hao, *Sui luan ji en*, 5b; Zha, *Renhai ji*, 113.

Li Shizhen's *Bencao gangmu*. Also following Li Shizhen, the editors located a poem by Du Fu 杜甫 (712–770 CE) titled "Getting Rid of Thornplants" (*chucao* 除草), in which the poet noted in particular that he was weeding *qian* 薂 grass from his garden:

Their poison is worse than bees or wasps,	其毒甚蜂蠆
They are so numerous they fill the roadside.[34]	其多彌道周

Because neither Wang Hao nor Zha Shenxing made the connection to the common nettle as described in Chinese literature when they encountered scorpion grass beyond the Great Wall, we can conclude that the analogy between these two plants was made only during the compilation process of *Guang Qunfang pu*. Li Shizhen's classification of nettles as a "poisonous grass" (*ducao* 毒草) clearly also influenced how scorpion grass was portrayed in *Guang Qunfang pu*.[35] The crossover of genres (pharmacopoeia, *leishu*, and catalogs) together with the highly coordinated compilation process at court caused the shifting depiction (and attendant domestication) of a plant from beyond the Great Wall.

Last, *Guang Qunfang pu*'s description of scorpion grass is only one possible version drawn from a dense and multivalent discourse circulating in and around Beijing at the time. For instance, the Manchu bannerman official Nalan Chang-an 納蘭常安 (1683–1748) identified scorpion grass with *qianma*, and confirmed its omnipresence in the southwestern province of Guizhou because he had served in office there. In the 1760s, a Hangzhou-based southern scholar named Zhao Xuemin 趙學敏 (fl. 1753–1803) drew from both Zha Shenxing's and

---

34. Wang Hao, *Yuding Peiwen*, vol. 847, 97:42b–43b (590, top and bottom panels). The character difference between *qian* 薂 and *qian* 蕁 did not figure here. See also Li Shizhen, *Bencao gangmu*, vol. 773, 17:58a (314, upper panel). We borrow Stephen Owen's translation of Du Fu's poem. See Owen, *Du Fu Shi*, vol. 4, 49.

35. For more context of the shifting connotations of *du* 毒 in Chinese medicine, see Yan Liu, *Healing with Poisons*. In Manchu, the word *horon* bears the double connotation of being both "potent, powerful" and "poisonous," analogous to Liu's discussion of early Chinese texts. Yet *horon* was tellingly not used to describe a plant like *gabtama* in the *Mirror*.

Chang-an's writings to describe the *qian* grass as a new entity not properly accounted for in previous pharmacopeias.[36] This northern nettle with Manchu and Mongolian names thus completed its journey into the Chinese record and found its kin both in older accounts and in distant regions.

Inner Asian Plants in Guang Qunfang pu *(2): Bush Cherries or Crabapples?*

A week after the imperial entourage crossed the Han Temur mountain range, Kangxi made a stop near Ulaan Khad (*Wulanhada* 烏蘭哈大; cf. Ma. Ulan hada < classical Mo. *ulayanqada* 'red cliff', whence also Ch. Chifeng 赤峰 idem, the current name of the city), where he went fishing by the rapidly running river surrounded by magnificent mountains. The Chinese court scholars were likewise told to practice fishing in front of the heir apparent's tent. As they fumbled with their fishing rods, a messenger from the emperor's entourage arrived at the scene and presented a plate piled full of freshly picked fruits, "complete with stems and branches" (*lian di dai zhi* 連蒂帶枝). In both Wang Hao's and Zha Shenxing's diaries, this freshly picked fruit "growing in bushes" (*congsheng* 叢生) in abundance amid the "thick forest" (*congmang* 叢莽) of Ulaan Khad was referred to as the "Ula crabapple" (*Wula nai* 烏喇奈; cf. Ula, one of the historic Jurchen tribes) but also by the alternative name *ouli* 歐李. Both similarly noted that the fruit had a "luminously red" (*wodan* 渥丹) color and sweet-sour taste, with a size larger than cherries but no bigger than plums and apricots. In addition, the fruit could also be made into preserves with honey.[37]

Presumably, the scholars recorded what they were told by the messenger, who acted as a surrogate for the emperor by bringing the actual fruits and relating the names and uses. In a pattern with which we are by now familiar, the Ula crabapple later appeared more or less along the same lines in both scholars' accounts. Wang Hao's somewhat private premonition against consuming too much of this fruit

---

36. Zhao Xuemin, *Bencao gangmu shiyi*, vol. 994, 5:65a–b (656, lower panel).
37. Wang Hao, *Sui luan ji en*, 7a; Zha, "Peilie biji" 228. Zha caught thirty fish on that day. Cf. Zha, *Renhai ji*, 114.

of "hot nature" (*xingre* 性熱), however, was a concern apparently not shared by Zha Shenxing.[38] Our inquiry could very well conclude here with this being yet another example of how imperial tours brought heretofore-unfamiliar species into a written Chinese-language discourse. Yet if we scrutinize the two names likely supplied by the emperor's messenger, we already see a bifurcating interpretation of the fruit between that of a crabapple (*nai* 柰) and a plum (*li* 李). The ontological ambivalence is further complicated by the fact that both names could have been phonological imitations of non-Chinese names for fruits documented in the Manchu *Mirror*. The Ula crabapple could well have been referring to the Manchu word *ulana* (< Mo. *ulan-a*, a variant of *ulayan-a* 'red currant'). If so, the Chinese syllable *nai* used by the Chinese diarist would be a phonetic transcription of the last syllable in the Manchu name, albeit one deliberately represented using a Chinese character that meant crabapple. Note that the original Mongolian name is related to *ulayan* 'red', a word that occured in the name of the town where the Zha and Wang encountered the fruit.

In the Manchu *Mirror*, *ulana* occurs as an alternative name for *mamugiya*, defined as follows:

> *mamugiya*. Name of a fruit. The color is red. It is like a cherry. The taste is sour but agreeable [or 'sweet']. Also called *ulana*. Also called *fulana*.
>
> *mamugiya, tubihe-i gebu, boco fulgiyan, ingtori adali, amtan jušuhuken bime jancuhūn. geli ulana sembi. geli fulana sembi.*[39]

The ground for understanding *ulana* as a kind of crabapple (*nai*) is thus already weakened by the fact that *ulana* was described in Manchu as being "like a cherry." The alternative name *ouli*, similarly, is also probably a transcription of the Manchu word *uli*, which likewise referred to a fruit. According to the Manchu *Mirror*, the nature of *uli* is

---

38. Wang Hao, *Yuding Peiwen*, vol. 846, 67:48a (797, upper panel).

39. *Han-i araha manju gisun*, 1708, 18:*tubihe-i hacin*:48b. The 1717 Manchu-Mongolian *Mirror* translated *mamugiya* as *ulan-a* in Mongolian (*Han-i araha manju gisun*, 1717, 18:*tubihe-i hacin*:119a). For the translation of the Mongolian, see Lessing et al., *Mongolian-English Dictionary*, 870–71.

still further complicated by the presence of a "Chinese" (*nikan*) variety of *uli*. First, the entry for the common *uli*:

> *uli*. Name of a fruit growing on trees. Smaller than haw in size. The color is deep red, the taste sour. In the "Nanshan you tai" 南山有臺 chapter of "Xiao ya" 小雅 of the *Poetry Classic*, it says: "On the Southern mountain there are *qi* willows" 南山有杞.[40]
>
> *uli, moo de banjire tubihe-i gebu umpu ci ajige boco fahala fulgiyan amtan jušuhun. ši ging ni siyoo ya-i nan šan io tai fiyelen de, julergi alin de uli bi sembi.*

Now the entry for the "Chinese *uli*":

> Chinese *uli*. The shape is like the crabapple, the taste slightly sour. The Chinese call it *huahong* [花紅 'the red of the flower']. It can be preserved in honey and eaten.
>
> *nikan uli, banin šag'o de adali, amtan majige jušuhun, nikasa huwa hung sembi. hibsu de gidafi jembi.*[41]

As discussed in chapter 2, quotes from Chinese literature, especially the Confucian classics, were frequently appended to definitions in the *Mirror*. The entry on the *uli* presents another example of this practice, whereby an ultimately Chinese quote helped anchor a Manchu word *cum* thing (here, *uli*) in the tradition of classical learning. The quote from the *Poetry Classic* is curious, however; *uli* is here called upon to stand in for quite different things.[42]

---

40. Translation from Karlgren (*Book of Odes*, 116–17), with a silent change of *k'i* into *qi*.

41. *Han-i araha manju gisun*, 1708, 18:*tubihe-i-hacin*:47b–48a.

42. In the passage quoted, *uli* is used to translate *qi* 杞. Bernhard Karlgren (1889–1978), whose translation we use here, treats the word as referring to a kind of willow. The reference of the Chinese character in the context of the *Poetry Classic* was obscure by late imperial times. The commentary in the edition from which the Shunzhi court translators worked when they produced the Manchu text quoted in the *Mirror* says that "the *qi* tree is like the Ailanthus (*chu*)" (*qishu ru chu* 杞樹如樗), a tree that is mentioned in the *Poetry Classic*'s poem "Qi yue" 七月 as a kind used for firewood. See *Shizhuan*

At the same time, the *Mirror* posits a clear difference from the "Chinese" *uli*, which tastes less sour, has a very stable vernacular name in Chinese (*huahong*), and belongs within the general kind of crabapples (Ma. *šag'o* < Ch. *shaguo* 沙果). Taken together, it is clear that what was described to the court scholars was some kind of hybrid version of *uli* that combined some characteristics of *nikan uli* (preserved in honey). All of a sudden, the ontological identity of "Ula crabapple" and *uli* becomes uncertain again. What was it that they saw and tasted out there by the forest of Ulaan Khad in 1703?

Going further back to earlier (and later) records by Chinese observers in the region, we realize that *uli* already caught Gao Shiqi's attention in 1683 simply as *wu-li* 烏立, with a different vernacular Chinese name *laoya yan* 老鴉眼 'the old crow's eye'.[43] Fang Shiji, writing from the western region of Manchuria in the 1710s, described *uli* (*ou lizi* 歐李子) as found in both Ningguta and Aigun.[44] Both Gao and Fang, however, seem to be describing a wilder version of bush cherries with soft stems and small leaves, small delicate white flowers, and purplish "round-like-pearl" (*yuan ru zhu* 圓如珠) fruits that resembled "small plums" (*xiaoli* 小李) with sour taste. More important, according to Gao Shiqi, *uli* was already made into filtered and concentrated juices for enjoyment in the 1680s. So it was indeed possible that the fruit picked from Ulaan Khad, which Zha Shenxing also described as growing on densely grown shrub rather than trees, was a red-colored *uli*

---

daquan, 78:9:53a (583, lower panel); Karlgren, *Book of Odes*, 99. On the Chinese edition used for the Manchu translation, see Tu, "On the Source Text." In this poem, the word is there translated into Manchu as *mukdehen* 'desiccated (or 'excrescent' [glossed as *furu moo*, with this meaning, in the commentary to the Manchu translation of "Qi yue"]) wood'. In the Manchu version of "Nanshan you tai," which is quoted in this entry from the *Mirror*, the *uli* is likewise compared to desiccated or excrescent wood (*uli moo mukdehen-i adali.*). See *Han-i araha Ši ging bithe*, chap. 9 (n.p.). Curiously, the kind of tree mentioned in the line of the poem that immediately follows is *li* 李 'plum', which the Manchu text gives as *foyoro* idem; a juxtaposition intended, perhaps, to indicate a similarity or complementarity between *uli* and plum? More curiously still, in "Qi yue," *uli* is used to translate *yu* 鬱 (*Shizhuan daquan*, 78:8:11b [541, lower panel]), a word that Karlgren leaves untranslated but specifies refers to a fruit from an eponymous tree.

43. Gao Shiqi, *Hu cong dongxun rilu*, appendix (*fulu*), 3a.
44. Fang Shiji, *Longsha jilüe*, 1418.

that should not be called a "crabapple" (*nai* 柰), which could properly speaking only be used in reference to the *nikan uli*.

A few decades later, one or the other of these fruits came to be featured in an important diplomatic encounter between the Qing emissary Tulišen (1667–1741) and the Torghut Mongol leader Ayuka Khan (1646–1724). In Tulišen's Manchu-language account, his embassy first met in person with Ayuka Khan on a summer day (KX 53/6/2, 1714/7/13). After delivering Kangxi's letter and engaging in friendly conversation with the Khan, Tulišen prepared for departure. Two days later, Ayuka Khan's consort hosted a banquet to honor the guests, and two "new Manchu" warriors in Tulišen's embassy demonstrated archery skills that impressed the hosts so much that Ayuka Khan sent for the two men again the next day. When the two archers, Gajartu and Mitio, brought their bows and arrows (all finely made gifts bestowed on them by Kangxi) to meet with Ayuka again (presumably without Tulišen present), Ayuka invited them to sit down in front of a plate full of fruits. According to Tulišen, Ayuka asked the two men while pointing to the *ulana* fruit on the plate, "Does the central state also have this fruit?" (*dulimbai gurun ere tubihe bio akūn seme fonjiha*). Answered Gajartu, "Our central state also has this kind of fruit" (*ere hacin-i tubihe meni dulimbai gurun de inu bi*). In the Chinese translation of Tulišen's text, which appeared along with the Manchu version in 1723, the name of the fruit is *ouli* 歐黎, this time written with a character meaning "pear" rather than "plum."[45] The text suggests that, from the Lower Volga River to the forests of Manchuria, the fact that Tulišen and the two "new Manchu" warriors could identify the fruit across such great distances helped buttress the Qing imperial diplomatic message. After all, Manchu rulership of "our central state" of China also encompassed the Mongol world, where this fruit was known. It is interesting that back in China, when a Chinese version of the record was produced, the identity of the fruit was once more in doubt, as the discrepancy between the Manchu and Chinese versions make it unclear just what fruit the Khan had been pointing at.

---

45. Tulišen and Chuang, *Man-Han Yiyu lu jiaozhu*, 137; Tulišen and Imanishi, *Kōchū Iikiroku*, 145.

Still another century later, Xi-qing 西清 (fl. 1806), learned bannerman and great-grandson of Ortai 鄂爾泰 (1677–1745), quoted from Tulišen's account in his chronicle on the history of the Amur region to reaffirm the existence of *ouli* in the area. He noted that Tulišen used the character for "pear" rather than "plum" and described it as a lovely fruit "the size of cherries, with the appearance and taste of plums" 如櫻桃，色味皆如李, especially good to enjoy with wine when preserved with honey. Dismissing a theory that associates the Chinese name *ouli* 歐李 to Northern Song scholar-official Ouyang Xiu 歐陽修 (1007–1072), Xi-qing presented another evocative thesis that *ouli* may in fact have been the same name as *yuli* 郁李 'bush cherry', a common plant in North China and long documented in the Chinese pharmacopoeia tradition.[46] Indeed, *ouli* 蕰梨 was recorded as a vernacular name for *yuli* as early as in the early fifteenth-century *Jiuhuang bencao* 救荒本草 (Materia medica for famine relief).[47] In 1735, Mingdo had in fact translated *uli* as *yuli* 郁李 and *duli* 杜李, while using *oulizi* 甌 (?; the print is a little unclear) 李子 as a translation of Ma. *mamugiya*.[48] The Manchu name *uli*, and the Manchurian bush cherries to which it refers—frequently confused with dwarf varieties of crabapples as they were—may have found its close kin in a homely shrub known across the North China plain through the attentive eyes of a Manchu bannerman stationed at the northernmost outpost of the Qing empire.

By cross-examining Manchu and Chinese sources, we have traced the unstable identity of *ulana*, *uli*, and *nikan uli* before and after the 1703 imperial tour and the sometimes arbitrary consolidation of these names in encyclopedic texts such as *Guang Qunfang pu*. If, as in the earlier example of *gabtama* and scorpion grass, the Chinese courtiers fused what they saw and heard from the tour with the encyclopedic research they conducted afterward, the confusion between a cherry-like shrub and more common garden varieties of crabapples can be attributed to Wang Hao's willingness to simply write down what he had heard without further investigation.

46. Xi-qing, *Heilongjiang waiji*, 8:3a. For more on Xi-qing's observations on Manchurian flora, see Rogaski, *Knowing Manchuria*, 168–77.
47. Zhu, *Jiuhuang bencao*, 7:30b (824, upper panel).
48. Mingdo, *Yin Han Qingwen jian*, 19:224a–b.

Inner Asian Plants in *Guang Qunfang pu* (3): A Broad-Leaf Tree

The last example of frontier species that appeared in *Guang Qunfang pu* involves a tall, broad-leaf, and versatile tree found in the mountains of Manchuria. During the 1703 imperial tour, Wang Hao described a tree he encountered there, with its foliage displaying full autumnal glory:

> And there is this kind of white *jia* tree. Its leaves are large like fans. When tender, the leaves can be steamed to make cold noodles. After frost, the leaves turn bright red like [those of the] maple. Its bark can be made into ropes and used as fishnets. In Ula, people use it to catch big fish.
> 又有一種白椵木，葉大如團扇，初生時可蒸冷淘，霜後則鮮赤如楓，其皮可治繩為魚網之用，烏喇網大魚常用之。[49]

The entry for *jia* tree, treated as a new addition in *Guang Qunfang pu*, resembles Wang Hao's account very closely but omits some details:

> (Addition:) The *jia* tree has very big leaves like round fans. Its bark can be used to make ropes as the major thread in fishnets. It is extremely strong.
> 〔增〕椵木葉最大，有類團扇，其皮可以當麻，取為魚網之綱，牢固殊常。[50]

Zha Shenxing, who also received "imperial instruction" (*shengyu* 聖諭) regarding this tree during the tour, had a similar entry in his diary. He confirmed important details such as the edible young leaves, the bright foliage color in autumn, the leaves being "as large as round fans" (*da ru tuanshan* 大如團扇), and the use of its bark to make ropes and fishnets. In addition, Zha recorded that the pliable and durable inner bark was furthermore used to make serpentine—a slow-burning cord

---

49. Wang Hao, *Sui luan ji en*, 5b.
50. Wang Hao, *Yuding Peiwen*, vol. 847, 81:26b (255, upper panel).

(*huosheng* 火繩)—in matchlock muskets.[51] Later, in his *biji* based on the diary entry, Zha added a few important reflections on his encounter with this tree: first, that this tree was "not found in the central state [of China]" (*Zhongguo suo wu* 中國所無). Second, Zha noted that although the leaves of this tree resemble that of tallow (*wujiu* 烏桕), the edges of its leaves were not smooth but jagged (*juchi* 矩齒). Last, although he heard this tree referred to as *duàn*, that pronunciation is an awkward fit for the character he uses, which appears in classical sources and is pronounced *jia* 椵 in modern Mandarin.[52] Here Zha Shenxing poses a conundrum of finding a stable Chinese name for the compellingly new presence of this tree to the classical corpus. What if the sound of its name—as relayed through the word of the emperor himself, no less—does not square with the glosses around the Chinese character one's learned colleagues have assigned for it?

The confusion between the two Chinese characters, *jia* and *duan*, already appeared in the first *Shengjing tongzhi*'s list of trees growing in the region. It does not help that the two characters were already entangled in the ancient lexicon *Erya*, where the character *jia* was in turn marked as a pomelo-like fruit tree, a hibiscus-like shrub, and a white poplar-like tree.[53] The gazetteer compilers sought a compromise by using the character *duan* to mark how the tree's name sounds locally and quoting the *Erya* to say that in fact what the region has is the poplar-like referent of the *jia* tree:

> It grows in mountains everywhere. A slightly reddish kind is called purple *jia*. It is firm and strong, and can be used to build houses and utensils. Another slightly yellowish kind is called chaff *jia*. Its texture is loose.

---

51. Zha, "Peilie biji," 224–25.
52. Zha, *Renhai ji*, 113.
53. This was likely responsible for the unfortunate rendering of the storied island Kado 椵島 near the Sino-Korean border, which was frequently translated as "Pomelo Island" due to the multiple referent of the character *jia*. For a list of polysemous characters that describe more than one kind of plant or animal, see Carr, "Linguistic Study," 528–31; for Carr's discussion on *jia*, see 531.

各山中皆有之。微紅者曰紫椵，堅實可為屋材及器皿。稍黃色者曰糠椵，其質鬆。⁵⁴

The exile's son Yang Bin also encountered the same tree during his time in Ningguta. Without the burden of reconciling it with the *Erya*, Yang simply refers to this tree as he heard it called as *duan*:

> *Duan* resembles gingko trees. It can be made into utensils; its bark can be used as shingles on the roof. When soaked in water for a long time, the bark can be made into ropes.
> 椵類銀杏，可爲器，其皮可爲瓦，浸水久之，可索綯。⁵⁵

As we can see, by simply going with the vernacular sound, Yang was able to describe the same tree without plunging into conundra posed by classical Chinese lexicography.

Another textual reason for insisting on the character *jia* to name this tree derives from the pharmacopoeia (*bencao*) tradition. The *bencao* pharmacopoeia since the late fifth century records an allegedly "Koryŏ" (*Gaoli* 高麗) ditty in praise of ginseng:

> The Koreans have praised ginseng as follows: "Three branches and five leaves / Back against sunlight and facing toward the shade / Those who seek my whereabouts / shall find me near the *jia* trees." The sound for the tree character is *jia*. It resembles parasol trees, being rather large in size.
> 高麗人作人參讚云：三椏五葉，背陽向陰。欲來求我，椵樹相尋。椵音賈，樹似桐，甚大。⁵⁶

---

54. *Shengjing tongzhi*, 21.8a. Cf. the more elaborate record in the *Qinding Shengjing tongzhi*, vol. 2, 27:11a (1311).
55. Yang, *Liubian jilüe*, 4:4a.
56. An excerpt from Tao Hongjing's 陶弘景 (456–536) commentary, quoted in Li Shizhen, *Bencao gangmu*, vol. 773, 12:13a (9, lower panel). Tao was comparing Koguryŏ ginseng with the kinds found in Shangdang (southern Shanxi) and also Paekje, which he considered superior to that in the Koguryŏ territory. *Shengjing tongzhi* refutes this rating based on the more recent valorization of Liaodong ginseng. The same ditty was

That *Guang Qunfang pu* also quotes extensively from the pharmacopoeia corpus, including this association of ginseng with the *jia* trees, may have been part of the reason for the persistent use of *jia* to name this tree encountered during the hunt, even if it sounded different.[57]

None of this would have posed a problem to this tree, which was known as *nunggele* in Manchu consistently across different sources. The Manchu *Mirror* describes it thus:[58]

Black bark, big leaves. The wood [lit. flesh of the tree] is fine and soft. It is used for carving flowery [patterns].

notho sahaliyan, abdaha amba, moo-i yali narhūn uhuken, ilha arame foloro colire bade baitalambi.[59]

In the court manual for performing ritual offerings, the broad *nunggele* leaves—freshly harvested in the early summer—were used to wrap steamed sticky rice balls with azuki bean fillings.[60] Shen Qiliang's dictionary from 1683 listed *nunggila* (read: *nunggile*) *moo* and *nunggele moo* with the translation *jiamu* 椴木 '*jia* tree', using the same Chinese character as *Shengjing tongzhi*.[61] (The *nunggele* tree later was rendered as *duan* 椴 in the *Ode to Mukden*, which we examine in more detail in chapter 6.) Over time, it was the vernacular name of *duan* that came to be used as the standard common name for this plant, a Northeast Asian kin to the linden tree (tilia) of Europe.

By directing their attention on certain characteristic plants beyond the Great Wall, the emperor and the court literati—who related their

---

also quoted by multiple literati authors, including Wang Shizhen. See Wang Shizhen, *Chibei outan*, vol. 2, 340 (no. 658).

57. Wang Hao, *Yuding Peiwen*, vol. 647, 93:8a–b (483, top panel).

58. Tsintsius's *Sravnitel'nyi slovar'* (vol. 1, 611) is not useful for this tree because only the Manchu word is listed.

59. *Han-i araha manju gisun*, 1708, 19:*orho-i hacin*:11b. Our translation has benefited from that in Imanishi, Tamura, and Satō's *Gotai Shinbunkan yakukai* (vol. 1, 861 [item 15170]) and Li Yanji's *Qingwen huishu* (2:32b [44, lower panel]).

60. This remains a local delicacy even today in parts of Liaoning. As Yang Bin noted, the similar leaves from Boluo trees were also used for the same purpose. See Yeh, *Manwen*, 398.

61. Hayata and Teramura, *Daishin zensho*, vol. 1, 52 (line 0351b1).

informal chats from the imperial excursions in writing—thus forged a new type of plant knowledge that helped bring the unfamiliar landscape of Manchuria into focus for readers at home. In the next section, we examine how these new creatures also found their way into the final version of *Gujin tushu jicheng,* thus preserving a key legacy of the Kangxi reign.

Frontier Flora in *Gujin tushu jicheng*

As discussed in chapter 3, Chen Menglei's first draft of this ambitious compilation already prominently featured plants and animals at the request by Chen's patron, Prince Yinchi. Rather similar to the kind of lexicographical encyclopedism displayed in *Guang Qunfang pu*, which was being compiled at the same time at the Kangxi court, Chen's encyclopedia divided plants and animals into several hundred minute entries rather than broadly defined "categories" (*lei*), a move that defied earlier conventions of *leishu*.

Earlier scholars maintained that the draft encyclopedia's fate was dramatically changed after Yinchi failed in his bid for imperial succession to his younger brother Yinzhen, who became the Yongzheng emperor in 1723 and immediately sent Chen Menglei back into exile. Yongzheng then ordered his own trusted court officials to take over the draft and eventually published it in 1728 as *Gujin tushu jicheng* in ten thousand fascicles, printed with copper movable type (with xylography for images) at the Hall of Martial Valor.[62] In this narrative, the focus is placed on the dramatic story of Chen Menglei and the encyclopedia being a ploy in court politics.

The publication of archival materials from the Imperial Household Department beginning in the 1990s, however, led to a modified consensus and new insights. In his recent study building on earlier works, Xiang Xuan 項旋 showed convincingly that a designated "Office" (*guan* 館) opened in the spring of 1716 (KX 55) with imperial permission, and that it was probably the Kangxi emperor who ordered the

---

62. For the history of movable type printing at the Qing court, see Weng, *Qingdai Neifu keshu yanjiu*, vol. 2, 254–60.

dramatic expansion of Chen Menglei's draft to add multiple "modern" (*jin* 今) sources, including *Guang Qunfang pu*. Chen Menglei was tasked to recruit more than sixty experienced scholars to work in said office, which likely was located in the vicinity of the Hall of Martial Valor, and that printing by movable type was largely completed by the time of the emperor's death in 1722. Furthermore, according to the memorials presented to Yongzheng by Jiang Tingxi, the newly appointed editor in chief following Chen's exile, Yongzheng explicitly ordered Jiang to not make substantive revisions to the "already printed" (*yishua* 已刷) portions of the text in order to avoid gossip that the new emperor was wantonly making alterations to his late father's last great scholarly project.[63] Hence we can conclude that the singular interest in plants and animals in the final layout of *Gujin tushu jicheng* to be a product of the late Kangxi reign that combined both Chen Menglei's faithful execution of Yinchi's orders and the new travel literature directly occasioned by Kangxi's hunting trips in the 1700s.

It is not our goal here to conduct a thorough analysis of the sections on plants and animals in *Gujin tushu jicheng*, but we want to highlight the influence of court compilations such as *Guang Qunfang pu* in the late 1700s on the final form of *Gujin tushu jicheng*, especially its overall presentation of natural erudition (*bowu* 博物). Iconic species and frontier flora encountered during the imperial hunting tours, still unknown to Chen Menglei when he compiled his first draft in the early 1700s, appeared in the finished imperial encyclopedia and were largely identical to the accounts given in *Guang Qunfang pu* via Wang Hao's hands. By now, we are familiar with several of them: the golden lotus flower of Mount Wutai; the luminous wood beyond the Great Wall; the Manchurian fruits *ilhamuke, fafaha,* and *misu hūsiha*; as well as the *duan* or *nunggele* trees, with their unstable references in classical Chinese lexicography. These all appeared in various sections in *Gujin tushu jicheng*.[64]

---

63. Xiang, *Qingdai* Gujin tushu jicheng, 55–83; Weng, *Qingdai Neifu keshu yanjiu*, vol. 2, 260–64.

64. For *ušarki / ula* 'crabapple', see Chen Menglei and Jiang, *Qinding Gujin tushu jicheng*, vol. 556 (17b, bottom panel; *juan* 313); for various flowers quoting *Guang qunfang pu*, see vol. 546 (10b, bottom panel; 11a, top panel; *juan* 183); for the "night luminous tree," see vol. 556 (23a, middle panel; *juan* 314); for entries quoting from the 1684 *Shengjing tongzhi*, including *ilhamuke, fafaha, misu hūsiha*, see vol. 556 (18a, top

This evidence of editorial intervention, then, tells us that the last great imperial encyclopedia owes its scope and coverage to Kangxi's conscious effort to edify frontier flora and fauna as imperial knowledge, and should thus not be understood as a single literati author's creation.

Furthermore, that Yongzheng appointed Jiang Tingxi as chief editor speaks to the new emperor's intention of continuing—rather than veering away from—Kangxi's agenda. As noted earlier, beginning in the 1700s, Jiang was being groomed in the same rarefied circle as Wang Hao and Zha Shenxing by accompanying the emperor in hunting tours beyond the Great Wall. As a capable and much sought-after artist, Jiang Tingxi was also instrumental in extending the imperial vision for botanical and zoological interest into a pictorial format, as others have shown in studies of his flower albums commissioned at court. Several of the European natural-historical images that the Jesuits brought to the attention of scholars at the Qing court were moreover duplicated in *Gujin tushu jicheng*, likely overseen by Jiang. In chapter 7, we examine new pictorial albums (called *pu* 'catalogs') of animals made at the Qing court that, we argue, should be considered as continuing the lexicographical encyclopedic tradition presented in *Guang Qunfang pu*, which unlike them was limited to plants.

Even though *Gujin tushu jicheng* is usually hailed as the pinnacle of the Chinese encyclopedic tradition (*leishu*), we also need to bear in mind how this book, together with other early eighteenth-century court compendia such as *Guang Qunfang pu*, subverted previous conventions by highlighting the sheer abundance and diversity of botanical life through rigorous and exhaustive textual research. Displacing the *leishu* genre's earlier primary function as a handy aid to literary composition, late Kangxi-era imperial scholarship did not just concern itself with lyricizing the elegant beings that pleased the literati gaze. Instead, it offered an encyclopedic vision that was both rigorously embedded in the classical tradition and open to local and vernacular additions, including from Manchu and other languages. The universality of this new encyclopedism was construed as the sum total of

---

panel; *juan* 313); for larch (Ch. *luoye song* 落葉松; Ma. *isi*) beyond the Great Wall, see vol. 547 (14a, middle panel; *juan* 197).

individual things rather than a cosmic vision of resonance across all ontological categories. In this sense, specific plants and animals played an outsized role in underpinning this new imperial encyclopedic project, simply by being the minuscule and diverse creatures they are, bearing many names in different tongues and registers. In part 3 of this book, we see how the Qianlong emperor used both *Guang Qunfang pu* and *Gujin tushu jicheng* as key references to revise and expand the Manchu lexicon in his new, Manchu-Chinese *Mirror*.

Interlude: *Gewu* with an Imperial Aura

So far, we have seen that the Kangxi emperor's remarks in *Guang Qunfang pu* on frontier flora and fauna were for the most part retold through the Chinese courtiers' writings, the exception being the occasional inclusion of some imperial poems. This relatively low-key timbre of the imperial voice, however, changed rapidly after *Guang Qunfang pu* was completed in 1708. As the studies by Catherine Jami and others have demonstrated, the emperor cultivated his own figure as a sagely ruler devoted to the Confucian precept of the "investigation of things" (*gewu* 格物).[65] Here, the subject figure who practices the Learning-of-the-Way precept of "the investigation of things" was no longer the individual literatus in his garden, but the emperor himself who governed All-under-Heaven and yet still had "leisure" to contemplate on the minutiae of plants and animals.[66] Frontier flora and fauna continued to play an important role in this newly glorified imperial persona, yet also created new challenges that called for a more systematic revision to the Chinese encyclopedic tradition.

This intensifying effort to consolidate the emperor's sagacious insight culminated in *Kangxi ji xia gewu bian*, which the Jesuit missionary Pierre-Martial Cibot (Han Guoying 韓國英, 1727–1780) referred to

---

65. Jami, "Investigating Things under Heaven." Earlier literature on Kangxi's role in the history of science largely focuses on his Western learning. See Jami, *Emperor's New Mathematics*; Han, *Kangxi huangdi*; Elman, *Their Own Terms*, 150–89.

66. We follow Elman, Duncan, and Ooms ("Rethinking 'Confucianism,'" 530) in using "Learning of the Way" rather than "Neo-Confucianism."

as the emperor's "observations on physics and natural history" (*Observations de Physique et d'Histoire naturelle*).[67] Kangxi's writings were anthologized every ten years beginning in the 1680s. The fourth and last installation, which covered his writings from 1711 through 1721 (KX 51–61), appeared in print posthumously in early 1733 (toward the end of YZ 10).[68] *Kangxi ji xia gewu bian* appeared as an integral part of this last installation, containing nearly a hundred short discussions on various topics including astronomy, meteorology, geography, philology, and—important for our purposes here—a large number of entries devoted to flora and fauna. Even though *Kangxi ji xia gewu bian* does not, overall, repeat what *Guang Qunfang pu* already covered, the relevant content to a large extent similarly derives from the emperor's extensive tours beyond the Great Wall, his military campaigns deep into Mongol territory to fight the Dzungars, and, as we will see, his extended summer stays at the new resort under construction at Rehe. In an observation of Mongolian seasonal migration customs, for example, Kangxi noted that

> Situations and circumstances beyond the border are hard to grasp by pure speculation. One must go there personally and then learn what is going on.
> 塞外情形不可臆度，必親履其境，然後能知之。[69]

The importance of travel and on-site observation was thus clearly articulated throughout *Kangxi ji xia gewu bian*.

To some extent, Kangxi's decision in the 1700s to build a permanent summer resort palace in Rehe—at a convenient site along the route of his hunting tours—represents an effort to provide a stable setting for accumulating and recording new knowledge derived from

---

67. Cibot, "Observations de Physique." On this translation, see Chen Shou-yi, "*Kangxi Ji xia gewu bian*"; Jami, "Western Learning and Imperial Scholarship," 159.

68. *Shengzu ren huangdi yuzhi wenji*, vol. 1298, "General table of contents" (*zong mulu* 總目錄), 1b–2a (1, lower panel; 2, upper panel). *Kangxi ji xia gewu bian* spans series (*ji* 集) 4, *juan* 26–31 of this work. On the editions and reception history, see Hiowan yei, Kangxi ji xia gewu bian *yizhu*, 1–4.

69. *Shengzu ren huangdi yuzhi wenji*, vol. 1299, ser. 4, 28:2a (581, lower panel).

"going personally" beyond the Great Wall. A rich literature has shed much light on the history of construction and management of the Rehe (now Chengde) summer resort, a unique landscape that art historian Cary Liu has called an "archive of power." Moreover, the Rehe site also witnessed many key political events unfolding away from the imperial capital, which provided historians with a different vantage point from which to retell a "new Qing imperial history."[70] In her recent work on *Kangxi ji xia gewu bian*, Catherine Jami highlighted the connection between imperial mobility and the construction of a new idea of the "Investigation of things."[71]

In the following sections, however, our focus is on the ways in which frontier animals and plants became introduced and domesticated in Rehe, continuing a trend in *Guang Qunfang pu*. Moreover, we are interested in how the discussion of such plants and animals in *Kangxi ji xia gewu bian* went beyond the lexicographical encyclopedism exhibited in the 1708 text and ventured toward a more integrated taxonomic vision that triangulated names recorded beyond the Great Wall with those found in earlier or contemporary Chinese records. The collection, which as mentioned was compiled after Kangxi's death, has clear links to the emperor's continued lecturing of his court officials on matters of natural history. We accordingly reference these late harangues of the aging emperor. The examples we use include wild cherries, so-called forest goji, a snow hare, the moose, a mammoth, and squirrels and bats standing in for headless monsters. Again, readers are advised to read select case studies or save the following for later.

Animals and Plants in Kangxi ji xia gewu bian (1): From Wilderness to Domestication

Many records in *Kangxi ji xia gewu bian* mention the transplantation of plants to the Rehe resort. One entry, for instance, concerns

---

70. See Dunnell and Millward, *New Qing Imperial History*; Cary Liu, "Archive of Power." The main palace halls including residential quarters were completed between 1708 and 1711.

71. Jami, "Investigating Things under Heaven," 75; on imperial travels beyond the Great Wall, 186–89.

178 CHAPTER FIVE

> A stalk of white millet that suddenly grew out of a hole in a tree in the region of Ula seven years ago. The locals used its seeds to grow a new crop, which soon yielded an abundant harvest that resulted in sweet-tasting grains with a gentle nature. Someone presented this millet to me. I ordered it to be planted in the mountain resort, and saw that its stalk, stems, leaves, and tassels were indeed twice as large as those of other varieties, and the seeds also matured faster. The grains are excellent for making steamed cakes with a clean white color like those made with sticky rice, with yet more delicate texture and fragrance.
> 七年前，烏喇地方樹孔中忽生白粟一科。土人以其子播穮，生生不已，遂盈畝頃。味既甘羙，性復柔和，有以此粟來獻者，朕命布植於山莊之內，莖幹葉穗較他種倍大，熟亦先時。作為糕餌，潔白如糯稻，而細膩香滑殆過之。[72]

The auspicious millet, which "suddenly appeared" in Manchurian wilderness, thus became domesticated as a symbol for the Qing civilizing project; the millet not only was similar to known Chinese grains but also surpassed them by virtue of its performance in yield, quality, and taste.

The transition from wilderness to domesticated setting, however, was always threatened with the potential loss of original rigor. Almost as if to mitigate this fear, one entry in *Kangxi ji xia gewu bian* mentioned the *qin-da-han* 秦達罕 (< Ma. *cindahan* < Mo. *činday-a[n]*), a large-sized hare living in the cold climate of the Khingan Mountains and now kept as a pet in the Rehe palace.[73] In the wild, the *cindahan*'s coat color resembles that of other hares (*tu* 兔) in spring and summer but changes to pure white toward the end of autumn. The female normally produces two litters in spring. According to Kangxi's observations, the *cindahan* that lived in Rehe was able to keep its shedding and reproductive cycles just as it did in its native habitat.[74] One might read the tale of this hare as an allegory of peaceful and nonthreatening domestication of frontier peoples into the fold of the Qing empire.

72. *Shengzu ren huangdi yuzhi wenji*, vol. 1299, ser. 4, 26:3b–4a (*juan* 353 in whole set, lower panel; 569, upper panel).
73. Rozycki, *Mongol Elements in Manchu*, 48.
74. *Shengzu ren huangdi yuzhi wenji*, vol. 1299, ser. 4, 26:11b (*juan* 353 in whole set; 572, lower panel); Cibot, "Observations de Physique," 464.

Some animals were not domesticated but described by Kangxi with a focus on their utility. The moose, or *kandahan*, which was described first in *Shengjing tongzhi* and subsequently in the Manchu *Mirror*, received treatment in *Kangxi ji xia gewu bian* as well (Kangxi transcribed the name as *kan-da-han* 堪達罕, a departure from the usage in the gazetteer). Kangxi described the physical features of the animal as similar to what was recorded in the *Mirror*, but added that it was found in "Soyolji and other areas" 索約爾濟等地方, referring to a mountain in the Khingan range.[75] Kangxi listed uses for parts of the moose's body, which makes his account similar to Lu Dian's in *Piya* and conversely different from both the gazetteer and the *Mirror* (see discussion in chapter 4). The moose's hooves had medical properties, according to the emperor, and its antlers—which already the gazetteer had compared to jade—he described as prized by contemporaries for making archery thumb-rings. Thumb-rings made from moose, an animal unique to the north, were superior to those made from ivory (*xianggu* 象骨). Kangxi did not give a Chinese name for the *kandahan*, nor did he make anything out of the Inner Asian character of the name.[76]

More directly related to the previous discussion on lexicographical encyclopedism, Kangxi in *Kangxi ji xia gewu bian* sometimes used Manchu names for plants to assert his authority over the supposedly new species growing beyond the Great Wall. At the emperor's order, a native Manchurian fruit tree was transplanted in large tracts all over the "thousand grove island" (*qian lin dao* 千林島) at the Rehe summer resort. The emperor described the fruit, which he referred to by its Manchu name *ying-e* 櫻額 (< Ma. *yengge*), as a smaller version of "wild black grapes" that were harvested in summer. The "fine fruit" (*jiaguo* 佳果) could be consumed fresh or made into dried powder transportable to faraway places.[77] What Kangxi did not mention is that Gao Shiqi already made a note of *yengge* in his Manchu word-list compiled in

---

75. Soyolji appears as a peak (*alin*) on the 1721 atlas of China, searchable at *Qing Maps* (https://qingmaps.org/).

76. *Shengzu ren huangdi yuzhi wenji*, vol. 1299, 27:91–b (578, lower panel). Hiowan yei, Kangxi ji xia gewu bian *yizhu*, 27, glosses *kandahan* as "reindeer" (*Rangifer tarandus*). See also Wu Weiping, "Qianlong huangdi yu yu banzhi," 99–101.

77. *Shengzu ren huangdi yuzhi wenji*, vol. 1299, ser. 4, 27:5a–5b (*juan* 355 in set; 576, lower panel); Cibot, "Observations de Physique," 471–72. We have studied the

1683, using the alternative Chinese characters *ying-wo* 英莪. Gao furthermore speculated, on the basis of hearsay, that the plant's name derived from what he called the "Parrot pass" (*Yingge guan* 鸚哥關), a place where it was found in abundance.[78] In records of the eighteenth century, this pass is called *yengge jase* (*jase* meaning "pass") in Manchu and—in a transcription of the Manchu—Ying'e 英額 pass in Chinese.[79] With the benefit of hindsight, we now know that the word *yengge* is well attested in Tungusic with the sense of wild cherry, also known as bird cherry (*Prunus avium*) because of its popularity with many kinds of birds.[80] Although Gao's association of *yengge* with parrots was clearly false, the use of the Manchu word as a loan in Chinese suggests that the plant had long been known among Chinese settlers.

We find an even more detailed account of some very similar fruits in the first *Shengjing tongzhi*, which, as discussed in chapter 4, was compiled at least partly on the basis of field investigations in the early 1680s. In the local product list of fruits, the gazetteer described a fruit named *chou lizi* 稠梨子 'dense pear'—likely a somewhat polished homophonous or near-homophonous version of what other early Qing Chinese sources had referred to as *chou lizi* 臭李子 'smelly plum'.[81] The gazetteer also mentioned the ways this fruit could be made into a dried powder used in a summer drink, which was cherished by the "local people" (*turen* 土人).[82] All in all, the Chinese sources present a clear

---

earlier edition of Kangxi's anthology and found this entry to be identical with the Siku quanshu edition. See Hiowan yei, *Yuzhi wenji*, ser. 4, 27:8a–9a (309, both panels).

78. Stary, "Manchu Word List," 583. Gao furthermore described it as bearing fruit "abundant like mulberries" 結實累累如桑椹 with a sweet taste.

79. Söderblom Saarela, "Public Inscriptions," 45.

80. Tsintsius, *Sravnitel'nyĭ slovar'*, vol. 1, 318.

81. Hayata and Teramura, *Daishin zensho*, vol. 1, 204 (line 1211b2): "wild grape; smelly plum; the name of a bird" 野葡萄。臭李子。鳥名. The unexpected mention of a bird is likely due to confusion with *yengguhe*, a kind of green or white parrot, which in turn may have come from the Chinese word *yingge* 鸚哥. See the definition of *yengguhe* in *Han-i araha manju gisun*, 1708, 19:*gasha-i hacin*:32a.

82. The gazetteer mentions under the entry for "grapes" (*putao* 葡萄) another variant called "mountain grapes" (*shan putao* 山葡萄) that were similarly small, dark-colored, had an astringent taste, and were cherished by the local people, whereas in the earlier dictionary by Shen Qiliang, "smelly plums" and "wild grapes" were two names for the same plant. Our conclusion is that the gazetteer may have duplicated

overlap with both the emperor's words in *Kangxi ji xia gewu bian* and the Manchu *Mirror* description of *yengge* as a "black-colored" and "astringently flavored" fruit that looked like small grapes (*boco sahaliyan, mucu de adalikan, amtan fekcuhun*).⁸³ What was achieved, then, by Kangxi in *Kangxi ji xia gewu bian* is thus not necessarily to introduce a species that had never been recorded in other languages before, but to elevate something that in vernacular Chinese was a none-too-impressive "smelly plum" into a "fine fruit" with a Manchu name that occupied a prominent spot in the newly built Rehe summer resort. The availability of the Manchu name *yengge*, newly rendered in two Chinese characters that suggested the fruit's morphological affinity with cherries, provided Kangxi with an important source to achieve this goal and showcase his *gewu* learning. In later Manchu lexicographical works, the equivalence of *yengge* as rendered by Kangxi and the vernacular Chinese equivalents of "mountain grapes" or "smelly plums" became firmly established.⁸⁴

We have so far described how the summer resort in Rehe came to be populated with species collected from beyond the Great Wall—an open space where such flora and fauna alike could flourish in a semi-domesticated setting. Once there, their places in the Manchu lexicon started to shift based on the new Chinese discourses also generated at court.

Animals and Plants in Kangxi ji xia gewu bian (2): Classical Lexicography and Taxonomic Speculations

It is one thing, as we have seen in *Shengjing tongzhi* and *Guang Qunfang pu*, to add frontier species with vernacular Chinese names or non-Chinese names to a compendium of encyclopedic scope. It is quite another, however, to try to triangulate species documented in earlier Chinese sources with those new names emerging from the empire's

---

two names of the same thing and made them into two separate entries. See *Shengjing tongzhi*, 21:11b.

83. *Han-i araha manju gisun*, 1708, 18:*tubihe-i hacin*:49a.

84. See, for instance, Li Yanji, *Qingwen huishu*, 10:9a (181, upper panel); Juntu, *Yi xue san guan*, 41a (137, lower panel).

new territories. Although the replacement of *chou lizi* with *yengge* may seem to be a relatively minor lexicographical move, Kangxi in *Kangxi ji xia gewu bian* also ventured to match names found in ancient Chinese sources to species in Manchuria. This attempt went even further than *Guang Qunfang pu*, which had already juxtaposed new entries for frontier flora with its exhaustive survey of earlier Chinese sources, and sometimes showed evidence of cross-examination as in the case of the scorpion grass and the use of the characters *jia* and *duan* to gloss a Manchurian tree. In *Kangxi ji xia gewu bian*, we detect a stronger desire on the part of Kangxi to venture further toward making permanent modifications to the Chinese lexicon based on both Manchu and Mongolian sources as well as his personal experience beyond the Great Wall.

One important motivation for Kangxi to seek out ancient Chinese wonders in distant lands is clearly articulated in *Kangxi ji xia gewu bian*:

> With Heaven and Earth being so capacious, there are so many wonders and marvels. What the Classics documented must have been based on evidence. Latter-day people have never known or seen them in their own eyes, therefore they cast doubt on them and do not believe their existence.
> 天地之大，奇異甚多，經典所載必有証據，後人因未親知實見，故疑而弗信。[85]

In the same paragraph, Kangxi then went on to discuss how an emissary he dispatched to Russia brought back evidence of an ancient wonder tale's (a bird cohabiting with a mouse) being real.[86]

Similarly, Kangxi anchored anecdotes about animals and plants from frontier regions to similar records found in earlier Chinese sources, many of which belonged to the tradition of documenting the "marvelous" (*yi* 異) or "divine" (*shen* 神). For example, *Kangxi ji xia gewu bian* included an account of a woolly animal carcass uncovered in the deeply frozen land of Russia, to wit, a mammoth. Kangxi had

---

85. *Shengzu ren huangdi yuzhi wenji*, vol. 1299, ser. 4, 26:3a–b (*juan* 357 in whole set; 568, bottom panel).
86. Cibot, "Observations de Physique," 470–71.

learned of the existence of this animal sometime before the spring of 1721, but probably before 1716. In that year (KX 55/3/13 or *yisi*, 1716/4/5 or 6), Kangxi had lectured his court on various events that proved stories from ancient literature right. One of them involved the *xi shu* 磎鼠 '*xi* rodent', which Kangxi had read about in *Yuanjian leihan*. This encyclopedia quoted *Shenyi jing* 神異經 (Classic of divine marvels), an "early medieval collection of fabulous tales" arranged by geographical area, which described the *xi* rodent as a huge, furry creature that lived deep beneath the ice in the frozen north.[87] Just as the existence in the present of this permanently frozen land had now been proven, so had the existence of the *xi* rodent.[88] Four years later (KX 60/3/*yichou*, 1721/3/31), Kangxi in another lecture specified just how the existence of permafrost and the *xi* rodent had been proven. It was Russian envoys to the Qing court who told the emperor of this animal. Its bones were reportedly like ivory (only slightly yellowish, according to Kangxi), and only after Kangxi had seen bracelets and combs made from it did he believe their story. This finding was then summarized in *Kangxi ji xia gewu bian*. Here, Kangxi furthermore referred to unnamed "character books" (*zishu* 字書); another indication of the lexicographical aspect of these investigations.[89]

In the record of Kangxi's first lecture, the creature was called *xi* rodent, but in the second lecture and in *Kangxi ji xia gewu bian*, it was called *fen shu* 鼢鼠, normally used to refer to the zokor.[90] This suggests

---

87. The term used is *mo-men-tuo-wa* 摩門橐窪, which is likely a feminine genitive form of the noun *mamont* in Russian. We thank Liya Xie for her help on this point. See *Shengzu ren huangdi yuzhi wenji*, vol. 1299, ser. 4, 30:8a–b (*juan* 361 in set; 595, bottom panel); Cibot, "Observations de Physique," 481. On *Shenyi jing*, see Knechtges and Chang, *Ancient and Early Medieval*, 860. The character *xi* 'stream' might be a variant of *xi* 谿, which *Shenyi jing* included as the first character of a binome referring to a beast in the shape of a dog. See *Yuding Kangxi zidian*, vol. 231, 29:4b (196, lower panel). Known recensions of *Shenyi jing* use different characters, but none of them *xi* 谿. See Wang Guoliang, *Shenyi jing yanjiu*, 100n2.

88. FHA, *Kangxi qiju zhu*, 2257; Wang Xianqian, *Donghua lu*, vol. 370, 97:4a–b (585, lower panel).

89. This entry is discussed in Liu Chao-ming, "Qingchu *Ji xia gewu bian*," 170.

90. The published *qiju zhu* or court diary does not extend to this date. In *Shengzu Ren huangdi shilu* (vol. 3, 291:14b [832, upper panel]) and Wang Xianqian's *Donghua lu* (vol. 370, *Kangxi* 107:4a [653, lower panel]), however, the character used is *fen*.

that the entry in *Kangxi ji xia gewu bian* was written up in conjunction with the 1721 lecture.

Another story that Kangxi related on the same occasion, about a beast with its eyes on its chest and its mouth by its navel—known to the Mongols and to one of his bodyguards, who had encountered it on patrol behind the (Dzunghar) front—similarly verified an account in *Shanhai jing*, but that story did not make it into *Kangxi ji xia gewu bian*.[91] The guardsmen reported that the Mongols called the creature *ob*. Some such creatures could allegedly fly, but the soldiers had not seen this kind. They just knew that the Mongols called this kind *jib*. These two words are listed in modern Mongolian dictionaries, which define them as "squirrel" and "bat."[92] Kangxi remarked that the Mongols called evil people *obtai* 'sly, shrewd, cunning' and *jibtei* 'treacherous, crafty', which to him proved that these animals were the headless creature known by different names in various editions of *Shanhai jing*, but referred to by Kangxi as *xingtian* 刑天, who made eyes out of his nipples and a mouth out of his navel after the emperor of spirits in the country of single arms cut off his head.[93]

A fish fossil that Kangxi encountered during his hunt with the Khorchin Mongols, in which the fish looked as if "still splashing the waves and swimming about" 猶作鼓浪游泳狀, in turn, was associated with similar accounts in various Chinese texts and encased in an inkstone case embellished by a *Songhua* stone, another fashionable product from Manchuria.[94]

Yet the urge to assert equivalence between ancient Chinese sources and modern-day reports may have at times prompted Kangxi to arrive at his conclusion too hastily. Notably, on examining the bright colors made from American cochineal presented to him by the Jesuits, Kangxi asserted that there should be "no doubt" (*wuyi* 無疑) that it must have been the same thing as the earlier insect-derived red pigment imported

91. Wang Xianqian, *Donghua lu*, vol. 370, 107:3a–4b (653, upper and lower panels).
92. Rikugunshō, *Mōkogo daijiten*, 291, 1511; Lessing et al., *Mongolian-English Dictionary*, 1047. These are different words than those known from the Manchu-Mongol *Mirror* and listed in Schlesinger's "Qing Invention of Nature" (71–72 [fig. 10]).
93. Lessing et al., *Mongolian-English Dictionary*, 598, 1048; Yuan, *Shanhai jing jiaozhu*, 214; Mathieu, *Étude sur la mythologie*, vol. 2, 399.
94. Named after the Sunggari River. See Ko, *Social Life*, 14.

since Tang and Song times from what was known as the Chenla kingdom of modern-day Cambodia.[95] We now know that the long-standing tradition of *Kerria lacca* production that was carried out in that country had really not been the same thing as the post-Columbian exploitation of cochineal.

It is of course not our intention here to judge Kangxi by our present-day truth claims of geographically defined biodiversity. Our point is simply to observe a prominent pattern in the ways in which Kangxi approached unusual plants and animals, especially the ones encountered beyond the Great Wall, namely, his deliberate triangulation of names in other languages with Chinese classical records. The meaning of the investigation of things for Kangxi involved more than offering lexicographical enrichment by introducing names from non-Chinese languages. Instead, it was likewise a case of lexicographical—and thereby also taxonomic—consolidation based on new insights gleaned from frontier species. In the parlance of modern taxonomy, Kangxi acted sometimes as a splitter, sometimes as a lumper.

One such example involves a shrub berry (*mumei* 木莓) named *pupan* 普盤. Identifying the berry with the character *qian* 藆 in the ancient dictionary *Erya*, and a related woody berry shrub in *bencao* pharmacopoeia, Kangxi tried to distinguish it from vine-grown berries commonly known in Chinese as *fupen* 覆盆. Going through the Chinese sources, he proposed a tripartite classification of berries into vine-grown ones, upright-growth shrub berries, and herbaceous berries that simply crawl on the ground.[96]

What was this fruit that had attracted the emperor's attention? *Shengjing tongzhi* of 1684 used two different characters for *pupan*, writing the word as 葡蟠. The unstable orthography suggests that the Chinese name was vernacular in origin. The gazetteer notably described this *pupan* as a "vine-grown" (*man sheng* 蔓生) fruit.[97] After *Kangxi ji xia gewu bian*, subsequent gazetteers of the Mukden region changed

---

95. See *Shengzu ren huangdi yuzhi wenji*, vol. 1299, ser. 4, 30:3a (593, top panel); see also Jami, "Investigating Things under Heaven," 197.

96. *Shengzu ren huangdi yuzhi wenji*, vol. 1299, ser. 4, 30:2a–b (*juan* 361 in whole set; 592, lower panel).

97. *Shengjing tongzhi*, 21:11b.

the name and description of the fruit in accordance with the imperial observation.[98]

It is more challenging, however, to identify this fruit with something in the Manchu lexicon. In the early Manchu-Chinese dictionary compiled by Shen Qiliang, the lexicographer listed a fruit called *una* (< a root meaning "to prickle") as the equivalent to *pupan* 普盻 and *pupan guo* -菓 '*pupan* fruit'.[99] Under *una*, Shen furthermore gave the translation *qi* 杞 'goji', quoting a line from the *Poetry Classic* where the *una* is used to translate *qi*.[100] Sangge's dictionary from 1700 translated the same Manchu word *una* as *pupan* 普盤.[101] The Manchu *Mirror* included the headword *una* ("red color, sweet taste" [*boco fulgiyan, amtan jancuhūn*]), under which it quoted the same line from the *Poetry Classic* that Shen Qiliang also used.[102] This identification of *una* with goji berries (*qi*) was then repeated in later compilations based on the Manchu *Mirror*,[103] sometimes while retaining the definition *pupan guo*.[104]

We can now see that *Kangxi ji xia gewu bian* effectively moved *una*—and its Chinese vernacular name *pupan*—away from the *Poetry Classic*'s reference to goji berries and instead realigned it with a different character in *Erya* as well as with the relevant descriptions of berries in the pharmacopoeia tradition. This intervention made it impossible for the *Expanded and Emended Manchu Mirror* in the 1770s to still use *qi* as the Chinese translation of *una*. Another vernacular name, *tuopan*

---

98. *Qinding Shengjing tongzhi*, vol. 2, 27:18a (1325).
99. Tsintsius, *Sravnitel'nyĭ slovar'*, vol. 2, 247.
100. Hayata and Teramura, *Daishin zensho*, vol. 1, 34 (line 030103), 36 (line 0306b3); Shen, *Da Qing quanshu*, 3:1a (57, lower panel), 6b (60, upper panel).
101. Sangge, *Man-Han leishu*, 24:1b.
102. *Han-i araha manju gisun*, 1708, 18:*tubihe-i hacin*:48b. We have benefited from the translations in Imanishi, Tamura, and Satō, *Gotai Shinbunkan yakukai* (vol. 1, 846 [items 14934, 14935]). The *Mirror* also included an entry after *una* that is *weji una* 'forest goji' ("the leaves are yellowish and shimmering, with a trunk like a shrub [or like sagebrush?]. It does not reach one *to* [the distance between thumb and middle finger on an outstretched hand] [in height]. It produces fruits like those of the *una*" [*abdaha sohokon bime gincihiyan, cikten suiha-i adali, emu to be dulerakū, una-i adali tubibe banjimbi*]).
103. Li Yangji, *Qingwen huishu*, 2:1a (29, upper panel), 12:28a (223, upper panel). See also Juntu, *Yi xue san guan*, 39b (137, upper panel).
104. Mingdo, *Yin Han Qingwen jian*, 19:224b.

guo 托盤果 'plate-top fruit', was chosen instead.[105] This example shows that it would be a mistake to conclude that *una* simply referred to two kinds of shrub-grown fruits. Instead, it suggests that the rationale by which Manchu names were connected to classical Chinese counterparts became more rigorous and precise beginning in the late Kangxi reign, notably after the first *Mirror* was completed.

Throughout *Kangxi ji xia gewu bian*, we see the emperor using the newly drawn connections between frontier observations and Chinese sources to make sweeping taxonomic claims. For instance, it included a long discourse defining the common features of willow (*liu* 柳) apart from populus (*yang* 楊), and crabapples (*nai*) from plums (*li*).[106] Again, Kangxi made several rushed judgments without substantiation, for example, by asserting that the name *linqin* 林檎 'forest birds fruit' referred to a kind of plum rather than a crabapple. Yet based on the detailed descriptions of *linqin* in Chinese sources, it is quite clear that the emperor was wrong.[107]

One last interesting example of lexicographical rearrangement—and taxonomic reconfiguration—across the division between Chinese and Manchu can be seen in Kangxi's discussion of *shanzha* 山楂 'haw'. Citing a Song-dynasty poem and the Song pharmacopoeia, Kangxi noted in *Kangxi ji xia gewu bian* that in the Shaanxi dialect, the fruit was called *wenpu* 楎桲, which closely resembles the Manchu name *umpu* in sound.[108] He speculated that the Manchu name from beyond the Great Wall may have derived from Shaanxi, or reversely from the far north to Shanxi.[109] What is implied in this claim is that this sweet and versatile fruit was already known and named several hundred years earlier in North China, and was most likely subsequently introduced

---

105. *Yuzhi zengding Qingwen jian*, vol. 233, 28:50b (174, lower panel).

106. *Shengzu ren huangdi yuzhi wenji*, vol. 1299, ser. 4, 30:6b (*juan* 361 in set; 594, lower panel); vol. 1299, ser. 4, 29:5a–b (*juan* 359 in set; 589, upper panel).

107. See, for instance, *Guang Qunfang pu*'s careful review of this term: Wang Hao, *Yuding Peiwen*, vol. 846, 57:5b–6a (594, bottom panel; 595, top panel).

108. *Han-i araha manju gisun*, 1708, 18:*tubihe-i hacin*:47a. Not listed in Schmidt's "Chinesische Elemente," and no cognates are found in Tsintsius's *Sravnitel'nyĭ slovar'* (vol. 2, 268).

109. *Shengzu ren huangdi yuzhi wenji*, vol. 1299, ser. 4, 30:7a (*juan* 361 in set; 595, upper panel).

to Manchuria along with the name by which it was currently known in that language.

It is important to note that such forced lexicographical equivalences across Chinese and Manchu also carried the risk of diminishing the ontological uniqueness of Manchurian nature. Notably, a few decades later, Kangxi's grandson Qianlong vehemently denied the validity of this claim. Starting in the next chapter, we see how the effort to invent new Manchu words for plants and animals also summoned forth a nativist tendency of abolishing this kind of fragile equivalences that were established toward the end of the Kangxi reign. In chapter 8, we discuss how Qianlong sought to bestow a different Chinese name on the haw, thus decoupling *umpu* from *wenpu*.

Conclusion

In this chapter, we have reviewed how the "court on horseback" in the 1700s and 1710s became a space that engendered imperially sanctioned new knowledge about plants and animals. In a manner that resonated with the "Houses of Experiment" in seventeenth-century England, the Chinese scholars beared witness to truth claims the emperor made en route in his excursions out of Beijing.[110] Unlike literati scholars isolated in their study discussing the investigation of things, from whom the imperial envoy Zhang Penghe distanced himself in the quote with which we opened part 2, Kangxi's ambulating lecture hall involved hands-on, personal experience. As we have seen, Kangxi kept up with his inquiry into flora and fauna across linguistic divides into the last years of his reign. By then, at home in the imperial capital or securely stationed at the new summer resort at Rehe, the emperor consistently summarized his musings as a higher form of *gewu*—the investigation of things.

110. In this sense, we might see Wang Hao and Zha Shenxing bearing an analogous role to the gentlemanly observers to experiments in seventeenth-century England as Steven Shapin described in "The House."

The problem with this grand imperial encyclopedism of the late Kangxi period, then, lies in the obvious challenge of reconciling lexicographical differences with ontological identity, or, to put it differently, how to match words to things and things to words, given the steep challenges of reconciling different vocabularies across space, time, and language. One immediate political challenge of this encyclopedic project lay in whether the Qing court was willing to uphold the uniqueness of Manchurian plants and animals while defending its ability to relate them to similar species documented in the earlier Chinese tradition. We turn to this question in the next chapter.

PART III

Neologisms for Plants and Animals at the Qianlong Court

FIG. 6.1. A miniature copy of the *Ode to Mukden* (title in Chinese) from the Qianlong emperor's curio collection. National Palace Museum, Taipei (photo by Söderblom Saarela).

CHAPTER SIX

Ode to Mukden *and the Manchu Homeland in Chinese Literary Form*

I was not a little embarrassed when I had to talk about animals; I expect to be so even more [now] that I have to talk about plants. I am no Botanist. Thus I will refrain from providing names solely on the basis of the Chinese or Tartar indications; this I leave to a Reader more able than me. I will content myself with relating what the Commentators & Dictionaries have to say, and I will give the plants the names they have in the original language, unless I know for certain their French names.

Je n'ai pas été peu embarrassé, lorsqu'il m'a fallu parler des animaux; je prévois que je vais l'être encore davantage en parlant des plantes. Je ne suis point Botaniste; ainsi je me garderai bien de donner des noms sur les seuls indices Chinois ou Tartares; je laisserai ce soin au Lecteur plus habile que moi; je me contenterai de rapporter ce que disent les Commentateurs & les Dictionnaires, et je donnerai aux plantes les noms qu'elles portent dans la langue originale, à moins que je ne sache très sûrement leurs noms françois.
—Joseph-Marie Amiot, first translator of the
Ode to Mukden, in 1769.[1]

I have endeavored, to the best of my ability, to provide European equivalents to the Manchu terms that the emperor uses in reference to the numerous natural products of the land of Mukden, and I have kept the Manchu

---

1. Amiot, *Éloge de la Ville de Moukden*, 267–68.

names only where I was unable to translate them with precision.

J'ai tâché, autant qu'il m'a été possible, de donner les équivalens européens aux termes mandchou qui désignent les nombreuses productions naturelles du pays de Moukden citées par l'empereur, et je n'ai conservé les noms mandchou que là où je n'avois pas le moyen de les traduire avec exactitude.
—Julius Klaproth, second translator of the
Ode to Mukden, in 1828.[2]

As discussed in chapter 5, Kangxi's multiple tours to Manchuria in the late seventeenth and early eighteenth centuries contributed to an increased interest in the flora and fauna of the region among Chinese literati writers. One of Kangxi's other tours—to Mount Wutai, which was located in China proper but had great significance in the Tibetan Buddhism popular among the Mongols—even inspired a Chinese rhapsody or ode to the golden lotus, a flower that also grew in Mongolia.

In this chapter, we continue our inquiries into the plurilingual scholarship on flora and fauna that had taken its form at court in Beijing but was also influenced by the exceptional status of Manchuria as the Qing imperial homeland. Shortly after succeeding the throne in 1735, Qianlong revived imperial touring with reference to his grandfather's travels, starting with the imperial hunt in 1741. Qianlong then used a hunt in 1743 as a springboard for traveling to the old Manchu capital and visiting imperial tombs in the area. Nominally, the *Ode to Mukden* that was occasioned by this trip was "written by the emperor" (Ch. *yuzhi* 御製, Ma. *han-i araha*), just like Kangxi's *Mirror of the Manchu Language*. As in the case of the *Mirror*, however, this appellation hid an effort whose collective and apparently undocumented nature has made largely anonymous. The poem was cast in the old Chinese

2. Klaproth, *Chrestomatie mandchou*, 238.

form of the *fu*, but the emperor understandably also published the poem in Manchu. It is thus the first text we discuss in this book so far that is explicitly bilingual in both Chinese and Manchu, and one that included long lists of Manchurian plants and animals.

The *Ode to Mukden* has rightfully attracted much attention among historians given the text's explicit reference to Manchu identity. Despite its title, the *Ode* was not written only about the Manchu original and secondary capital of Mukden (today's Shenyang), but also about "Greater Mukden," a notion that predated and largely overlapped with that of Manchuria.[3] The poem, Mark Elliott writes, "is a tribute to the entire region . . . whose cragged peaks, wild forests, and fertile plains were unbelievably dense with life." The *Ode* included lists of animals, birds, plants, and trees: "The list of fish and mollusks goes on for twenty lines."[4] Michael Chang has shown that the mid-Qing emperors' extensive touring were acts of governance, albeit justified as the fulfillment of filial obligations to the imperial forbears. According to him, Qianlong used the Mukden trip to strengthen Manchu imperial power by portraying it as following also "more distant *ethnic* predecessors"—the erstwhile Jurchen rulers of Manchuria.[5]

In this chapter, we pay close attention to the fact that the *Ode* went through a two-stage revision process, having first been published in 1743 and then appearing in a revised edition in 1748. Specifically, we note how the revised edition altered the way that the Manchu text of the *Ode* translated Chinese terms, including in numerous instances animals and—especially—plants. In sum, it is our contention here that the revised translations quietly launched Qianlong's project to revise the Manchu lexicon, which had been initiated a year earlier but had not yet yielded anything in print. This project of aligning the Manchu lexicon with that of Chinese through the replacement of loans by means of neologisms later grew to encompass plants and animals from outside the empire as well. That story is told in later chapters. We focus here on the *Ode* as a Chinese literary work written on the basis of Chinese

---

3. Elliott, "Limits of Tartary," 614.
4. Elliott, "Limits of Tartary," 615–16.
5. Chang, *Court on Horseback*, 94–95.

texts—but one that also carried the paramount task of enumerating Manchurian natural products in Manchu language at the same time.

This chapter focuses on the lists of plants and animals found in the *Ode*. We first describe the genre of the *fu* to which Qianlong's poem belonged. Our point here is that aggrandizing *fu* about imperial capitals had a long history; that extensive cataloging—including of the natural environment—was integral to the genre; and that by the Qing period it generally involved research in encyclopedic genres, notably including the collections of treatises known as local gazetteers.

The second section of the chapter focuses on the *Ode*'s relationship to *Shengjing tongzhi*. This gazetteer, introduced in chapter 4, appeared in several editions before the writing of the *Ode*, and furnished it with much of its plant and animal vocabulary. In terms of its description of natural products, *Shengjing tongzhi* was a Chinese text ultimately based on knowledge first articulated in Manchu and other Tungusic languages. This Chinese text was then used as a source for the *Ode*, which once more rendered parts of that vocabulary back into Manchu.

Third, we briefly situate the *Ode* in relation to earlier examples of Manchu cataloging of plants and animals and Chinese *fu* on the natural products of Mukden.

Fourth and finally, we turn to the *Ode* itself and survey what it says about the flora and fauna of the old capital and its vast hinterland. The book was distributed to serving and aspiring civil officials and highly placed bannermen, but it also reached Europe in a French translation a few decades after its publication. Joseph-Marie Amiot, a Jesuit in Beijing, did the translation. Amiot's edition of the *Ode* added another language to an already plurilingual text. Yet it did more as well. Amiot consorted with Manchu aristocrats and benefited from access to both European scholarly literature and the bilingual dictionaries that his missionary brethren had compiled in China. Most important, Amiot brought a product of court scholarship, ignored in Qianlong's poem itself, to bear on the *Ode*: the Manchu *Mirror* of 1708. Amiot used the Manchu reference work to identify plants and animals mentioned in the poem. In his footnotes, the two texts were brought together for the first time. We reference Amiot's translation in our discussion of the *Ode* because it shows another facet of the scholarship that was carried out in Beijing in several languages.

The Chinese Genre of the *fu*

The genre of the *fu* to which Qianlong's new text belonged had ancient and imperial roots; doubtless a reason the emperor considered it an appropriate form for his praise of the Manchu homeland. The *fu* 'rhapsody' or 'ode' originated as poetry recitations that synthesized earlier genres at the court of the Han empire in antiquity.[6] Unsurprisingly, as a kind of oration at the center of power, the *fu* has links to early rhetoric.[7] Indeed, the circumstances of the presentation of *fu* in the Han period are clearer than the specific literary characteristics and origins of the early texts grouped under this rubric.

The word **pah* 賦, whence *fù*, originally meant "to give, to contribute." In Han texts, the character used for this word was also used to write the unrelated but similar word **pha* 敷 > *fū* 'to spread out, extensively', cognate to **pâh* 布 > *bù* 'to spread out; display, announce, disperse'.[8] It was perhaps in this sense that *fu* "came to be applied in the pre-Han period to the 'presentation' of poems at the courts of kings and lords."[9]

However, early in its history, critics believed that this acceptation of the character *fu* 賦 referred to characteristics of the poems thus "presented." The character *fu* occurs as one of the "six Principles of Poetry (*liu yi* 六義) first enumerated in *Zhou li* 周禮 (Rites of Zhou) and repeated and explained in the Great Preface (*Daxu* 大序) to the Mao version" of the *Poetry Classic*.[10] The preface glossed the character as "to display and to enumerate."[11] Literature classified as *fu* thus came to be seen as having an exhaustive and encyclopedic listing as a central characteristic. Although unsustainable as an explanation for why early *fu* carry this name, the theory derived credence from the fact that "*fu* became the name of a type of declamatory poem commonly

---

6. Williams, "Introduction: The Rhapsodic Imagination," 2.
7. Knechtges, *Han Rhapsody*, 5–6.
8. Schuessler, *ABC Etymological Dictionary*, 173, 245. Reconstructed forms are Schuessler's "Minimal Old Chinese" forms. For the use of the character *fu* 賦 for *fu* 敷, see Karlgren, "Loan Characters," 84 (item 306); Knechtges, *Han Rhapsody*, 12.
9. Knechtges, *Han Rhapsody*, 12.
10. Rusk, *Critics and Commentators*, 7.
11. Su, "Introduction," 20.

characterized by a mixture of prose and verse, relatively long lines . . . , parallelism, elaborate description, dialogue, extensive cataloguing, and difficult language."[12] The genre "is notorious for its long catalogues of animal, plant, fish, and mineral names, many of which are often only vaguely identified even by the most learned commentators."[13] The "poems on things," a subcategory of rhapsodies, evidence this kind of cataloging. In the eastern Han, things covered in them included fruits, cicadas, horses, and different kinds of birds.[14] As discussed in chapter 4, the Kangxi emperor wrote several "poems on things" during his tours north of the Great Wall.[15] The encyclopedism of such pieces is one of a single subject, not an aspiration to comprehensive knowledge of all fields.

Rhapsodies on a more expansive subject might contain several exhaustive descriptions. Notable examples are the *fu* on the early imperial capitals. The historian Ban Gu's 班固 (32–92 CE) "Liang du fu" 兩都賦 (Two capitals rhapsody; 58–75 CE) praised the establishment of the eastern Han.[16] Zhang Heng's 張衡 (78–139 CE) "Er jing fu" 二京賦 (Two metropolises rhapsody)—a more critical poem—used a more recherché vocabulary.[17] Zhang's rhapsody included plants such as "Wood sorrel, nutgrass, eulalia, wool grass, | Vetch, brake, iris, | Arthraxon, frittilary, carex, | Hollyhock" and the elusive *huaiyang*, which neither the commentators nor the translator David Knechtges could identify.[18] Qianlong, who assumed both the role of emperor and court poet, like Ban praised his dynasty's capital and perpetuated the tradition of using rare vocabulary for the products of the natural environment.

Qianlong's poem made overt references to the classical Han rhapsodies on capital cities but not to the many *fu* written during the intervening centuries. Notably, many odes to Beijing had been written

12. Knechtges, *Han Rhapsody*, 13.
13. Knechtges, "Problems of Translating," 329.
14. Guo and Xu, *Zhongguo cifu fazhan shi*, 184.
15. See also Lai, "Qianlong chao *Niaopu*."
16. Guo and Xu, *Zhongguo cifu fazhan shi*, 153; Xiao and Knechtges, *Wen xuan, or Selections*, 1:93–180.
17. Guo and Xu, *Zhongguo cifu fazhan shi*, 158–61.
18. Xiao and Knechtges, *Wen xuan, or Selections*, 1:208–9.

during the preceding Ming period but received no mention by the Manchu emperor or his scholarly staff.[19]

Not just the long cumulation of *fu* over the centuries made writing such a poem a different matter in 1743 than it had been in the first century CE. A writer in the Qing period benefited from a mature print culture. The writer of odes could mine several encyclopedic genres for material. The mid-Qing literatus Yuan Mei 袁枚 (1716–1798) wrote perceptively that

> they had no local gazetteers (*zhishu*) and no encyclopedias (*leishu*) in antiquity. For that reason, when [Zuo Si 左思, ca. 250–305 CE] in the "Three Capitals Rhapsody" and [Ban Gu] in "Two Capitals Rhapsody" wanted to express the beauty of customs, landscapes, and natural products, no matter the mountain, river, or plant, bird, beast, fish, and insect, they had to resort to the most thorough investigations and comprehensive inquiries. It was a precise and concentrated effort. For that reason, [these rhapsodies] took three years to complete. . . . Nowadays our gazetteers and encyclopedias are fine and complete, so that if Ban and Zuo had been born into our age and then written these rhapsodies, they could have written them up after only a few days of browsing and excerpting.
>
> …古無志書，又無類書，是以《三都》、《兩京》，欲敘風土物產之美，山則某某，水則某某，草木、鳥獸、蟲魚則某某，必加窮搜博訪，精心致思之功。是以三年乃成。…今志書、類書、美矣、備矣，使班、左生于今日，再作此賦，不過翻擷數日，立可成篇。[20]

When the Qianlong emperor in 1743 mobilized his scholar-officials to write a rhapsody on the old Manchu capital, they proceeded as Yuan described. To "display and enumerate" the plants and animals of Mukden, they flipped through *Shengjing tongzhi*. This book was revised several times following its initial publication in 1684. The edition printed in 1736, the year Qianlong ascended to the throne, included new natural products not in the first edition. In several instances, the editors noted that newly added plants were harvested by the Imperial Household Department or submitted to it in tribute, which shows the

---

19. Guo and Xu, *Zhongguo cifu fazhan shi*, 24.
20. Yuan's preface in Pu, *Lidai fuhua jiaozheng*, 3.

importance of this exchange for knowledge production.[21] Many of the new plants and animals in the revised gazetteer did not have any descriptions at all, however, so the reader was left at a loss as to what they referred. A bilingual text, even without definitions, would have provided an additional layer of information. The gazetteer came close to doing that in the few instances that the lemmatized name was a Chinese transcription of a Manchu or otherwise Tungusic name, with an alternative Chinese name or "vernacular appellation" (*su hu* 俗呼) provided within the definition.[22] Even in its revised form, *Shengjing tongzhi* was thus a problematic source for a text on the nature of Manchuria such as Qianlong's *Ode*.

Shengjing tongzhi and the *Ode*

Despite its problems, *Shengjing tongzhi* was the main source for the names of plants and animals included in the *Ode*. The *Ode* was the first book the court published that juxtaposed Manchu and Chinese vocabulary for flora and fauna; as mentioned, the court had not yet published a bilingual Manchu-Chinese dictionary. The *Ode*, like other rhapsodies, relied heavily on *Shengjing tongzhi* in its notes on plants and animals—as would be expected in Qing times. Indeed, the trip to Mukden that occasioned the writing of the *Ode* also prompted Qianlong to commission a revision of *Shengjing tongzhi*, which he found unsatisfactory.[23] Curiously, the *Ode*'s commentators quoted from both the 1684 and the 1736 editions of *Shengjing tongzhi*, but without distinguishing the two for the benefit of the reader.

The notes on plants and animals, and the quotes from the gazetteer they included, were translated into Manchu in the Manchu version of the *Ode*. Thus knowledge that had originated in Manchu or in one of the Tungusic languages of northern Manchuria and then been rendered in Chinese in the gazetteer was turned back into Manchu once

---

21. E.g., *cuocao* 矬草 and *hua* 樺 in *Qinding Shengjing tongzhi* (vol. 2, 27:7b [1304], 11a [1311]).
22. *Qinding Shengjing tongzhi*, vol. 2, 27:18a (1325).
23. Zhang Yichi and Liu, "Qingdai 'da yi tong,'" 163.

more. In that sense, the *Ode* represents a step in the integration of Manchu and Chinese terminologies of the natural world. Yet the Manchu element in this bilingual edifice remained weak. Not once did the compilers quote the one product of court scholarship that actually explained plants and animals in Manchu: the *Mirror* of 1708. It was not until Amiot translated the *Ode* into French in the 1760s that the *Mirror*'s definitions were brought back into the discourse on the environment of the northeast.

Manchu Cataloging of the Natural World and Chinese *fu* on Mukden before the *Ode*

New as the *Ode to Mukden* was as a literary creation, it was neither the first time that elements of the natural world had been cataloged in Manchu, nor the first time that a *fu* was written about the natural products of Mukden. This section discusses a few examples of Manchu-language cataloging and the Manchu literatus and official Nalan Chang-an's rhapsodies on Mukden, the literary texts that most closely resemble the *Ode*.

In terms of Manchu-language texts, parts of the Manchu *Mirror* actually to some extent read like the kind of catalogs a reader might expect in a *fu*. Not only was its contents divided into chapters on semantic grounds, but within sections, words tended to be grouped further. Other texts predating the *Ode* more directly resemble a *fu*, however. We discuss two, one of which was written in French but listed Manchu vocabulary.

The first text is the Manchu-Chinese primer *Lianzhu ji* 聯珠集 (Collection of stringed pearls) from 1699, which was published as a supplement to the second edition of *Broadly Collected Complete Text in the Standard Script* (titled *Tongwen huiji* 同文彙集 [Collection in the standard script]) in 1700 and then in a trilingual edition, with Mongolian added, in 1728.[24] *Lianzhu ji* was originally a Chinese text composed

---

24. Liu Shun in the preface called the book "a volume for instructing children" (Ma. *juse tacibure emu debtelin*; Ch. *xun tong . . . yi ce* 訓童 . . . 一冊): Zhang Tianji, *Lianzhu ji*, preface, 1b; Söderblom Saarela, "Manchu and the Study," 184. We have not

by one Zhang Tianqi 張天祈 (n.d.). Zhang's student Liu Shun 劉順 (fl. 1687–1700) was impressed by the text and thought that "if translated, it would be like the pulses and grains, cloth and silk of those who study Manchu texts" (Ma. *erebe ubaliyambuci manju bithe tacire ursei turi jeku, boso suje-i [adali] ombi dere seme*; Ch. 譯而出之，誠清書家之菽粟、布棉也). Liu thus prepared a Manchu version.[25]

Lianzhu ji included, among other things, names of insects, animals, fish, and plants. These were gathered in their respective categories (Ma. *hacin*; Ch. *lei*) in different parts of the text. Because of the nature of the text as a Chinese composition that made no reference whatsoever to Manchuria, the natural environment it described was that of China proper, with many Chinese localities mentioned by name. Even though Liu had the text corrected by "famous (Manchu and Chinese) scholars" (Ma. *gebungge baksi de dasabufi*; Ch. 正諸大方滿漢), many words for plants and animals, or the products made from them, were not translated into Manchu in the text. They were instead left in Chinese transcription.

Among the aquatic creatures, Zhang and Liu listed, for example, carp (Ma. *mujuhu*; Ch. *li* 鯉), turtle (Ma. *aihūma*; Ch. *bie* 鱉), and eel (Ma. *horo*; Ch. *man* 鰻), but also several kinds of fish given only in Chinese transcription in the Manchu text, such as yellow croaker (*hūwang ioi* [< *huangyu* 黃魚] *nimaha*), mackerel (*cing ioi* [< *qingyu* 鯖魚] *nimaha*), and perch (*lu ioi* [< *luyu* 鱸魚] *nimaha*). Similarly, among the vegetables, they listed the edible bracken (Ma. *fuktala*; Ch. *jue* 蕨), celery (Ma. *gintala*; Ch. *qin* 芹), and Chinese chives (Ma. *senggule*; Ch. *jiu* 韭), but also several Chinese regional specialties that were predictably given in Chinese transcription, as well as the Sichuan pepper (Ma. *hūwa jiyoo* [< Ch. *huajiao* 花椒]; Ch. *songjiao* 菘椒), a borderline case because the Manchu word was a Chinese loan (*huajiao*), but not a transcription of the Chinese word used in this specific text (*songjiao*).[26] In its use of Chinese transcriptions instead of Manchu words, *Lianzhu ji* was similar to the later *Ode to Mukden*.

---

seen the trilingual edition, which Gabelentz described ("Mandschu-Bücher," 540–41).

25. Zhang Tianqi, *Lianzhu ji*, preface, 1b–2a.
26. Zhang Tianqi, *Lianzhu ji*, 14b–15a.

Whereas Zhang and Liu's text gave an impression (wrong, to be sure) of a Manchu language lacking in terms for plants and animals, their contemporary Dominique Parrenin (1665–1741), Jesuit in China, praised the richness of the Manchu vocabulary in this semantic field. In a letter sent to France in 1723, Parrenin gave the example of the Manchu words for dogs. He probably gleaned these words from Kangxi's *Mirror*, but presented them in a narrative form reminiscent of the cataloging of Chinese *fu*.[27]

Parrenin, who knew Manchu very well, wrote that Manchu had little use for paraphrases and circumlocutions because Manchu had relatively short words that could be used in their stead. "This is readily apparent," he wrote, "when speaking about domestic or wild, winged or aquatic animals" (*C'est ce qui se voit aisément, quand il s'agit de parler des animaux domestiques ou sauvages, volatiles ou aquatiques*). He then entered into a canine rhapsody of sorts: "I have chosen the [example] of the dog; among all domestic animals, it provides the fewest terms in their language, and still they have many more than we do" (*Je choisis celui du chien; c'est celui de tous les animaux domestiques qui fournit le moins de terms dans leur langue, et ils en ont cependant beaucoup plus que nous*): "You want to say that a dog has abundant and long hair on its ears and tail? The word *taiha* will suffice. That it has a big and long muzzle, a similar-looking tail, big ears, and drooping cheeks? The sole word *yolo* says all that" (*Voulez-vous dire qu'un chien a le poil des oreilles et de la queue fort long et bien fourni? le mot* taiha *suffit; qu'il a le museau long et gros, la queue de même, les oreilles grandes, les lèvres pendantes? Le seul mot* yolo *dit tout cela*), and so on. Parrenin listed a good dozen words, and further strengthened his point by claiming that the Manchus had twenty times as many words for horses as they did for dogs.[28]

The words *taiha* and *yolo*—as well as *beserei*, a mix between a *taiha* and a common house dog, which Parrenin mentioned next—appeared in this order in Kangxi's *Mirror*.[29] In a sense, Parrenin was transforming the reference work into a literary text somewhat like a *fu*. Later,

27. Cf. Stary, "Linguistic and Cultural Limits," 292–93.
28. Parrenin, "Lettre du Père Parennin," 225–26.
29. *Han-i araha manju gisun*, 1708, 20:*eiten ujima hacin*:2a.

after the publication of the *Ode to Mukden*, a Manchu scholar took this further and summarized the whole *Mirror* similarly.[30] There was, then, a tendency to narrate the word lists of dictionaries to display or even celebrate the richness of the lexicon. The words Parrenin listed, however, did not refer to animal species as much as to difference in appearance, sex, or age. The Manchu version of Qianlong's *Ode* actually struggled a bit even to keep up with the very rich horse-related vocabulary that literary Chinese could produce (resorting at times to compounds, which Parrenin asserted Manchu did not need).[31] However, Qianlong's translators faced a more striking challenge when they attempted to represent the species or kinds of plants and animals that the Chinese text mentioned.

Earlier examples of listing elements of the natural world in Manchu or in reference to Manchu thus exist. Yet texts more closely resembling the *Ode to Mukden* are found, by contrast, in Chinese literature. The Manchu writer Chang-an held a low-pressure post as vice president of the Ministry of War in the secondary Qing capital of Mukden for half a year in 1740 (QL 4/latter half of 12–QL 5/intercalary 6). He used his ample spare time to write about "greater Mukden." Chang-an wrote a "Shengjing wuchan fu" 盛京物產賦 (Rhapsody on the natural products of Mukden), one on its "Gourds and Fruits" ("Shengjing guaguo fu" 盛京瓜果賦), one on "Vegetables" ("Shengjing shucai fu" 盛京蔬菜賦), and more focused rhapsodies on the region's muskmelon (*xianggua* 香瓜, whence Ma. *hengke*), pine nuts (*songzi* 松子), and more. The rhapsodies, all in Chinese, listed numerous plants, animals, fish, and birds.[32] Chang-an mentioned the estates managed by the Imperial Household Department and the products submitted to Beijing in tribute.[33] We consider it likely that at least some of the individuals involved in the writing of the *Ode to Mukden* knew about Chang-an's pieces.

---

30. I.e., Juntu, *Yi xue san guan*.
31. E.g., *kara* (< Mo. *qar-a* 'black') for Ch. *li* 驪 'black horse' but *kara alha* for *yu* 騟 'black horse with white hind parts' (*Han-i araha Mukden-i fu bithe*, 43a [Chinese], 54b [Manchu]; Rozycki, *Mongol Elements in Manchu*, 133).
32. An, "Nalan Chang-an yu," 100–101.
33. Chang-an, *Shenshui sanchun ji*, 233:7:4a–b (395–96).

A Manchu *Ode*?

The *Ode to Mukden* was published simultaneously in Chinese and Manchu. Zhang Yichi 張一弛 has shown that the court circulated it fairly widely thereafter. In the capital, it was distributed to members of the elite through the educational and cultural establishment. In the provinces, it was sent to high-ranking civil and military officials through the palace memorial system. Chang-an, by then the governor of Zhejiang, received a copy this way. He was one of the officials in the lower Yangzi who asked to reprint the book to use it to instruct the gentry through the local government schools there. Chang-an was a bannerman and it is reasonable to assume that he was sent a Manchu copy of the *Ode*, but it is far from certain that the Manchu version was reprinted in the south.[34] The government at multiple levels clearly made efforts to disseminate the work, however.[35]

Although the Chinese characteristics of the *Ode* have always been obvious, no consensus has been reached in the scholarship of the status of the Manchu version. As far as we know, no one has claimed that the Chinese text was translated from the Manchu, but a resistance is noticeable in the literature to treat the Manchu text as a translation from a Chinese original. However, the publication history and the *Ode*'s coexistence within the otherwise largely Chinese poetical oeuvre of Qianlong and, most important, the characteristics of the Manchu text itself very strongly suggest that the poem was conceptualized and first written up in Chinese. This section discusses the circumstances of the text's publication and some of its characteristics. In the following section, we turn to the words for plants and animals.

Qianlong returned from his Manchurian tour after November 16 (QL 8/10/1). According to the *Shilu* 實錄 (Veritable records), the emperor "stayed over in Mukden on this day, imperially created [i.e, commissioned] the *Ode to Mukden* and prefaced it" 是日，駐蹕盛京，御

---

34. Consider, for example, that stele erected by the government in the south could forego the Manchu versions that accompanied the Chinese text on stele in the north. See Söderblom Saarela, "Public Inscriptions," 38.
35. Zhang Yichi, "Yuzhi *Shengjing fu*," 42–44.

製《盛京賦》,并序.³⁶ *Qiju zhuce* 起居注冊 (Register of rise and repose), the imperial diary, makes no mention of the poem's being commissioned on this day. The poem is only mentioned on December 10 (QL 8/10/25), when the Empress Dowager Xiaosheng Xian (1692–1777)—who accompanied Qianlong on the tour and indeed justified Qianlong's presence—entered the Meridian Gate at the palace in Beijing and was greeted by officials recounting the emperor's filial accomplishments.³⁷

The *Ode* was apparently printed the same year (QL 8), in which two months remained when the imperial retinue returned to Beijing. Two Manchu editions and one Chinese edition exist that have been thought to have been published that year. One of the Manchu editions, however, carries the words *fujurun* 'ode' and *šutucin* 'preface', whereas the other has the Chinese loanwords *fu* and *sioi,* respectively. The first pair are neologisms introduced on December 12, 1748 (QL 13/10/22), to which this edition must therefore date.³⁸

The other Manchu edition and the Chinese edition, however, were probably published either in late 1743 or in 1744. Two of the undersigned officials died in 1745, which would appear to be a terminus ante quem for the book's initial publication. On June 11 of that year (QL 10/5/12), the Imperial Household workshop was making a case for the book's printing blocks, presumably after they had been used in the production of the book.³⁹ Publication ought therefore to have taken place in 1744 or the first half of 1745.

The reprint of the poem with the introduction of neologisms in 1748 or thereabouts coincided with the publication of an edition printed in thirty-two newly made Manchu seal scripts (without commentary).⁴⁰

36. *Gaozong Chun huangdi shilu*, vol. 3, 596 (202:5b). We have to read *yuzhi* 御製 and *xu* 序 as verbs here given that they are linked by *bing* 并.

37. Chang, *Court on Horseback*, 98; *Qianlong qi qiju*, vol. 11, QL 8/10/second half, 17.

38. Zhang Hong, Cheng, and Tong, "Qianlong chao 'Qinding Xin Qingyu' (san)," 54.

39. *Qinggong Neiwu fu*, vol. 13, 649.

40. I.e., *Qianlong yuzhi*. On the seal scripts, see Stary, *Die chinesischen und mandschurischen Zierschriften*. Furthermore, Lin Shih-hsuan 林士鉉 has drawn attention to a manuscript edition with ten seal scripts held at the University of

A probable scenario is that an edition with ten seal scripts was produced and presented to the throne, upon which Chinese loanwords were detected, and the text revised. The second edition in standard script ought thus to have been printed after this event because it incorporates the corrections. Amiot sent his translation to France in 1769.[41] Amiot's translation enjoyed a readership in Europe that extended beyond the circle of Sinologists.[42]

The text of the *Ode* was changed between editions, with great consequences for our understanding of how it was written (we will return to this issue). We do not know precisely when and how the *Ode* was first written. It certainly counts among the products of court scholarship churned out with great speed, much like the expansion of *Qunfang pu*. Although the emperor is presumed to have written the main text of the poem, several high officials declared in a colophon to the *Ode* that they had written the substantial commentary that provided definitions of words and identified literary allusions.

The undersigned were the grand councillors Ortai, Zhang Tingyu 張廷玉 (1672–1755), and Xu Ben 徐本 (1683–1747), who were some of the most powerful people in the realm; the ministry presidents Ren Lanzhi 任蘭枝 (1677–1746) and Zhang Zhao 張照 (1691–1745); senior president of the Censorate Liu Tongxun 劉統勳 (1700–1773); ministry vice presidents Liang Shizheng 梁詩正 (1697–1763), Wang Youdun 汪由敦 (d. 1758),

---

Washington Library and at the Liaoning Provincial Library, which contains yellow slips pasted over some words, with apparent corrections of the underlying text. It is unclear what the corrections represent, however. They are not numerous enough to represent the replacement of Chinese loanwords with neologisms. Neologisms such as *miyahūtu* (which we discuss presently), for example, are already in the text of the University of Washington copy (*Han-i araha Mukden-i fu bithe*, 2:23). See Lin Shih-hsuan, "Huang yi pei du," 65; Lin Shih-hsuan, "Meiguo," 380–81.

41. Cf. Stary, "'L'Ode di Mukden' dell'imperatore Ch'ien-lung," 237; Jin, "Manwen shi *Shengjing fu*," 71. The "hymn" (Ch. *song* 頌) was included in the first edition of the *Ode*. Amiot saw the first edition, given that he noted which of the Manchu words were new in the revised translation. The discussion of the editions Amiot used in Long ("Qian Deming fayi," 169–70) is valuable, but Long does not distinguish between the 1743 and the 1748 Manchu editions with commentary. *Pace* Crossley, "Introduction to the Qing," 22n31, we have found no evidence that Amiot was working from a presumed English translation.

42. Etō, *Kenryū gyosei "Seikyō fu."*

and Qian Chenqun 錢陳群 (1690–1774); junior vice commissioners Ji Huang 嵇璜 (1710 [?]–1794) and Zhang Ruoai 張若靄 (1713–1746); and the subexpositor Guan-bao 觀保 (1711 [?]–1776).[43] The undersigned were Chinese save for Ortai and Guan-bao. Yet Zhang Tingyu, Liu Tongxun, Qian Chenqun, and Ji Huang had studied Manchu in the Hanlin Academy after passing the civil service examinations. Zhang and Liu, at least, were good at Manchu. They worked on Manchu scholarly projects, taught the language to their juniors, or graded their Manchu exams.[44]

Despite what the evident Manchu skills of the commentarial team would make one believe, the *fu*'s commentary was written entirely in reference to the poem's Chinese text. To a large extent, the commentary was literary and lexicographical, quoting earlier rhapsodies in which a certain word occurred, or defining a character using glosses from dictionaries. Such information referred specifically to the Chinese text but, except for most remarks on character variants, it was translated and entered into the Manchu text.

Information drawn from texts such as *Shengjing tongzhi* or *Bencao gangmu*, which tended toward the encyclopedic rather than the lexicographic, was less obviously out of place in a Manchu commentary. The absence of any discussion on how Manchu vocabulary aligned with that of Chinese and of any comments on what exactly certain words referred to, however, meant that the appropriateness of these notes in reference to the Manchu vocabulary was not demonstrated. It is not obvious that the commentators even ever saw the Manchu text.

The poem itself was, as mentioned, ostensibly written by Qianlong. It is probably true that he contributed to it, even wrote it at least in part.

---

43. For the position of senior president of the Censorate, see Brunnert and Hagelstrom, *Present Day Political Organization*, 76 (item 207a). The Chinese title of the junior vice commissioners is *you tongzheng* 右通政 (Ma. *ici ergi alifi hafumbure hafan*), not listed at the expected place in Brunnert and Hagelstrom (*Present Day Political Organization*, 484 [item 928]). Hucker (*Dictionary of Official Titles*, 553, item 7467), gives "vice commissioner" for *tongzheng*. Gimm (*Kaiser Qianlong*, 49) writes that the editors were Ortai, Prince Yinlu 胤祿 (1695–1767), and eleven others. Xu ("Qianlong *Shengjing fu*," 128) writes that Fuheng 傅恆 (1721–1770) and colleagues edited the blockprinted edition.

44. Huang, "'Qingshu shujishi' kaoxi," 125, 129.

If one goes by the number of poems published in Qianlong's name, the emperor "was by far the most prolific poet in Chinese history" (he published more than forty-two thousand pieces).[45] If it is true that he often spent part of the afternoon writing poetry, Mark Elliott writes, then "this would suggest that Qianlong may indeed have authored most of the verse ascribed to him; or at least that the essential idea behind many, if not all, poems was quite likely his."[46] The *Ode* was thus probably Qianlong's idea too, even if the commentators might have polished it as they combed through the classical literature that they then cited in the notes.

If Qianlong wrote or drafted the *Ode*, did he do so in Chinese or Manchu, or both? Qianlong's engagement with poetry, the genre of the *Ode*, and—most crucially—the text itself and the history of its revision all suggest that the poem was at heart a Chinese composition.

Qianlong's poetry is overwhelmingly written in Chinese and in Chinese genres. The Manchus certainly had poetry, however. Manchu poems, or ballads, commonly carry *ucun* 'song' in their title. The *Ode*, however, was not called an *ucun*. In the first printing of the text, it was simply called a *fu*, as in Chinese. This was later changed to the neologism *fujurun*.

In Manchu-Chinese discourse of the late seventeenth and early eighteenth centuries, Chinese *fu* existed in proximity not to *ucun*, but to *irgebun* 'poem, poetry'. This word was primarily used in reference to Chinese poetry or translations of it. It is probably an early loan from a Mongolic language, but the semantic development of the word in Manchu occurred under the influence of Chinese.[47] Shen Qiliang's Manchu-Chinese dictionary from 1683 glossed *irgebun* using three of the "six Principles of Poetry"—including *fu*—as mentioned, and the word *yong* 咏, another ancient term referring to poetic recitation or song. The corresponding verb *irgebumbi* Shen translated as *fu shi* 賦詩 'to compose poetry', where *fu* is the verb.[48]

---

45. Fang Chao-ying, "Hung-Li," 371.
46. Elliott, *Emperor Qianlong*, 111.
47. José Andrés Alonso de la Fuente, personal communication (2020). Cf. Tsintsius, *Sravnitel'nyĭ slovar'*, 326, s.v. ИРГЭБУ-.
48. Hayata and Teramura, *Daishin zensho*, vol. 1, 28 (line 0236a2).

Shen thus associated Chinese *fu* with Manchu *irgebun*. He also associated *irgebun* with a word that in the hands of Qianlong's court scholars later yielded a new Manchu word for "ode, rhapsody." As a gloss to the word *fujurungga* 'elegant, refined', Shen offered a translated line from an essay by the Tang poet Li Bai 李白 (701–762 CE) (where it translates Chinese *ya* 雅) in which *irgebun* (translating *jiazuo* 佳作) occurs.[49] On the basis of this word *fujurungga*, Qianlong's scholars coined *fujurun* to translate Chinese *fu* (the new word conveniently had *fu* as its first syllable).[50] Thus it appears that the vocabulary for talking about poetry in Manchu was a product of the encounter with the Chinese literary tradition.

With a new Manchu word for "ode" thus invented, *irgebun* came to be used specifically as the Manchu translation of Chinese *shi*, both in the sense of regularized, Tang-style verse (*lüshi* 律詩) and in reference to the *Poetry Classic*. Bilingual *shi* poems published in Qianlong's name were accordingly called *irgebun* in Manchu. However, Qianlong also wrote bilingual *ucun* in which the Chinese version was called a *ge* 歌 'song'. Judging by Giovanni Stary's and Tatiana Pang's respective studies of one of Qianlong's *irgebun* and one of his *ucun*, however, their meter is similar.[51] Qianlong's Manchu poetry—a far more limited output than his Chinese compositions—thus had one foot in the Chinese tradition. Indeed, if Amiot is to be believed, Qianlong on one occasion composed a Manchu hymn to prove to his high officials that Manchu, and not only Chinese, was a medium fit for poetry.[52] Yet the emperor's Manchu poetry did not (and could not easily) represent the metrical conventions of Chinese verse. Instead, the *ucun* and *irgebun* exhibit formal similarities with the popular Manchu ballads. Their verse scheme has Mongol precedents.[53]

---

49. Hayata and Teramura, *Daishin zensho*, vol. 1, 247 (lines 1428a5–b1).

50. Ch'oe ("Manmun ŭro ssyŏjin Kŏllongje," 98) writes that the word *fujurun* is composed of *fu* plus the "derivative suffix" *-jurun*. However, standard dictionaries include no words with this suffix other than *fujurun*, so this explanation cannot be sustained.

51. Stary, "Mandschurische Miszellen," 83; Pang, "Three Versions of a Poem," 90.

52. Amiot to Bertin, Beijing, May 11, 1779, in Amiot, *Hymne Tartare-Mantchou*, vi.

53. Stary, "Fundamental Principles," 80–81.

Curiously, the *Ode to Mukden* had a different poetic form. "Most of the poem is written in prose," Giovanni Stary writes, "according to the classical rules of composition in the genre of the *fu*, but at irregular intervals it contains a considerable number of passages clearly set in verse according to criteria never (or only rarely) used in other Manchu lyrical works" (*la maggior parte del poema è scritta in prosa, secondo le regole classiche della composizione del genere fu, ma ad intervalli irregolari contiene un numero notevole di brani chiaramente versificati secondo criteri mai [o solo raramente] applicati in altri componimenti lirici mancesi*).[54] The Manchu version of the *Ode*, then, has literary qualities unrelated to the Chinese poem. This becomes all the more clear when we consider that a Manchu version of the poem that is but a transposition of the contents of the Chinese actually does exist. Later in the eighteenth century, the *fu* was included in another work of court scholarship, which was written in Chinese and translated into Manchu. This Manchu version, unlike the original, is reportedly entirely in prose and thus a new translation from the Chinese.[55]

The literary qualities of the Manchu text have bearing on the question of how the text was written. Clearly, it was more than a routine translation like those carried out in the Manchu-Chinese Translation Office (Nei Fanshu Fang 內翻書房) in the Grand Secretariat.[56] Qianlong probably had a hand in crafting the Manchu version, or at least in approving it. On the same trip that witnessed the writing of the *Ode*, Qianlong criticized the Manchu translation of funerary orations read at the ancestral tombs, so clearly he was paying attention to such things during the period.[57]

The area in which the two *fu* differ most starkly is in the register of their vocabulary. Manchu did not have the rich and historically layered vocabulary of Chinese. Manchu was a vernacular language in the first half of the eighteenth century, and the *fu* written in it was accordingly of a different register than the Chinese text. Vernacular *fu*, however, had precedents. In the Tang period, when *fu* writing was quite

---

54. Stary, "'L'Ode di Mukden' dell'imperatore Ch'ien-lung," 237.
55. Stary, "'L'Ode di Mukden' di Qianlong," 1096.
56. On this institution, see Zhao Zhiqiang, "Lun Qingdai de Nei Fanshu Fang."
57. *Qianlong qi qiju*, vol. 11, 10 (QL 8/9/21; 1743/11/6).

common, a kind of "*fu* in common speech" (*su fu* 俗賦) appeared that made "abundant use of vernacular and oral elements in its diction and syntax."[58] A Manchu *fu* was thus certainly a radical break with the classical Chinese rhapsody, but not unique.

Jean-Pierre Abel Rémusat (1788–1832) noted the difference in lexical register between the Manchu and the Chinese versions of the poem.[59] "Written in Manchu and Chinese," Rémusat wrote, the *Ode* "differs greatly between the two languages":

> In Chinese, it is a perpetual cento, a heap of the most difficult, the most recherché, the most sublime expressions that are to be found among the poets of antiquity; in this form, the poem is unintelligible without the help of a commentary. In Manchu, by contrast, its style is simple, and although the two texts are both original compositions, the Tartar is extremely easy to understand, something which cannot be explained without entering in great detail into the spirit of the two languages.
>
> *c'est l'Eloge de la ville de Moukden, composé en chinois et en mandchou, et fort different dans l'une de ces langues de ce qu'il est dans l'autre. En chinois, c'est un centon perpétuel, un amas des expressions les plus difficiles, les plus recherchées, les plus sublimes qui se trouvent dans les anciens poètes : sous cette forme le poème est inintelligible sans le secours d'un commentaire. En mandchou, au contraire, le style en est simple, et quoique ces deux textes soient tous deux originaux, le tartare est extrêmement facile à entendre, fait qui ne pourrait s'expliquer qu'en entrant dans de grands details sur le genie des deux langues.*[60]

The two texts of the poem were indeed fundamentally different. Several commentators have relied on this fact to argue that the *Ode* was a Manchu work. Amiot, who knew the Manchu text intimately, considered it the original, and preferred it over the Chinese because he believed it "expressed more naturally the ideas of the Author, who is Manchu" (*j'ai cru qu'il exprimoit plus naturellemeny les idées de l'Auteur*

---

58. Kroll, "Significance of the *fu*," 94.
59. Rémusat's name is spelled following Pino, "RÉMUSAT (Abel-Rémusat)."
60. Rémusat, *Nouveaux mélanges asiatiques*, 59.

qui est Mantchou).⁶¹ Pamela Crossley calls it "one of the last literary compositions in Manchu."⁶² She sees in it two voices, one Manchu and one Chinese.⁶³ Giovanni Stary, as mentioned, stresses that "the Manchu version is not a calque on a hypothetical Chinese original" (*la versione mancese non è una traduzione stereotipa di un'ipotizzata versione originale cinese*).⁶⁴ We believe, on the contrary, that the *Ode to Mukden* was originally a Chinese text, composed on the basis of Chinese literature that included the Chinese-language *Shengjing tongzhi*. This text was then translated into Manchu, but with some difficulty. This becomes especially clear when we consider the words for flora and fauna, to which we now turn.

Plants and Animals in the *Ode to Mukden*

The selection of words for plants and animals included in the *Ode* shows that the text was written in Chinese and subsequently translated into Manchu. The *Ode* presents a vision of nature in Manchuria entirely based on the knowledge that had been produced on the region within the Chinese-language discourse that we saw emerge in parts 1 and 2 of this book. Even though the form in which this Chinese knowledge was codified, notably the several editions of *Shengjing tongzhi*, relied on fieldwork and field reports that had once existed in Manchu, translating the Chinese words back into Manchu in the context of the *Ode* was not easy. The translators quite simply did not have the Manchu words for what the poem presented as the natural products of the Qing homeland. This was especially true for the plants. When the Manchu text was revised for the new edition of 1748, words that in 1743 had simply been transcribed from the Chinese were replaced with neologisms. The court scholars thus created a Manchu language capable of describing the Manchurian nature that they were already able to write about in Chinese.

61. Amiot, *Éloge de la Ville*, 212.
62. Crossley, "Introduction to the Qing," 22. See also Crossley and Rawski, "Profile of the Manchu Language," 95.
63. Crossley, *Translucent Mirror*, 269.
64. Stary, "'L'Ode di Mukden' dell'imperatore Ch'ien-lung," 236.

The creation of a Manchu vocabulary for these plants (perhaps a recreation, in some cases, given that vernacular names might very well have existed in Manchuria before their erasure in the Chinese-language gazetteer) was the first step toward the integration of natural historical discourse in the two languages that was later carried out at the Qing court.

Although the Chinese language afforded the authors of the *Ode* a way to describe a great variety of plants, it also limited what natural products could be included in the poem. Neither of the editions of *Shengjing tongzhi* had been able to fully transpose the Manchurian flora into Chinese. Some words were left untranslated and unglossed. These words were not entered into the Chinese text of the *Ode* and thus were not present in the Manchu version either, even though the words were available in that language. We discuss these two characteristics of the *Ode*'s vocabulary for the natural world, both of which resulted from the poem's origin as a Chinese text.

We first discuss the Chinese words for which Manchu had no equivalents. In the quoted passage, Rémusat noted the difference that rich, historically layered vocabulary made for the Chinese *fu*—a genre that harnessed this very quality. He rightly noted that the Manchu text was simpler. Yet he strangely downplayed the difficulty of identifying the referents of many of the Manchu words for plants and animals, something that both Amiot and Rémusat's colleague Julius Klaproth remarked on.

The plants and animals in the *Ode to Mukden* were concentrated to five lists of twenty-six terrestrial animals, thirty-one birds (or winged creatures), thirty-two plants (flowers, grasses, and trees), thirty-five aquatic creatures, and thirty-seven agricultural crops (grains, pulses), in that order. These numbers are subject to debate given that some words, even in the middle of the lists, appear generic. We have excluded generic terms that introduce a list (e.g., Ch. *gu* 穀; Ma. *jeku* 'grain'), but included those that appear within lists (e.g., Ch. *shu* 蔬; Ma. *sogi* 'vegetable'). The text, moreover, includes a list of horses of various colors (just mentioned) and isolated words for animals (dragon, phoenix). Notably absent are insects (only the mosquito gets a passing mention). We concentrate on the five lists of flora and fauna.

The writers and translators had no problem finding Manchu equivalents of most of the words in these lists. Naturally, the tiger, leopard, wolf, or fox were not difficult to identify in either language, nor were birds and fish such as the swallow, magpie, eel, or plants such as radish, scallion, or peach. However, the translators could not come up with a Manchu word for a creature such as the jellyfish and as many as seventeen or so plants.

Transcriptions from Chinese in the Manchu Ode

Whenever the translators could not come up with a Manchu translation, they transliterated the Chinese and often appended a generic term such as *orho* 'grass, herb' or *moo* 'wood, tree'. For example, the Chinese poem contained a list of reeds that the Manchu language did not accommodate. But then again, how could it? The inflated number of words in the Chinese poem in reality hid a smaller number of plants; here, abundance was a linguistic fiction. The Chinese text listed both *lu* 蘆 and *wei* 葦, and *xiao* 蕭 and *di* 荻, but then quoted *Erya*'s and its subcommentary's equation of *lu* with *wei* and *xiao* with *di*.[65] Although *luwei* was a compound in Mandarin, *xiaodi* was not, so the translators treated them as separate words.

Three of the words—*lu*, *wei*, and *di*—were also listed in both editions of *Shengjing tongzhi* alongside vernacular names such as *zhangmao* 章茅 (referent unknown) and *huangbei cao* 黃背草 'Themeda triandra'.[66] The gazetteer in these instances reads as if its editor tried to compensate for the lack of up-to-date information by including single-character entries provided with only the most general philological glosses. The editors of the 1736 edition actually vowed to reduce the philological verbiage, yet kept a word like *di* 荻.[67] In fact, they even made it more out of place by removing the note that "when the locals

---

65. Fèvre and Métailié (*Dictionnaire Ricci des plantes*, 465) similarly presents *wei* as a synonym of *lu*.
66. Fèvre and Métailié, *Dictionnaire Ricci des plantes*, 199.
67. *Qinding Shengjing tongzhi*, vol. 1, 23 (*fanli*:8a).

thatch houses they often use it, calling it *di*" 土人結茅屋多用之，呼為荻.[68] Probably the locals in fact used a vernacular name that was not even Chinese, but had at some point been rendered as *di* during the translation process that underlay the writing of the gazetteer. Perhaps because the description was unclear without a mention of the original name, the 1736 edition of the gazetteer removed it (while keeping the similar notes that followed the vernacular names *zhangmao* and *huangbei cao*, presumably because the notes were more accurate in reference to these names) and only kept the philological remark "*Erya* calls it *xiao*" 《爾雅》謂之蕭 for *di*—not much to work with for the editors and translators of the *Ode to Mukden*.[69]

In the first Manchu edition of the *Ode*, *lu* was translated as *ulhū* 'common reed', defined in Kangxi's *Mirror* as a plant that "grows in places covered in water. At the joints, its center is hollow. Used to make curtains and mats" (*muke noho bade banjimbi. jalan de niyaman kumdu, hida derhi arambi*).[70] As for *wei*, it was translated as *darhūwa* 'elephant grass', which in the *Mirror* was described as "a kind of *ulhū*. The color differs, it can be red or white. Its ears are white" (*ulhū-i duwali. boco fulgiyan šanggiyan adali akū, suihe šanggiyan*).[71] Furthermore, *xiao* was translated as *ficakū orho*, which the *Mirror* described as "similar to *darhūwa*, but shorter. Grows in high places in the mountains" (*darhūwa-i adali, darhūwa ci fangkala, alin kuru bade banjimbi*).[72] Finally, *di* was not translated but instead transcribed as *di orho* '*di* grass', a made-up word that was naturally not found in the *Mirror*.[73] In the 1748 revision of the Manchu *Ode*, *di orho* was changed to *darhūwa orho* '*darhūwa* grass', which might be a calque on a presumed Chinese *lucao* 蘆草 'reed canary grass, lady's-laces' (not attested in the text), a plant with similar characteristics as *Miscanthus sacchariflorus* to which the word *di* can refer.[74] The translators did not have the Manchu synonyms at hand

68. *Shengjing tongzhi*, vol. 8, 21:6a.
69. *Shengjing tongzhi*, vol. 8, 21:6a; *Qinding Shengjing tongzhi*, vol. 2, 1306 (27:8b).
70. *Han-i araha manju gisun*, 1708, 19:*orho-i hacin*:3b.
71. *Han-i araha manju gisun*, 1708, 19:*orho-i hacin*:3b.
72. *Han-i araha manju gisun*, 1708, 19:*orho-i hacin*:3b.
73. *Han-i araha Mukden-i fu bithe*, 25b (Manchu).
74. *Han i araha Mukden-i fujurun bithe*, 25b; Fèvre and Métailié, *Dictionnaire Ricci des plantes*, 102, 286.

that would enable them to translate the Chinese original, which benefited from a vocabulary rich in synonyms that had accumulated over many centuries.

Furthermore, the translators used transcriptions for several other plants in the first edition of the *Ode*. For example, *ni* 蘱 (early Mandarin pronunciation, today read *yi*), 'ladies' tresses, pearl-twist', an orchid growing on prairies and in mountain forests, was transcribed as *ni orho* '*ni* grass' and then changed to *bulha orho* 'polychrome grass'.[75] This orchid comes in many colors, so it was an appropriately descriptive name.[76] Another example is found in the treatment of *mugin* 木槿 (today read *mujin*) 'hibiscus' (perhaps *Hibiscus syriacus*).[77] This word was first transcribed as *mu gin moo* '*mugin* tree' and then changed to *mooyen ilha* (with 'tree' changed to 'flower').[78]

Finally—but more examples could be mentioned—*shanglu* 商陸 'India pokeberry' was first transcribed as *šang lu* and then changed to *fiyelesu*.[79] Amiot's difficulty in identifying this last plant is an indication that it was not a household name in Beijing at the time. Amiot had found in a "Dictionary"—surely referring to a manuscript Chinese-European language dictionary compiled by the Jesuits in Beijing—a translation of *shanglu* as "Phitolaca" (i.e., *phytolacca*, still the current name), but he could not locate the word in any of the (European) "Botanists" that he had handy.[80]

---

75. For *ni*, see Morrison (*Dictionary of the Chinese Language*, vol. 2, 67), which gives the pronunciation *neïh*. The character is not listed in Pulleyblank, *Lexicon of Reconstructed Pronunciation*. Giles (*Chinese-English Dictionary*, 1634) gives the pronunciation *yi*, glosses it as *Spiranthes australis*, and notes that it is "a common orchid throughout China." English translation and description of habitat from Fèvre and Métailié (*Dictionnaire Ricci des plantes*, 347, s.v. *pán long shēn*; *Han-i araha Mukden-i fu bithe*, 21a [Chinese], 25b [Manchu]; *Han i araha Mukden-i fujurun bithe*, 25b).

76. Morrison, *Dictionary of the Chinese Language*, vol. 2, 67.

77. Pulleyblank, *Lexicon of Reconstructed Pronunciation*, 156; Fèvre and Métailié, *Dictionnaire Ricci des plantes*, 327.

78. *Han-i araha Mukden-i fu bithe*, 21a (Chinese), 26a (Manchu); *Han-i araha Mukden-i fujurun bithe*, 26a.

79. Fèvre and Métailié, *Dictionnaire Ricci des plantes*, 403; *Han-i araha Mukden-i fu bithe*, 21b (Chinese), 26b (Manchu); *Han-i araha Mukden-i fujurun bithe*, 26b.

80. Amiot, *Éloge de la Ville*, 273.

The names of some animals, moreover, were also transcribed rather than translated in the *Ode*. A creature called *haizha* 海蛇 'sea jellyfish' (the noun adjunct *hai* 'sea' was actually only provided in the commentary, not the main text of the *Ode*) was transcribed as *hai je* in the first edition of the *Ode* and changed to *sangguji* in the second. Amiot apparently did not know what it referred to—he transcribed the Manchu name in his translation—and neither did Klaproth, who called it a "sea snake" (*serpent de mer*).[81]

The first version of the Manchu *Ode* thus contained a number of transcriptions from Chinese. For some words, the translators were evidently unable to come up with Manchu equivalents, as in the case of the reeds. Yet in other cases, the Chinese transcription is better understood as a loanword in Manchu that was in actual use and not a hapax legomenon produced by the translation process. Loans are seen in Imperial Household archival documents of the period, and the word for Sichuan pepper, mentioned in the context of *Lianzhu ji*, is another example of the same phenomenon.

Words Invented and Retracted

Such would also be the case with a word of Chinese origin that was used in the 1743 Manchu edition, replaced in 1748, but ultimately not adopted in the court's official Manchu standard once it was promulgated. The Chinese version of the *Ode* included an animal that it called *si* 麆. This animal was listed already in *Shengjing tongzhi*, which quoted earlier literature: "Big ones are called 'elk' (*mi*), thus they are a kind of elk. The flavor of its flesh is inferior to that of the Chinese water deer (*zhang*). Its hide can be made into shoes" 大者曰麋則麋類也。肉味不及麞。其皮可作履舄.[82] The definition was in the 1736 edition of the gazetteer changed to "a kind of muntjac deer (*jun*) with long fur and

---

81. *Han-i araha Mukden-i fu bithe*, 24a (Chinese), 30b (Manchu); *Han-i araha Mukden-i fujurun bithe*, 30b; Amiot, *Éloge de la Ville*, 288–89n117; Klaproth, *Chrestomatie mandchou*, 256.

82. *Shengjing tongzhi*, 21:20b.

dog feet, hide suitable for shoes" 麕類，毛長，犬足，皮堪履舄.[83] This description was quoted in the commentary to the *Ode*.[84]

The 1743 edition of the Manchu *Ode* rendered this animal as *gi buhū*, where *gi* comes from early Mandarin or dialectal *ki* 麂 (now *ji*), perhaps meaning "Red muntjac."[85] Kangxi's Manchu *Mirror* defined *gi buhū* as "small body, similar to the Siberian roe deer [or Chinese water deer?] with feet similar to those of the musk deer. The horns on its head grows in the place of the roe deer" (*beye ajige, sirga de adali, bethe miyahū adali, uju-i weihe gio de bade banjimbi*).[86]

Presumably because the first element of *gi buhū* is a loan from the Chinese, the word was replaced in the 1748 edition of the *Ode*. The editors replaced it with the neologism *miyahūtu*, formed on the basis of *miyahū* 'musk deer'. This latter word was not included anywhere in the *Ode*, but Kangxi's Manchu *Mirror* defined it as "like the roe deer [or Chinese water deer?] (*sirga*, cf. Mo. *sirga* 'isabelline, greyish yellow') but smaller, blackish color.[87] There is a lump at its navel, which is the musk. It is also called *mikcan*" (*sirga de adali bime ajige, boco sahahūn, erei ulenggu de dalgan bi, uthai jarin inu. geli mikcan sembi*).[88] Amiot noted that the word *miyahūtu* was new and that it had replaced *gi buhū*. On the basis of the word's derivation, he believed it was a kind of musk deer (although he called it a civet [*civette*], which usually does not refer to a deer but to other animals that secrete a strong scent).[89] For whatever reason, however, the word *miyahūtu* was not used in the place of *gi buhū* in the later *Catalog of Beasts* (for which see chapter 7) nor in

---

83. *Qinding Shengjing tongzhi*, vol. 2, 27:30b (1350).

84. In the Manchu version, "A kind of roe deer [or Chinese water deer?], long fur, dog's feet, the hide can be made into boots and shoes" (*gi buhū, sirga-i duwali, funiyehe golmin, indahūn-i bethe, sukū be gūlha sabu araci ombi*). *Han-i araha Mukden-i fu bithe*, 21a (Manchu).

85. For the pronunciation, see Pulleyblank, *Lexicon of Reconstructed Pronunciation*, 140. For the identification of *ji* with the Red muntjac, see Yu and Zhang, *Qinggong shoupu*, 408 (item 30).

86. *Han-i araha manju gisun*, 1708, 19:*gurgu-i hacin*:46a.

87. Doerfer, *Mongolo-Tungusica*, 100, item 302.

88. *Han-i araha manju gisun*, 1708, 19:*gurgu-hacin (jai)*:46a.

89. Amiot, *Éloge de la Ville*, 251.

later official dictionaries.[90] The editors who revised the Manchu *Ode* had probably been asked to weed out Sinicisms, but in this case, the original word was ultimately considered acceptable as proper Manchu.[91] The reform of the lexicon was thus not an altogether straightforward project.

Absence of Manchu Names in the Chinese Text

The preceding paragraphs discuss examples of Chinese names of plants and animals that were represented in Manchu either as Chinese loans, or not at all, in which case the translators chose to transcribe the Chinese word and use it as such in the Manchu text. These examples

---

90. For *gi buhū* in the *Catalog of Beasts*, see Chuang, Shoupu *Manwen*, vol. 1, 203.

91. The translators' treatment of the different kinds of deer and the words used to refer to them suggests something of the rushed character of the work. In the commentary for the string *lu zhang pao si* 鹿麞麅麆, *Shengjing zhi* 盛京志, referring to some edition of the gazetteer of greater Mukden, was quoted several times. The notes for *zhang* and *pao* were compiled into one, with information relative to the two characters linked by *you* 又 'furthermore'. Here they quote the 1684 edition. Yet immediately the commentators again quote *Shengjing zhi* in reference to *si*, where they provide the definition from the 1736 edition. This suggests that before the commentary was compiled, the words in the main text of the poem were allocated to different commentators, who did not all use the same edition of the gazetteer when preparing their notes.

The notes on the deer are interesting in another regard as well. Contrary to the practice elsewhere in the text, when translating the comment on the Chinese *pao* 麅 'Siberian roe deer', the translators did not omit the Chinese paleographical remarks included in the 1684 *Shengjing tongzhi* and carried over into the Chinese commentary to the *Ode*. As a word, *pao* was not the problem. It was a colloquial Chinese word that Shen Qiliang knew translated to the Manchu *gio* (with cognates in other Tungusic languages and in Mongolian). The 1684 *Shengjing tongzhi*, however, added the note that "[the word] is also written as *pao* 麃." The two characters pronounced *pao* are variants and have no relevance outside a Chinese-language context. The translators of the commentary tried to render the distinction in Manchu as well, however. They translated the Chinese *yi zuo pao* 一作麃 as *geli emu gebu gūran sembi* 'another name is *gūran*', where *gūran* translates the graphic variant of *pao*. Shen Qiliang translated *gūran* as "the rope on a pouch; a tie string" 荷包上繩子。繫子; but he also listed the compound *gūran giyoo* (i.e., *giyo*) as *jiaopao* 角麅 'horned Siberian deer'. Thus two character variants became two words in the Manchu version. See Hayata and Teramura, *Daishin zensho*, vol. 1, 71 (0506b2–3), 221 (1313b1); Rozycki, *Mongol Elements in Manchu*, 89; see also Amiot, *Éloge de la Ville*, 251.

show that the *Ode* was first and foremost a Chinese text that could be translated into Manchu only with difficulty. That the poem originated as a Chinese text had another consequence, however; natural products for which Manchu—but not Chinese—words existed were precluded from entering the poem.

The *Ode to Mukden* was written to celebrate the greatness of Manchuria in its specificity. A remarkable house such as that of the Aisin Gioro could only have emerged from a remarkable place. It was indeed a place quite different from China proper, and thus, as we showed in part 2, the Chinese language at times did not have the words to talk about it. Thus Gao Shiqi's and Yang Bin's travel diaries, *Shengjing tongzhi* and *Guang Qunfang pu*, all listed local plants and animals that had no agreed-upon Chinese names, only Manchu names that these Chinese-language texts transcribed into Chinese characters and glossed with various levels of detail. The *Ode*, however, did not include a single such name, in neither the Chinese transcription nor the native Manchu orthography. Thus plants such as the kind of "cherry" (*yingtao* 櫻桃) called *mi-sun-wu-shi-ha* 米孫烏什哈, that is, *misu hūsiha*, which the 1736 edition of the *Shengjing tongzhi* listed as a product from Ningguta but did not translate, were absent from the *Ode to Mukden*.[92] Yet, like other plants that the gazetteer treated similarly, this plant was obviously characteristic of the natural world of Manchuria, which is why the gazetteer that listed the region's natural products took note of it.

It is possible, but perhaps not very likely, that the *Ode*'s authors were unaware of these words because they did not know Manchu well enough. Alternatively, they might have believed that obviously non-Chinese words were unsuitable in a poem in the *fu* genre. Had the authors been working on the basis of Manchu-language sources, they would probably have included a word such as *misu hūsiha*, and the Manchurian nature described in the poem would accordingly have appeared differently. Writing the *Ode* as a Chinese text based on Chinese sources and generic conventions—in the context of a still incompletely integrated bilingual discourse on flora and fauna—meant that the totality of the knowledge that actually did exist in court circles in

92. *Qinding Shengjing tongzhi*, vol. 2, 1325 (27:18a).

Beijing regarding the flora and fauna in the region could not be represented in the *fu*.

Establishing Equivalences

In some instances, the translators opted for transcribing the sounds of a Chinese word even though a Manchu word with a similar meaning was available to competent users of Manchu. One likely explanation for this is that early Manchu dictionaries were ambiguous or contradictory regarding the definition and, especially, the translation of the names for plants and animals. Shen Qiliang left a number of words in this category untranslated in his dictionary. Furthermore, Kangxi's *Mirror* was monolingual, which often left room for interpretation, and other dictionaries offered varying translations.

The problematic nature of their sources notwithstanding, the translators managed to establish equivalences that would stand the test of time. This happened, at least, in the case of the Siberian salmon. The Chinese version of the *Ode* mentioned a fish called *xilin* 細鱗 'narrow-scaled', a descriptive name that the editors probably got from *Shengjing tongzhi*, which listed a *xilin yu* 細鱗魚 'narrow-scaled fish', a fish that Kangxi had once offered to his Chinese officials as a delicacy when on tour in Manchuria.[93] The translators of the *Ode* matched this word to *niomošon*. Jonathan Schlesinger noted that *niomošon* had earlier been left untranslated in the Chinese version of Tulišen's Manchu account of the author's journey to Russia.[94] Now, however, it was matched to *xilin*.[95] In 1683, Shen Qiliang had given *xilin yu* 細鱗魚 as one of three translations of *secu nimaha*,[96] which the *Mirror* treated as a different fish.[97] The referents of all of these words were thus not

---

93. *Shengjing tongzhi*, 21:23a; *Qinding Shengjing tongzhi*, vol. 2, 27:32a (1353); Lai, "Qianlong chao *Niaopu*," 21.

94. Schlesinger, *World Trimmed with Fur*, 34–35.

95. *Han-i araha Mukden-i fu bithe*, 24a (Chinese), 30a (Manchu); *Han-i araha Mukden-i fujurun bithe*, 30a.

96. Hayata and Teramura, *Daishin zensho*, vol. 1, 102 (0644a3).

97. *Han-i araha manju gisun*, 1708, 20:*birai nimaha-i hacin*:31a; see also 20:*mederi nimaha-i hacin*:37b, where the sea fish *sohoco* is said to be "somewhat similar to the *secu*" (*secu de adalikan*). This is the entire entry.

necessarily the same in all instances. In any case, the identification made in the *Ode* was upheld in Qianlong's Manchu-Chinese *Mirror*, even after a preparatory vocabulary had given *niomošon* a different Chinese translation.

A possible case of the translators in the first round resisting settling on an actual Manchu word in favor of a transcription from Chinese is found in the treatment of *cang* 鶬, which the *Ode* used in reference to what *Shengjing tongzhi* called *cangji* 鶬雞. The word was not new or unknown, but what it referred to in the gazetteer was not obvious. Li Shizhen in *Bencao gangmu* had said that *cangji* was a word used by southerners (*nanren* 南人) to refer to a bird known by various other names around China.[98] The gazetteer described it as a kind of large water fowl, big as a crane, with a long neck and legs and red cheeks.[99] The commentary to the *Ode* quoted this definition (with *bulehen* used for "crane" in the Manchu version).[100] The translators did not know a corresponding Manchu word; they transliterated *cang* as *tsang gasha* '*cang* bird'. In the revised translation, this was replaced by *kūrca*, a variant of what Kangxi's *Mirror* spelled as *kūrcan* 'hooded crane'. The *Mirror* defined it as "like the black crane but grey in color, it makes a lot of sound when it flies" (*yacin bulehen de adali bime fulenggi boco, deyere de jilgan ambula*).[101]

It is not clear that the *cangji* of the gazetteer (or of *Bencao gangmu*) was the hooded crane. It is possible that the translators of the *Ode* did not know of a bird that fit the description quoted in it, in which case its later identification with *kūrca~kūrcan* might have been motivated more by a desire to replace Chinese transcriptions with properly Manchu words than a wish to accurately render the Chinese text in Manchu. Alternatively, the identification of *cang* with *kūrca~kūrcan* was correct (but who is to tell?) and the word simply escaped the translators when they first transposed the poem into Manchu. It seems that in 1748, in any case, the translators neglected to consult the *Mirror*, given that they did not use the normative spelling of the word.

98. Li Shizhen, *Bencao gangmu*, vol. 774, 47:3b (348, lower panel).
99. *Shengjing tongzhi*, 21:16a; *Qinding Shengjing tongzhi*, vol. 2, 27:26b (1342).
100. *Han-i araha Mukden-i fu bithe*, 19a (Chinese), 23b (Manchu).
101. *Han-i araha manju gisun*, 1708, 19:*gasha-i hacin*:25b.

Another example of the translators initially opting for a transcription, only to have it replaced later with an existing Manchu word, is found in the treatment of *bianxu* 萹蓄 'knotweed'. The first translation of the *Ode* rendered this as *biyan hioi*, the expected transliteration of the two Chinese syllables.[102] This invites the question whether or not there was a Manchu word for knotweed. The *Mirror* included the word *banda hara*, which it defined as "string-like, growing by spreading out along the ground" (*na de dedume banjiha suihe ohongge be*).[103] The description captures a characteristic feature of knotweed, but it is less than unambiguous. For that reason, perhaps, Li Yanji's Manchu-Chinese dictionary, which was based on the *Mirror*, as well as Daigu's Manchu-Chinese dictionary, both similarly described the plant in Chinese without giving a corresponding Chinese name.[104] During the reform of the lexicon, however, the two words were equated. In the revised Manchu translation of the *Ode to Mukden*, *banda hara* was used to replace *biyan hioi* as the word for knotweed, an identification upheld in later official dictionaries.[105] Thus an equivalence was created.

Yet instances such as this hint at the difficulty of matching words both with each other and with things. When Amiot compared the Chinese and Manchu texts with the help of dictionaries, including the *Mirror*, some of these problems became apparent.

The Ode, the Mirror, and the Challenge of Forging a Bilingual Vocabulary

As noted, the use of a nonstandard spelling in the revised Manchu *Ode* raises the question whether the translators even consulted Kangxi's *Imperially Commissioned Mirror of the Manchu Language* in their search for equivalents of the Chinese words for plants and animals.

---

102. *Han-i araha Mukden-i fu bithe*, 21b–22a (Chinese), 26b–27a (Manchu); Fèvre and Métailié, *Dictionnaire Ricci des plantes*, 34.
103. *Han-i araha manju gisun*, 1708, 19:*orho-i hacin*:1b–2a.
104. Li Yanji, *Qingwen huishu*, 3:26b (70, lower panel); Daigu, *Manju gisun-i yongkiyame*, 6:33b.
105. *Han-i araha Mukden-i fujurun bithe*, 26b.

The first scholar to bring the *Mirror* to bear on the understanding of the *Ode* was Amiot, who translated the revised version in Beijing in 1769.

A Beijing resident, Amiot was a friend of the Qianlong emperor's cousin Prince Hongwu 弘旿 (1743–1811) and thus had access to a circle of educated individuals.[106] Amiot accordingly "interviewed" (*interrogés*) both Chinese and Manchus in connection with the *Ode*.[107] Yet when it came to plants and animals, he mostly quoted from the *Ode*'s commentary. Sometimes he resorted to the *Mirror of the Manchu Language*, which he used to get a sense of what tree the editors might have meant by Ch. *duanmu* 椴木 and Ma. *nunggele* 'Tilia', a common tree in Manchuria.[108] Amiot was unable to identify it but quoted "the description that they use in their Dictionary" (*la description qu'ils en font dans leur Dictionnaire*), meaning the *Mirror*.[109] Amiot even used the *Mirror* to provide synonyms for Manchu words used in the text: He noted that *mujuhu*, which the *Ode* used for 'carp', had a synonym in *hartakū*, whose definition Amiot quoted from the dictionary.[110] He relied on the *Mirror* to define several other kinds of fish as well.[111]

On two occasions, Amiot hit upon contradictions between the reference works in several languages that he had at his disposal. In the case of Ch. *zun* 鱒 and Ma. *jelu* 'trout' (?), he noted, the *Mirror* did not accord with an unnamed Chinese dictionary. In this case, however, he ascribed the discrepancy to the fact that the Chinese dictionary was compiled by European missionaries, who might have misunderstood their Chinese source text.[112] Thereby Amiot saved the Qing books from criticism.

Amiot more explicitly acknowledged the difficulty in reconciling the plurilingual sources at his disposal with regard to Ch. *li* 鱧 and its

---

106. Statman, "Forgotten Friendship."
107. Amiot, *Éloge de la Ville*, 225.
108. Fèvre and Métailié, *Dictionnaire Ricci des plantes*, 114, s.v. *duàn* 椴, acceptation 2.
109. Amiot, *Éloge de la Ville*, 722.
110. Amiot, *Éloge de la Ville*, 279.
111. Amiot, *Éloge de la Ville*, 284n105, 285n109.
112. Amiot, *Éloge de la Ville*, 280–81.

Manchu translation *hūwara*. Zhang Tianqi's and Liu Shun's *Lianzhu ji* used *hūwara* to translate *li*, and the *Mirror* presented it as a synonym of *horo*.[113] This last word was in turn used to translate *li* in the *Poetry Classic* and, as we saw, *Lianzhu ji* used it as a translation for "eel."[114] Amiot thought that *hūwara* ought to refer to a moray eel, lamprey, remora, or an electric ray of some sort, but noted that "the Dictionaries are not in accord regarding *hūwara*" (*Les Dictionnaires ne s'accordent pas entr'eux sur le* Houara).[115] Amiot thus glimpsed the problems that faced anyone who wanted to reconcile the Manchu dictionaries with a multilayered Chinese vocabulary for flora and fauna. The creation of a standard bilingual vocabulary through the publication of bi- and multilingual dictionaries from the early 1770s onward did not resolve these problems as much as hide them.

Conclusion

The *Ode to Mukden* occupies a special place in the history of court scholarship of the eighteenth century. As a bilingual rhapsody, it is rare. It was chosen as the text to showcase the newly invented Manchu seal scripts, a striking product of court art that still awaits comprehensive scholarly treatment. Importantly for our purposes, the *Ode* was the first officially published text to attempt to align Manchu and Chinese discourses on the natural world. In numerous instances, especially in the case of plants, the translators were unable to come up with actual Manchu equivalents.

The first edition of the poem preceded Qianlong's Manchu-language reform project by a few years, but when the time came to prepare the calligraphic edition in 1748, the decision to weed out Chinese transcriptions from Manchu texts had already begun. Scholars at court thus revisited the Manchu text of the *Ode* and replaced the difficult-to-translate terms with new coinages. Yet by replacing

---

113. Zhang Tianqi, *Lianzhu ji*, 14b.
114. *Han-i araha manju gisun*, 1708, 20:*birai nimaha-i hacin*:33a; Gabelentz, *Sse-schu, Schu-king, Schi-king*, 254; Legge, *Chinese Classics*, vol. 4, 269.
115. Amiot, *Éloge de la Ville*, 283.

the old transcriptions, the editors of the revised edition unwittingly laid bare the difficulties that educated bilinguals in Beijing had in talking about the natural world, even that of the Manchu homeland, in Manchu. The words used by the ethnically and linguistically diverse communities that inhabited "greater Mukden" had already been hidden from view in the Chinese-only *Shengjing tongzhi*, on which the editors relied so heavily. The *Mirror of the Manchu Language* ought to have been available as a resource, and it is not impossible that it was consulted during the translation of the *Ode*. Yet if it were, it left little trace in the final product, and as mentioned, within the imperial rhapsody are even indications that its translators did not rely on the *Mirror* as a guideline for spelling.

To our knowledge, the *Ode to Mukden* has not received much attention in the sizable literature on Qianlong's reform of the Manchu lexicon. In this literature, this reform has often been traced to late in 1747 (QL 12/10/14), when the emperor approved a list of new Manchu names for offices in the central government.[116] Much of the government apparatus had a Chinese history that predated the Manchus, and replacing Chinese names with Manchu ones had been done before, when the young Manchu state expanded in the 1630s.[117] The scale of Qianlong's reform of administrative terms was remarkable, but it was arguably not qualitatively different from earlier terminological overhauls.[118] Qianlong's reform of the Manchu lexicon subsequently expanded, however, and turned into something quite different from the language care evidenced by his predecessors.

In this chapter, we have shown that the expansion of the reform beyond official terminology is evident in the second version of the *Ode to Mukden*, which was prepared in 1748. All the Manchu names for plants and animals that in the original edition had been apparently ad hoc transcriptions of the Chinese were given new Manchu names in the new edition. The *Ode to Mukden* included a rich description of Manchurian nature based on *Shengjing tongzhi*. The reliance on Chinese sources for this book about the Manchu homeland led to a work

---

116. Zhang Hong, Cheng, and Tong, "Qianlong chao," 1994, 68.
117. Kicentai, *Teikoku o tsukutta*, 95–100.
118. Tong and Guan, "Qianlong chao," 37.

in which the Manchu text was secondary to the Chinese and the translators were unable to find Manchu equivalents for many of the names for animals (such as the jellyfish, as we have seen) and numerous plants. It is thus the first book that sought to bring the Manchu language into accord with the Chinese as a resource for talking about the natural world. At the same time, one can also deduce that the urge of naming Manchurian flora and fauna in Manchu should be acknowledged as an early precedent that triggered the invention of more neologisms later on at the Qianlong court.

The timing of the second edition is curious, as it was undertaken during the tumultuous year that saw Qianlong lose both a son and the empress, Lady Fuca (1712–1748).[119] Precipitated by these events or not, the year has been treated as a turning point in the Qianlong reign, one in which, Gao Wangling 高王凌 (1950–2018) remarked, the court's idealistic and activist statecraft withered into a kind of *mission civilisatrice* that was quite hollow in comparison.[120] We do not have to accept this bleak assessment of the latter part of the Qianlong reign to note court involvement in culture and scholarship, including in Manchu-language discourse on the natural world, is much more evident from around the time of the second edition of the *Ode to Mukden*.

At the time that the *Ode* was published, the court had not yet published a bilingual Manchu-Chinese dictionary. The *Mirror* from 1708 was in Manchu only, and the bilingual edition of the *Mirror* from 1717 translated the headwords not into Chinese but instead into Mongolian (a revised version appeared in the same year as the *Ode*).[121] But a normative dictionary that offered translations into Chinese did not yet exist. Indeed, in the intervening years between the composition of the *Ode to Mukden* and the publication of the Manchu-Chinese *Mirror* in 1772, many new Manchu words were coined in order to designate flora and fauna from near and far, past and present. A reading of Qianlong's encyclopedic poem reveals it as a key point in the history of court-sponsored writing about the natural world, midway between

---

119. Elliott, *Emperor Qianlong*, 41–44.
120. Gao, *Qianlong shisan nian*, 157.
121. Söderblom Saarela, "Manchu and the Study," 287–88.

Kangxi's encyclopedic and monoglot Manchu *Mirror* and Qianlong's Manchu-Chinese expanded edition, which incorporated the new words. In the next chapter, we return to the Qianlong court to discuss a series of subsequent projects and the accompanying backstage deliberations in the 1750s and 1760s that laid the groundwork for the Manchu-Chinese *Mirror*.

CHAPTER SEVEN

Qianlong's Manchu-Language Reform and Natural-Historical Philology

In the previous chapter, we discussed the replacement of Chinese loanwords for plants and animals with Manchu neologisms in the *Ode to Mukden*. The revision of the *Ode* in the late 1740s was an early example of Qianlong's care for the Manchu language being reflected in books produced at court. The introduction of the new words for flora and fauna shows the centrality of the description of the natural world in this process. Soon thereafter, with the initiation of the project to revise and expand the Manchu *Mirror* in 1750, the efforts to change Manchu started to converge toward a more systematic attempt to add neologisms to the Manchu language. The goal was twofold: to purify and to enrich the lexicon. Even though the ideological drive behind Qianlong's language reform has inspired much commentary in the literature, our research fleshes out the routinized philological labor behind the creation of neologisms during the 1750s and 1760s within Qing officialdom. It was in the writing of bilingual legends to illustrated catalogs (albums), and in inner-court documents produced in response to requests from the throne or high officials that the meaning of Manchu words for plants and animals was clarified and new words invented.

This chapter begins with a general overview of how new words to name plants and animals were coined and reviewed in the Qing bureaucracy. We then examine the continued relevance of the genre of "catalogs" (*pu*), now used in reference to illustrated albums, to the practice of lexicographical encyclopedism. Although the illustrated *Catalog of Birds* and *Catalog of Beasts* are best known today for their eclectic blending of Chinese and Western painting techniques, we suggest in the first half of this chapter that the bilingual legends that accompanied the creatures depicted there are best understood as a continuation of the late Kangxi-era encyclopedism that we discussed in parts 1 and 2. The remainder of this chapter uses bilingual documents archived at the Grand Council to highlight the search for proper Manchu neologisms that was carried out at court, as well as the mixed results that came of it. Importantly, in these documents, we begin to see how this politically driven effort to achieve bilingual parity likewise started to affect the Chinese lexicon. Not only were Manchu words for plants and animals created at court, but Chinese zoological and botanical terminology was standardized as well.

Neologisms in Qianlong's Manchu-Language Reform

Qianlong's Manchu-language reform, commonly dated to 1747, was related to the plurilingual scholarship produced at court. Indeed, a defining characteristic of the Manchu-language scholarship carried out at the Qianlong court was its plurilingualism. The revived project of compiling language *Mirrors* shows this clearly. Several *Mirrors* were published under Qianlong's supervision, and all of them contained at least one other language in addition to Manchu. Plurilingual scholarship at the Qianlong court was carried out with a heightened awareness of its inherent complexities, which was reflected in the reform of the Manchu lexicon, but official editorial projects involving several languages began earlier.

A bilingual Manchu-Mongolian edition of the *Mirror* had already been produced under Kangxi, but afterward Manchu court lexicography lay dormant for many years. After a quarter-century hiatus, the Qianlong emperor resumed work on the *Mirrors*. As true of Qianlong's

Manchu translations of the Confucian classics (some new, some revised), there is no reason to believe that Qianlong or his court from the outset envisioned a series of *Mirrors* that included an increasing number of languages.[1] The timing of the new editions of the *Mirror* was uneven, but nevertheless commonalities are unmistakable.

Qianlong's first *Mirror* was a new edition, published in 1743, of Kangxi's Manchu-Mongol *Mirror*. Qianlong ordered the Mongolian text in this book transcribed into the Manchu script in order to disambiguate the Mongolian orthography. In the preface, Qianlong assumed the perspective of a ruler of a multilingual empire whose languages had to be brought in alignment for the purpose of government.[2] An attention to phonetic detail and an acknowledgment, even celebration, of the multilingualism of the empire are thus discernable in this work. But indications that Qianlong sought to reform and enrich the Manchu language to encompass within itself the diverse world that these languages reflected are not yet evident. He was not yet coining new Manchu words.

The creation of new words at court—albeit not Manchu words—strictly speaking began with Kangxi's Manchu-Mongolian *Mirror*. The publication of this work, titled *Han-i araha manju monggo gisun-i buleku bithe | Qayan-i bičigsen manju mongyol-u gen-ü toli bičig* (Imperially commissioned Mirror of the Manchu and Mongolian languages), in 1717 represented the *Mirror*'s leap into plurilinguality. The work was headed by an official from the Mongol banners and the team included Buddhist clerics and instructors of Tibetan. Work began in 1710.[3] The new book celebrated Qing ascendancy over the eastern Mongols, but as a reference work it was directed primarily at the Mongols within the Eight Banners, who after several generations in the Manchu military allegedly had a poor grasp of their ancestral language.[4]

The editorial team translated both headwords and definitions in the original *Mirror* into Mongolian. Kangxi ordered the translators to

---

1. Regarding the classics, see Yeh, "Man-Han hebi," 16.
2. *Han-i araha manju monggo gisun*, 21:2b–3a (pagination of the preface dated QL 8/4/11); Chunhua, *Qingdai Man-*, 117.
3. Chunhua, *Qingdai Man-*, 112.
4. See Murakami, *Shinchō no Mōko kijin*, 18–22.

proceed like the compilers of the original *Mirror*, and "ask among the old men of the Eight Banners and among knowledgeable Mongols, write down their answers, and present your findings in list form as the work progresses" (*jakūn gūsai fe sakda, sara monggoso de fonjifi arakini bahara be tuwame emu udu afaha durun weilefi tuwabume wesimbu*).[5]

The task was not easy because, as the compilers explained, "one cannot translate this book in the way that one would translate ordinary books; [in the case of the *Mirror*,] the meaning of every word, every character has to be made the same" (*ere bithe be an-i bithe ubaliyambure adali ubaliyambuci ojorakū, urunakū emu gisun emu hergen sehe seme tere gūnin be emu obume*). The translators did not know how to render many Manchu words in Mongolian, so they asked not only within the Eight Banners, but also among the various Mongol representatives who periodically visited the court in Beijing.[6] Yet even so, for many terms simply no word was available in Mongolian. This led to the first official creation of new vocabulary, which the court later engaged in for both Manchu and Chinese as well.

Despite the translators' professed goal of establishing the Mongolian equivalences through interviews with Mongol bannermen and visiting dignitaries, in many cases the translators evidently simply transcribed the Manchu names into the Mongol script, with some changes to accommodate Mongolian pronunciation. Linguists in the twentieth century sought to disentangle these words from the Mongolian lexicon as a whole by checking which of the *Mirror*'s words were attested as actually used in the vernacular Mongolian dialects, but the words have otherwise received little attention.[7] These words were

---

5. Original Manchu text reprinted (typeset) in Rémusat ("Notice sur le dictionnaire intitulé," 25). Rémusat's translation is on page 37. The text also exists in a Mongolian version that we do not transcribe here.

6. Rémusat, "Notice sur le dictionnaire intitulé," 27–29.

7. For example, Doerfer, *Mongolo-Tungusica*, 246–54. Doerfer ("Terms for Aquatic Animals") checks for the presence of words for aquatic animals found in the pentaglot in the Mongolian dialects to ascertain whether the words were ever in actual use. By contrast, linguistic and anthropological studies of how the Mongols named and classified animals focus on terrestrial animals with a greater presence in Mongol society: Poppe, "On Some Mongolian Names"; Hasbagan and Chen, "Cultural Importance of Animals"; Meserve, "Expanded Role of Mongolian."

silently coined, and no sources have come to light to indicate that they were influential in Mongolian-language natural history. We omit them from the present discussion, which focuses on the Qianlong period.

The project of coining new Manchu words, begun in the 1740s, continued in the 1750s and 1760s with many new words for flora and fauna. The pictorial *Catalog of Birds* and *Catalog of Beasts*, both completed in 1761, contained exotic creatures from far outside the empire's borders. In part, these creatures were drawn from European sources that had been introduced to China by the Jesuits and then influenced *Gujin tushu jicheng*.

The ways new Manchu words were created at the Qianlong court have certain parallels in the Chinese tradition of language study, but also relied on Mongolian morphology. The Chinese parallels can be discerned from a characteristic of many of the new words first noted by Erwin von Zach (1872–1942) in 1900. Von Zach wrote that among the neologisms, "the numerous synonyms of an animal genus have the same endings" (*die zahlreichen Synonyma einer Thiergattung gleiche Ausgänge besitzen*). In the names for peacocks, for example, "the word *tojin* 'peacock' [a loanword in Manchu; cognates exist in, e.g., Persian and Mongolian] enters into combinations with Manchu words as well as Chinese loanwords in which its first syllable disappears without a trace" (*Wir sehen also das Wort 'tojin' sowohl mit Mandschuwörtern, als auch chinesischen Lehnwörtern Verbindungen eingehen, wobei seine erste Silbe spurlos verschwindet*), yielding words that all end in *-jin*.[8] Von Zach gave examples including *jujin*, *kundujin*, and *molojin*, all of which occur in the *Catalog of Birds*.[9] These words can be characterized as portmanteaus, or blends, such as the English word "smog," which is formed from "smoke" and "fog."[10]

Von Zach had initially come across the new words not in the *Catalog of Birds*, but instead in Hans Conon von der Gabelentz's (1807–1874) Manchu-German dictionary from 1864. When von Zach later looked them up in Qianlong's Manchu-Chinese *Expanded and Emended*

---

8. Tsintsius, *Sravnitel'nyĭ slovar'*, vol. 2, 191.
9. Zach, "Über Wortzusammensetzungen," 242; Chuang, Niaopu *Manwen*, vol. 1, 25.
10. We owe this comparison to Andreas Hoelzl.

Mirror, he was able to show that they were based on corresponding Chinese expressions. Yet von Zach did not make anything more of the history of these imperial neologisms or the history of the derivational techniques used to form them.[11] Erich Hauer (1878–1936) later revealed many more words similarly composed, as did, much more recently, Chuang Chi-fa 莊吉發.[12] Both have shed much light on Qianlong's lexical creations.

The process of coining new words by making blends probably suggested itself to the language reformers because of their familiarity with Chinese scholarly and artistic practice. The method used in Qianlong's plurilingual scholarly works (not limited to the *Mirrors*) to transcribe the sound of Manchu and other words using Chinese characters bore some semblance to the blends.[13] The central concept of Chinese character-based transcription was the Chinese word *qiè* 切 'to press close together, fit together'. In the system used in Qianlong court scholarship, initials, medials, and finals of Chinese syllables, each represented by a Chinese character, were "pressed close" to form the sound of a single syllable in Manchu or another language.[14] The Manchu blends were not single syllables like the units spelled by the Chinese character–based transcription system, yet they were formed by truncating other comparable units and fitting them to form new ones.

The Chinese linguistic tradition contained other techniques that might have been at the back of the reformers' minds, called "dissecting graphs" (*chaizi* 拆字) or "character splitting" or "graphosection" (*xizi* 析字) depending on the area of application. Describing such a technique, Wolfgang Behr writes that it "involves the analysis of complex characters into their constituent elements, or even into single strokes, and the semantic reassociation of these parts to unrelated lexical items."[15] Thus both on the level of sound and of writing were late imperial

11. Zach, "Über Wortzusammensetzungen," 242–43. See also Zach, *Lexicographische Beiträge*, 3:96–107.

12. Hauer, *Handwörterbuch der Mandschusprache*, 1952; Chuang, "Xiangxing huiyi."

13. We owe this realization to José Andrés Alonso de la Fuente.

14. Söderblom Saarela, "Alphabets *avant la lettre*," 248–50; *Early Modern Travels*, 158.

15. Behr, "Interstices of Representation," 306. See also Führer, "Seers and Jesters."

intellectuals familiar with splitting words and reassembling them to form new ones.

In addition to the creation of blends, perhaps under the influence of Chinese practices, the language reformers apparently relied on Mongolian morphology. B·Sod has noted that a number of Manchu neologisms, the names of several mythical animals among them, end in *-tu*, and that this ending is probably a loan from Mongolian. Such words include *sabintu* 'unicorn'. Unlike in the case of the Mongolian words invented as by-products of the court's Manchu lexicography, words in *-tu* are attested in the Mongolian vernaculars. The ending is found in the Khorchin dialect (the Khorchin being some of the earliest allies of the Manchus), where its "function is identical with how it is used in Manchu" 功能与满语完全相同.[16] In both cases, it is a nominalizing suffix.[17] It thus appears that the Manchus formed new words on Mongolian patterns.

As had been the case with Kangxi Manchu-Mongol *Mirror* of 1717, Qianlong's plurilingual *Mirrors* inadvertently created many new words for the newly included languages. For example, in reference to the section on birds in the *Pentaglot*, which included translations into Uighur or eastern Turki, Edward Denison Ross (1871–1940) wrote in 1909 that "in many cases the Turki seems merely a translation (sometimes a mistranslation) of the Manchu or the Chinese, and in all too many cases the Turki name simply represents the briefest possible summary of the Manchu definition."[18] Doerfer made similar remarks in relation to the names of aquatic animals in this book, noting that the Mongolian words were Manchu transcriptions, calques, or—as in the case of the "foreign creature" the *sangguji* 'jellyfish' (which made its first appearance in the second edition of the *Ode to Mukden*)—paraphrases (*dalai-yin mögersü* 'sea cartilage' in this case).[19]

---

16. On the Khorchin as allies, see Weiers, "Mandschu-Khortsin Bund von 1626."
17. B·Sod, "Cong *Wuti Qingwen jian*," 40.
18. Ross, *Polyglot List of Birds*, 255.
19. Doerfer moreover mentions a fourth category—mistakes—which we have excluded from the enumeration. Doerfer, "Terms for Aquatic Animals," 193–94. See also Corff et al., *Auf kaiserlichen Befehl*, vol. 2, 982 (item 4497.2); Lessing et al., *Mongolian-English Dictionary*, 545 (s.v. *møgeresy[n]*). On Mongolian neologisms introduced from Manchu, see Tsai, "Qing qianqi Manzhou," 241.

Much like Mongolian and Turki in the *Pentaglot*, some Chinese words were created by translation in the *Expanded and Emended Mirror* of 1772. Unlike Qianlong's Manchu neologisms, however, the inventions in languages such as Mongolian and Turki were not, as far as we can tell, discussed at court, but appear to be ad hoc by-products of the compilation of the dictionaries in question. The Chinese words are an exception. The court carefully discussed and documented Chinese words for certain plants, which left its trail of archival documents, as we show in the remainder of this chapter.

Relative to words in other languages, Qianlong's reform of the Manchu vocabulary involved much more explicit discussion, even outside the court itself. In terms of the wording used in public inscriptions, field officials took part by submitting suggestions to the throne.[20] The circulation of the new words to the bureaucracy can be gauged in an examination of documents sent between the Imperial Household Department in Beijing to its counterpart in Mukden, which Zhang Hong, Cheng Dakun, and Tong Yonggong have carried out. Their research showed that names for plants (sugarcane, apple, Chinese crabapple, pomegranate) were being revised as of late in 1747 and early 1748 (QL 13/11).[21] Animals soon followed, with *ihasi* 'rhinoceros', among others, introduced on March 11, 1749 (QL 14/1/23).[22]

It is not always obvious what made the court think that a new word was needed in the first place. In cases where a word already existed in Chinese but not in Manchu other than as a Chinese loan, the motivation for coining a new Manchu word is clear. Yet in some cases, the announcement sent to Mukden did not contain a corresponding, existing Chinese term. For example, on November 14, 1762 (QL 27/9/29), it was announced that "by imperial edict, [the word for] green seedless grape has been fixed as *kišimiši*" (*faha akū niowanggiyan mucu be hesei kišimiši seme toktobuha*).[23] The document did not specify, but *kišimiši*

---

20. Söderblom Saarela, "Public Inscriptions."
21. Zhang Hong, Cheng, and Tong, "Qianlong chao 'Qinding Xin Qingyu' (san)," 54.
22. Zhang Hong, Cheng, and Tong, "Qianlong chao 'Qinding Xin Qingyu' (san)," 57–58.
23. Zhang Hong, Cheng, and Tong, "Qianlong chao 'Qinding Xin Qingyu' (ba)," 87.

was a transcription—circuitous, perhaps—of the Turki *kīšmīš*. Qianlong the following year (1763, *guiwei*, i.e., QL 28) wrote a Chinese poem on this grape titled *qishimishi* 奇石蜜食 'marvelous stone honeyed food', in which he noted that the word was from Turki (*huiyu* 回語 'Muslim language') and discussed the fruit's history in Central Asia.[24] The Turki word *kīšmīš* (or *kishmish*) was indeed later given as the equivalent of Manchu *kišimiši* in the pentaglot *Mirror*, but just as there had been no Chinese word included in the document that announced the new Manchu coinage, so no Chinese name was given in that dictionary or in the Manchu-Chinese *Mirror* on which it was based.[25] In the imperial *Mirrors*, *kišimiši* was instead translated paraphrastically as "seedless green grape" 無子綠葡萄.[26]

Manuscript collections of new words were compiled, and one was even printed privately in order to provide a handy reference work for new words.[27] Yunggui 永貴 (1706–1783) printed a collection of words and phrases in 1748 or 1749, when he was lieutenant-governor of Zhejiang.[28] Yunggui's list of phrases (focused not on flora and fauna, but

---

24. *Qinding Huangyu Xiyu tuzhi*, 251:43:12a (820, top panel). We thank Arina Mikhalevskaya for drawing our attention to this source.

25. Eric Schluessel helped us research the use of this word in Turkic-language sources. In most varieties, the word means "raisin," not the fresh fruit of the grape (Bābur, *Bābur-nāma [vaqāyi']*, 455; Räsänen, *Versuch eines etymologischen Wörterbuchs*, 272). However, Gunnar Jarring (1907–2002), on the basis of fieldwork in southern Xinjiang in the 1930s, defines *čišmiš~kišmiš* (the palatalized form represents the pronunciation in Kashgar) as "small, sweet grapes, of the size of peas, without kernels, well-known all over Asia" (Jarring, *An Eastern Turki*, 74). Corff et al., *Auf kaiserlichen Befehl*, vol. 2, 1050 (item 4730.4).

26. *Yuzhi zengding Qingwen jian*, 1782, vol. 233, supplement, 3:9b (383, lower panel). When we consider that the Manchu word is ultimately a transcription of a Turki word, *kišimiši* is curious in that it appears to follow the Chinese-character form more closely than it follows the original. The Manchu word has vowels inserted in two instances to prevent consonant clusters, as would be demanded by the phonotactics of Mandarin Chinese. It might have been that Qianlong or his secretaries took the Chinese transcription into account when coining the Manchu name, but we should note that Manchu transcriptions of foreign words often took on this Chinese-looking form. On this point, see Söderblom Saarela, *Early Modern Travels*, 147–48.

27. Chunhua, *Qingdai*, 154–58 (items 116–119).

28. Yunggui, *Qingwen jian wai xinyu*. The date 1748–1749 is based on dates mentioned in the collection and the time of Yunggui's tenure in Zhejiang. For

on government institutions) was noted as "sent out by the Grand Secretariat" (*neige chaochu* 內閣抄出). It was thus evidently intended to help officials and clerical staff in the provinces to keep up with the changing terminology. The new words sent out by the Grand Secretariat apparently circulated within officialdom, as at least one attentive user of Manchu made notes of when new words were announced in this way.

Not all new words are attested in these sources, nor were all words so announced later entered into the revised *Mirror*. Some of the words simply came too late: *deyengge nimaha* 'flying fish' (for Ch. *feiyu* 飛魚 idem), announced in 1771 (QL 36), is one example.[29] As we discuss presently, court documents that were not made public at the time similarly proposed words that eventually did not make it into the official standard.

Such was the case with some of the new words created during the translation or retranslation of the Confucian classics into Manchu. The revised translation of the *Poetry Classic*, a book that included many obscure terms for flora and fauna, was finished in 1768 (QL 33).[30] Some new words used in this work, such as *dartaha* 'ephemera' (for Ch. *fuyou* 蜉蝣), *agada moo* 'tamarisk' (for Ch. *cheng* 檉), *iceku orho* 'madder' (lit. pigment herb, for Ch. *rulü* 茹藘—introduced as a translation for *qiancao* 茜草 'madder' in 1761 [QL 27/7/8]), *kartukū* 'barbel' (for Ch. *jiayu* 嘉魚), and *jodorho* 'water plantain' (for Ch. *xu* 蕍) were for reasons unknown subsequently not included in the Manchu-Chinese *Mirror*.[31] Nor was *kirho* or *guruki* included, despite in two separate poems in the *Classic* being used used to translate *qi* 芑 'common

---

Yunggui's position, see Brunnert and Hagelstrom, *Present Day Political Organization*, 405 (item 825).

29. Ihing, *Qingwen buhui*, 5:18a. In cases where the new word did not end up in the *Expanded and Emended Mirror*, Ihing noted that they had been announced by the Grand Secretariat on such and such a date.

30. Yeh, "Man-Han hebi," 15.

31. Zhang, Cheng, and Tong, "Qianlong chao," 2000, 31. The English gloss barbel follows Legge's translation of the Chinese (*Chinese Classics*, vol. 4, 270). Water plantain follows Fèvre and Métailié (*Dictionnaire Ricci des plantes*, 571, s.v. *zé xiè*). Cf. Legge, *Chinese Classics*, vol. 4, 165. *Yuzhi fanyi Shijing*, 3:92b (511, lower panel, for *dartaha*) and 6:24b (618, upper panel, for *agada moo*), 3:16b (473, upper panel, for *iceku orho*), 3:21a (475, lower panel, idem), 4:20a (532, upper panel, for *kartukū*), and 3:38b

sowthistle, milkweed, hare's lettuce'.[32] Incidentally, that the same Chinese word—treated as the same word by exegetes of the *Poetry Classic*—was translated in two ways in the Qianlong revision of the Manchu translation suggests that the translation work was not entirely coordinated.[33]

In most of the documents sent to the Mukden Imperial Household Department we have examined, Manchu words were introduced as the equivalents of Chinese words. Rarely was a Manchu word introduced on its own without a Chinese translation (as in the case of *kišimiši*). Here we see how the desire toward linguistic parity, especially vis-à-vis the classical Chinese lexicon, really drove the lion's share of the language reform conducted at the Qianlong court.

(484, upper panel, for *jodorho*). Ihing (*Qingwen buhui*, 1:14a, 1:37b, 2:37a, 5:10b, and 7:8b) says that the words originate in the *Poetry Classic*.

Relatedly, it is worth pointing out that the *Expanded and Emended Mirror* lists three names of birds that it says originate in the "*kanjur* canon" (*g'anjur nomun*) or "Buddhist canon" (*fucihi nomun*). These are not words drawn from the Manchu version of the Buddhist scriptures, however, but translations of Chinese words that occur in the legends to the *Catalog of Birds*. Thus *molojin*, an alternative word for the peacock, is a translation of *moyouluo* 摩由邏 (or 羅 < Sanskrit *mayūra*), which occurs in a quote from *Bencao gangmu* (in turn perhaps lifted from *Gujin tushu jicheng*), where it is presented as a name for the peacock used in "Sanskrit writings" (*Fan shu* 梵書). Similarly, *joni*, a word for the magpie, translated Ch. *chuni* 芻尼, corrupted in the *Expanded and Emended Mirror* to *chumao* 除毛, and *šaruk* translated *sheluo* 舍羅 'myna' (< Sanskrit *śārikā*). See Li Shizhen, *Bencao gangmu*, vol. 774, 49:18a (398, lower panel); Chuang, Niaopu Manwen, vol. 1, 45, 69, 69n49, vol. 2, 77; *Yuzhi zengding*, vol. 233, supplement, 4:7b (408, lower panel) and 4:26b (418, upper panel) and 4:32b (421, upper panel); Corff et al., *Auf kaiserlichen Befehl*, vol. 2, 1071–72 (4811.4), 1081 (item 4857.1), 1084 (item 4870.4).

32. *Yuzhi fanyi Shijing*, 185:5:62 (539, upper panel) and 6:38a (625, upper panel); Ihing, *Qingwen buhui*, 7:41a (for *kirho*; we could not find *guruki*). For the English translation of *qi*, see Fèvre and Métailié, *Dictionnaire Ricci des plantes*, 255, s.v. *kǔ jù cái*.

33. Legge, *Chinese Classics*, vol. 4, 284, 463: "white millet"; Karlgren, *Book of Odes*, 122, 199: "k'i plant(s)."

Neologisms in Pictorial Albums ("Catalogs") of Flora and Fauna

Before we proceed to discuss the creation of bilingual pictorial albums of birds and beasts made at the Qianlong court during the 1760s, it is necessary to investigate whether any plans were ever made to make bilingual albums of flowers, fruits, and trees. Indeed, one might even wonder why none existed for marine animals. Paintings of sea creatures were evidently available at the Qing court. The compilers of Kangxi's original Manchu *Mirror* might even have been aware of such images. That would be one explanation for the *Mirror*'s curious note regarding "sea fish," that such creatures were numerous and that their local names differed, presumably referring to their names in various languages. An encounter with European natural histories or Chinese albums made on their basis with Jesuit participation might very well have prompted this comment. Yet with one exception of a smaller album of fragrant Manchurian plants with Manchu and Mongolian captions that we discuss presently, we have not been able to establish any link between the production of pictures on plants and animals on the one hand, and the systematic creation of new Manchu words for fish, flowers, and trees on the other.

Perhaps it is no longer productive to ask why the Qing court did not produce a complete set of illustrated natural-historical works, as it seemed poised to do after the 1760s. What we see here instead, we suggest, is a clear trajectory of evolving consideration with what the genre of *pu* consisted of and what it could be made to accomplish. In chapter 3, we translated this term as "catalog" to better describe *Guang Qunfang pu*, a textual compilation that added more encyclopedic dimensions to the genre of *pu* but included no pictures. We also traced the emergence at the late Kangxi court (1710s–1720s) of pictorial portrayals of plants and animals by polymathic officials such as Jiang Tingxi. These portrayals were largely contemporaneous with the encyclopedic expansion of the genre of the *pu*. We suggest that Jiang's role as the chief editor of *Gujin tushu jicheng* in the 1720s, a true encyclopedia with both texts and images, can be seen as a fusion of those two trends of scholarship and art. In a sense, the pictorial (and textual) *pu* of birds and beasts of the 1760s were really part of the same larger project as the pictureless *pu* of plants in the late 1700s, with the fusion between textual and pictorial elements accomplished via *Gujin tushu jicheng* in the 1720s. The more pertinent question, then, becomes why it was important to make the captions bilingual in the 1760s and how this insistence on lexicographical parity between Manchu and Chinese affected this new kind of court-based encyclopedic documentation of flora and fauna?

Pictorial Albums of Manchurian Herbs and Foreign Flowers

Let us explore this implicit shift from one kind of *pu* to another through the earliest known example of the latter kind: a compact pictorial bilingual album of four Manchurian plants known by their fragrance. Titled *Jiachan jianxin* 嘉產薦馨 (Auspicious products of commendable fragrance), this slim album was attributed to the court painter Yu Xing 余省 (1692–1767 or later), who apprenticed with Jiang Tingxi and adhered to the latter's style of rendering plants. Yu would have produced the album in or around 1747 (QL 12). The two extant copies were likely initially housed in Beijing and Mukden respectively (the Mukden copy has since been returned to Beijing and the copy originally housed there is now in Taipei). Four plants—or, more likely, dried specimens of the plants sent to Beijing from various regions of Manchuria—were depicted in pictures and matched with captions in Chinese (right to left) and Manchu and Mongolian (left to right; see fig. 7.1).

Wang Zhao used a modern naturalist's eye to identify all four plants and matched them with extant species found in the northeast today. He also located memorials that allowed him to convincingly link the procurement of the herbs to Qianlong's effort to overhaul and standardize Manchu ritual practices at court in 1747, an effort that coincided with the preparation and compilation of an illustrated compendium.[34] Likewise with a naturalist's quest for precision, Wang found much to fault in the textual elements of the herb album. Why, he asked, did the album match these Manchurian plants with Chinese names drawn from older

---

34. See Yeh, *Manwen "Qinding Manzho,"* 15–21.

FIG. 7.1. Album leaf depicting *ayan hiyan* (Ch. *yunxiang*) in Yu Xing, *Jiachan jianxin* (1736–1748). Digital Archive, National Palace Museum, Taipei. https://digitalarchive.npm.gov.tw/Painting/Content?pid=9715&Dept=P.

Chinese literature, a corpus in which most of the sources were in fact describing fragrant herbs growing in the far south?[35]

Yet if we consider the description of those four Manchurian plants in light of late Kangxi *gewu* practices, discussed in chapter 5, the granting of southern Chinese plant names to Manchurian species does not seem so out of place. After all, this was what the Kangxi emperor had been doing throughout his *Kangxi ji xia gewu bian*, that is, to locate earlier Chinese records relevant for understanding the new frontiers of the Qing empire. The more interesting feature of this album on "auspicious products of commendable fragrance," which Wang did not discuss, however, is that all four Manchu names it included were mentioned in the 1708 Manchu *Mirror*. The captions in Manchu and Chinese were also closely based on the *Mirror*'s descriptions and included

35. Wang Zhao, "Dixiang qingfen," 64–75; see also Wang Zhao, "Zhong shen zhi xiang."

only a handful of discrepancies. Most notably, the 1708 *Mirror* listed *sengkiri hiyan* as simply another name for *ayan hiyan*, which it describes in this passage:

> The stem is like the stem of the pear-leafed crabapple. The leaves are smaller than the leaves of *niyanci hiyan* and fine and thin. It grows in thickets in marshy land. Another kind of *ayan hiyan* is found in places where a cliff larch has grown.[36] Its stem spreads out across the rock and sticks to it. Its leaves are like pine needles but shorter. The color of the seeds is like that of ripe [or black] grapes.[37] Their shape is as big as that of the bird cherry. These two kinds are also called *sengkiri hiyan*.
>
> cikten mamugiya-i cikten-i adali, abdaha, niyanci hiyan-i abdaha ci ajige bime, narhūn nekeliyen. lebenggi bade fuldun fuldun-i banjimbi. geli emu hacin-i ayan hiyan, hada-i isi banjiha bade bi. cikten sireneme wehe de latume banjimbi. abdaha, sata-i adali bime foholon. use-i boco boroko mucu-i adali, yengge-i gese amba. ere juwe hacin be geli sengkiri hiyan sembi.[38]

In the 1747 album, however, *ayan hiyan* and *sengkiri hiyan* (as well as their Mongolian translations *sülü* 'incense' [sic] and *qarabur* 'blackish' [sic]) were used as two distinct names referring to the two different plants that were originally both interchangeably called *ayan hiyan* and *sengkiri hiyan* in the 1708 *Mirror*.[39] The reinterpretation of synonyms as separate entities reflects the court's greater effort to establish a one-to-one correspondence between nomenclature and taxonomy in Manchu—that is to say, one name corresponds to one thing. In chapter 8, we trace how this approach to the Manchu lexicon was further elaborated and systematized in the *Expanded and Emended Manchu Mirror*.

---

36. Cf. *Han-i araha manju gisun*, 1717, 19:*orho-i hacin*:3a: *qada-yin qaryai* 'cliff larch.'

37. Cf. Li Yanji, *Qingwen huishu*, 1:8a (7, upper panel): *hei putao* 黑葡萄 'black cherries.'

38. *Han-i araha manju gisun*, 1708, 19:*orho-i hacin*:2a.

39. See Kowalewski, *Dictionnaire mongol-russe-français*, vol. 2, 833; Lessing et al., *Mongolian-English Dictionary*, 743, 932.

In sum, this fine herb album—as well as the immense effort the Qianlong court threw behind the procurement of these Manchurian plants—represents a continuity with the encyclopedic lexicographical work carried out during the Kangxi reign and described in earlier chapters. It remained deeply connected to the Manchu *Mirror* and the Chinese encyclopedic works that discussed Manchurian flora in order to connect them with an earlier classical literature in Chinese. Ongoing research by Lin Shih-hsuan 林士鉉, from which we profited when writing this section, will when published present a comprehensive analysis of the Manchu and Chinese names used for these plants.[40]

Yet the album also discloses redefined priorities new to the Qianlong court, namely, the much more extensive use of paintings, which here served the ideological purpose of enhancing the iconic status of these plants associated with the Manchu homeland. The album did so in a move very similar to that seen in the proliferation of golden lotus flower pictures in the late Kangxi reign, but the *Jiachan jianxin* album was even further removed from Buddhist iconography than the golden lotus pictures had been. Our purpose here, however, is not to embark on a full analysis of the pictorial conventions, which has been done admirably by others, but rather to highlight similar lexicographical rearrangements that were instrumental in the making of pictorial albums of birds and beasts as well.

Pictorial Albums ("Catalogs") of Birds and Beasts

In large part, the new words for terrestrial fauna and fowl in the Manchu-Chinese *Mirror* originated in the translated Manchu legends to the pictures collected in two albums finished at court in 1761. The Manchu-Chinese albums of birds and beasts were, like some of the other albums made in the palace workshop, based in part on images that originated in illustrated post-Renaissance European natural histories. Natural and wondrous creatures seen or heard about during the expansion of European empires and trading networks across the globe appeared as engraved images in scholarly works such as Conrad Gessner's (1516–1565) *Historia animalium* from the 1550s. These books reached the Jesuit

40. Lin Shih-hsuan, "Qianlong nianjian duiyu Dongbei."

libraries in China, where they incited the interest of the Kangxi emperor and influenced several Chinese-language publications, including *Gujin tushu jicheng*. This book in turn provided much of the textual material later included in the Manchu-Chinese albums. It likewise provided information on plants that, as far as we know, were never included in any album but later appeared with Manchu names in Qianlong's revised *Mirror*, which we discuss in chapter 8.

As heavily politicized imperial cultural products, the albums of beasts and birds were related to *Huang Qing zhigong tu* 皇清職貢圖 (Album of tributary peoples of the august Qing), a work with an obvious ideological intent.[41] Work on the three albums under Qianlong imperial auspices was in all cases begun in 1750 (QL 15), although many of the paintings originated in the courts of Qianlong's predecessors.[42] The Chinese and Manchu texts for the albums were produced after 1757. All three albums—both text and images—were completed by the end of 1761 (QL 26).[43]

Both of the images and the text in the *Catalog of Beasts*, one of the albums, relied primarily on text and images from *Gujin tushu jicheng*. The 180 animal illustrations in the *Catalog of Beasts* are subdivided into 107 auspicious animals (59 percent of the total), 61 terrestrial animals endemic to the Qing territory (34 percent of the total), and 12 animals from foreign countries (*yiguo shou* 異國獸, 7 percent of the total).[44] The last were, like the rest of the album, drawn from *Gujin tushu jicheng*, and ultimately derived largely from Gessner via Verbiest. Lai Yu-chih 賴毓芝 has interestingly showed that Qianlong's court artists probably consulted Gessner in the original as well, which indicates that realism was an important concern in the making of this album.[45] The 360 birds (sometimes males and females were presented separately) depicted in the sister work *Catalog of Birds*, by

---

41. Hostetler, *Qing Colonial Enterprise*, 41–49. On this work, see also Walravens, "'Tribute-Bearers in Manchu and Chinese'"; Walravens, "Das Huang Ch'ing."
42. Yanagisawa, "Kokyū Hakubutsuin," 20–22; Lai Yu-chih, "Qing gong dui Ouzhou," 13–14; Greenberg, "Taxonomy of Empire."
43. Lai Yu-chih, "Qing gong dui Ouzhou," 11–12.
44. Zou, "*Shoupu*," 142. The number 180 is from Zou. Cf. Lai Yu-chih, "Domesticating the Global," 125, who gives 183.
45. Lai Yu-chih, "Qing gong dui Ouzhou," 27–28.

contrast, might have been based almost in its entirety on paintings made in the late Kangxi period, probably on the basis of real-world observation.[46] As expected from such a work, it included depictions of numerous animals not previously described in Manchu. These animals were given new names, sometimes on the basis of their physical or behavioral characteristics. The opossum, a New World animal, for example, was named *sumaltu* on the basis of *sumala* 'pocket'.[47]

It is by virtue of the presence of Manchu-language legends that the albums are important for the issues under scrutiny here. The use of Manchu in the legends of natural-historical paintings appears to be a Qianlong-era phenomenon; the Kangxi-period pictures did not, to our knowledge, contain any Manchu text. Lai Yu-chih has shown, however, that some Kangxi-era paintings of Inner Asian birds carried Chinese names that were loans from Manchu.[48] In the Qianlong-era album, as the editors added Manchu legends they also removed the Manchu loans in Chinese and replaced them with Chinese-sounding names.

The legends of the images in the Qianlong-era albums on birds and beasts, by contrast, in most cases linked the creatures described to the Chinese scholarly tradition. Based on *Gujin tushu jicheng*, they included quotes from classical Chinese literature (including *Erya*). The Manchu legends were in general translations of the Chinese. In that sense, these natural-historical texts resembled the *Ode to Mukden*,

46. Lai Yu-chih, "Qianlong chao *Niaopu*." This paper revises Lai's earlier explanation of the history of the *Album of Birds*. She had previously proposed that Qianlong's Grand Council's collection of specimens (live birds or bird "skins" [*pi* 皮]) might have been related to the paintings in the album. We expect that she will sort out the role played by such specimens in her forthcoming work, which was not available to us at the time of writing. For the role of specimens, cf. Lai Yu-chih, "Images, Knowledge and Empire," 80; Yu, "Huangdi de bowutu," 97; Yanagisawa, "Kokyū Hakubutsuin," 19 (regarding males and females). The number 360 is from Greenberg's "Taxonomy of Empire." Yanagisawa gives the number 361.

47. On the opossum in *Shoupu* and its European sources, Lai Yu-chih, "Qing gong dui Ouzhou," 24–28. The etymology is from Walravens, "Konrad Gessner," 91n14. Note that Walravens, writing in the 1970s, did not know of the *Album of Beasts* but made the remark in reference to Ihing's later dictionary based on the Manchu-Chinese *Mirror* (which in this case presented a word that originated in the *Album*). We return to this point. Regarding *sumaltu*, see also Chuang, Shoupu *Manwen*, vol. 2, 49.

48. Lai Yu-chih, "Qianlong chao *Niaopu*," 20.

which also was a Chinese text in character and origin. There is at least one exception, however, as we discuss presently.

The Manchu text in some entries follows the Chinese texts so closely, while still observing the principles of Qianlong's language reform, that the result borders on the ridiculous. In the legend to the image of the black-naped oriole, many synonyms for the headword *huangli* 黃鸝 (Ma. *galin cecike*) are given.[49] The synonyms are originally Chinese; many of them are even presented as Chinese regionalisms. Yet they are nevertheless translated into Manchu. Thus when the Chinese text, quoting from Chinese literature, says that "the people of [the ancient kingdom of] Qi call it *boshu* 搏黍," the Manchu text has "the people of Shandong [i.e., the Qing province roughly corresponding to the ancient state of Qi] call it *tulin cecike* 'small *tulin* bird'" (*šandong ba-i niyalma tulin cecike sembi*), where *tulin* is a neologism that to our knowledge is only attested in this phrase.[50] Neither the ancient people of Qi nor the inhabitants of eighteenth-century Shandong ever used the phrase *tulin cecike* to refer to the black-naped oriole, so it is not clear what purpose the Manchu text is here intended to serve.

Lai has shown that some northern birds that in the Kangxi-era monolingual album had been referred to using vernacular Chinese names, or even transcriptions of Manchu names, were given classical Chinese names in the *Catalog of Birds* and provided with quotes from Chinese literature.[51] Even though some legends had such origins, many of the birds listed in Kangxi's Manchu *Mirror* were not mentioned in the Manchu legends in the *Catalog of Birds*. Yet a certain amount of Manchu-language research evidently went into some legends, beyond coining Manchu names for the birds and beasts that either had none or had previously been referred to using Chinese loanwords. The legend that accompanied the picture of the jay (Ma. *isha*; Ch. *songya* 松鴉), notably, cited Kangxi's Manchu *Mirror*, which as we know did

---

49. Yanagisawa identifies *huangli* with *Oriolus chinensis* ("Kokyū Hakubutsuin," 36). The name is still current.

50. Chuang, *Niaopu Manwen*, vol. 4, 123.

51. Lai Yu-chih, "Qianlong chao *Niaopu*."

not include any Chinese characters.[52] The reason was evidently that this bird "lives in the mountains outside the border and is not to be seen in the interior" (Ch. 生於塞外山中，非內地所產; Ma. *jasei tule alin de tucimbi, dorgi bade banjihangge waka*) for which reason no records in the Chinese tradition were available to consult.

That being the case, certain characteristics of the legend only make sense if we assume that the authors of the text consulted the *Mirror*, wrote their legend in Chinese, then had it translated into Manchu by someone else. The *Mirror* said that the jay was "somewhat similar to the paradise flycatcher, but smaller and with a tail that is short and mottled" (*baibula de adalikan, beye majige ajige, uncehen foholon alha*).[53] The paradise flycatcher, in turn, was described as somewhat similar to the magpie, with a long and ash-colored tail.[54] Regarding the jay, the *Mirror* interestingly remarked that this bird is "avaricious when it comes to food, [thus] very avaricious, bad people are compared to the jay" (*be de doosi, umesi doosi ehe niyalma be, isha de duibulembi*).[55] The Chinese legend accompanying the picture of the jay in the *Catalog of Birds*, however, stated that "the *Mirror of the Manchu Language* calls it *yi-si-ha* [< *isha*], and says that this bird's avarice for food is like the insatiable avarice of people, which is why it has received this name" 《清文鑑》名「伊思哈」，謂此鳥貪食如人貪婪，故得此名. This was evidently not what the *Mirror* said, which Tsai Ming-che, who has discussed this legend, also pointed out.[56] The Manchu legend in the *Catalog* translated the sentence rather than reproduce the different sentence actually found in the *Mirror* (*manju gisun-i buleku bithede, isha seme gebulehengge, ere gasha jetere de doosi, niyalmai doosi gamji-i adali ofi, tuttu duibuleme uttu gebulehebi sehebi*).

Furthermore, whereas the Chinese legend said that this "pine crow"—which is what the Chinese name *songya* translates to—was "shaped like a crow" (*zhuang ru ya* 狀如鴉), the Manchu had "shaped like a duck" (*isha-i arbun niyehe de adali*). Chuang Chi-fa, who published

---

52. The identification with the jay (*garrulus glandarius*) is from Yanagisawa, "Kokyū Hakubutsuin," 35.
53. *Han-i araha manju gisun*, 1708, 19:*cecike-i hacin*:33a.
54. *Han-i araha manju gisun*, 1708, 19:*cecike-i hacin*:32a.
55. *Han-i araha manju gisun*, 1708, 19:*cecike-i hacin*:33a.
56. Tsai, "Qing qianqi Manzhou," 243.

the Manchu and Chinese text of the album, remarks on the basis of this discrepancy that the authors of the *Catalog*'s Chinese text mistakenly wrote *ya* 鴉 'crow' rather than the homophonous *ya* 鴨 'duck'. We find it more likely that the Manchu translator associated the Chinese character for "crow" with the homophonous one for "duck"—perhaps someone read the Chinese text out loud to him—and translated it accordingly. Had he read Kangxi's Manchu *Mirror*, he would have known that the jay looked far more like a crow—which today's taxonomy groups in the family *Corvidae* along with the magpie—than a duck.[57]

The Mirror *and the* Catalog of Beasts: *Differing Descriptions of Rams and Wethers*

The case of the jay shows that the *Mirror of the Manchu Language* was at least on occasion consulted for its natural-historical contents by the authors of the legends to the *Catalog of Birds*, just as it was by translators at the Grand Council in animal-related matters about this time. By contrast, evidence relating to the *Catalog of Beasts* is mixed. The translators of the *Catalog of Beasts* evidently consulted the *Mirror*, but did not follow it to the letter. We discuss an example involving sheep.

An undated and bare-bones Manchu memorial found among the Grand Council's file copies shows that Kangxi's *Mirror* was used at court to resolve terminological issues relating to sheep. By virtue of its placement in the archive, the memorial in question has been dated to 1760 (QL 25/4), which coincides with the writing of the Manchu-Chinese legends to the albums of birds and beasts. The existence of the memorial suggests that even though the reform of the Manchu lexicon in terms of flora and fauna involved the introduction of many new words, it did not mean that the existing vocabulary posed no difficulties even for competent users.

A juxtaposition of this memorial with the albums shows that usage and translation practice was not uniform. More precisely, when talking about sheep, the *Catalog of Beasts* used words in a way that was at odds with Kangxi's *Mirror* and with translations of routine administrative

---

57. Chuang, Niaopu *Manwen*, vol. 2, 213.

documents—in this case one of Qianlong's edicts—in the Grand Council. The memorial reads in full as follows:

> Searching in the *Mirror*, it says that a castrated male sheep is called *buka* 'wether', an intact male sheep is called *kūca* 'ram'. Neutered rams are used for provisions for the army, intact rams are used for breeding. Following this, we changed it on the edict and respectfully memorialize to show it [to you].
>
> *buleku bithede baicaci, aktalaha haha honin be buka, aktalahakū haha honin be kūca sembi, coohai urse de kunesun obure de, buka baitalame, fusen gaire de kūca baitalame ofi, ere songkoi hese de halame arafi gingguleme tuwabume wesimbuhe.*[58]

The memorial was concerned with sheep. It is thus understandable that its paraphrase of the *Mirror* left out a part of the definition of *buka* that did not concern sheep directly. Yet in fact, although the *Mirror*, as the memorial noted, defined *kūca* as an "intact male sheep" (*aktalahakū haha honin*)—not a goat—it also stated that "male sheep (*honin*) and male goats (*niman*) are both called *buka*" (*haha honin, haha niman be gemu buka sembi*).[59] One word seemingly referred to a subset of goats and sheep, the other only to a subset of sheep.

Although *honin* 'sheep' (Ch. *yang* 羊) was used as a heading for a legend in the contemporary *Catalog of Beasts*, *niman* 'goat' was not. In the album, *honin* is depicted as a white, long haired sheep with long horns. The Manchu legend states that "male sheep are called *buka*... male black goats are called *kūca*" (*haha ningge be buka sembi... sahaliyan niman-i haha ningge be kūca sembi*).[60] The word we translate as "goat" here, *niman*, was defined in the Kangxi *Mirror* as sheep with

---

58. Manchu palace memorial file copy, 03-0178-1818-011, FHA.
59. See *Han-i araha manju gisun*, 1708, 20:*eiten ujima hacin*:1b.
60. Chuang, Shoupu *Manwen*, vol. 2, 349 (image on 310). The word *buka* has cognates in several Turkic languages and in Mongolian, where it means "bull"; a semantic change must have taken place as it was borrowed, or later. See Sinor, "Some Altaic Names," 320.

"long hair and a short tail" (*funiyehe golmin, uncehen ajige*), which sounds very much like the sheep painted in the album.[61]

Whatever the referent of *niman*, the definitions of *buka* and *kūca* given in the *Catalog of Beasts* are very different from those given in Kangxi's *Mirror*. In one case, *buka* is a castrated male versus simply a male in the other, and *kūca* has a much more restricted meaning in the *Catalog* that only partially overlaps with that given in the *Mirror*. The reason for the discrepancy might have been that the translators felt compelled to represent the terminological distinctions made in the Chinese legend, which said that "males [are called] *fen* . . . black mountain sheep [are called] *li*" 牡羒 . . . 黑羖䍽, using obscure vocabulary and ancient phrasing. If so, accuracy might have mattered less than fidelity to the classical Chinese to the Manchu translators.

Alternatively, the discussion and varying use of these words in court documents suggest either some unclarity regarding their meaning, or that normative usage (represented by the original *Mirror*) and actual usage differed. Tellingly, in Qianlong's later *Expanded and Emended Mirror*, the definitions of these words were revised.

In addition to the albums mentioned here, which had bilingual inscriptions, Qianlong commissioned other paintings with animals as subject matter. Series of paintings of horses included Manchu and Mongolian legends alongside Chinese, and paintings of peacocks were associated with Chinese imperial poems, some of which were inscribed onto the paintings.[62] Like other court art, the paintings and poetry had strong political overtones (horses were tribute items from Inner Asia, and peacocks were associated with the newly conquered Western regions). Both have a relationship to the Manchu-Chinese *Mirror* but not to natural history. The horses were not so much different kinds as different individuals that Qianlong had received. The interesting Manchu-Mongol-Chinese vocabulary, included in subsequent *Mirrors*, that

---

61. See *Han-i araha manju gisun*, 1708, 20:*eiten ujima hacin*:1b.
62. For Qianlong's poetry on peacocks and the paintings he commissioned, see Kleutghen, *Imperial Illusions*, chap. 4.

developed to describe them concerned their appearance (e.g., color), but not the identification of different breeds.[63]

The peacock paintings and poems, by contrast, included only Chinese-language material, but the vocabulary referred to kinds rather than individuals. Accordingly, the supplement to Qianlong's Manchu-Chinese *Mirror* was enriched with newly invented names for varieties of peacock. Thus the bilingual—and later multilingual—*Mirrors* produced from the 1770s onward had ties to both art and literature produced at court that built on late Kangxi-era natural erudition but with a heightened sense of the political symbolism carried by animals.

Grand Council Documents on Plants and Animals

Most of the names included in the *pu* albums of birds and beasts made it into the Manchu-Chinese *Mirror*. Yet even more animal names were coined that did not stem from these albums. Archival documents show that animal names were discussed, revised, and invented at court beyond what can be ascertained by reading the finished albums. In other words, Manchu neologisms were invented not only for iconic plants and animals but also at a much more mundane level of lexicographical standardization. We note four examples: a bird, an insect, various deer, and a berry. In many cases, establishing a new word involved research in the Chinese scholarly literature. Again, readers should feel free to read the case studies in any order they like, or skipping along to the end of the chapter.

That happened also in the case of a gourd that already had a Manchu name. In this instance, the investigation apparently centered entirely on the plant itself, not on its nomenclature. There is even an indication that it went beyond the consultation of scholarly literature to include the study of a specimen of the plant. The document on this gourd is the last of the Grand Council documents that we discuss before

---

63. On the albums of horses and their relationship to the Manchu *Mirrors*, see Lin Shih-hsuan, "Qianlong shidai de gongma." On the colors of horses in Manchu, see Uray-Kőhalmi, "Die Farbbezeichnungen der Pferde"; Shinneman, "Horse Colors."

turning to the *Expanded and Emended Mirror of the Manchu Language* itself in the next chapter.

New Words Discussed at the Grand Council
(1): A Bird

First, an anonymous memorial with an imperial rescript dated February 28, 1754 (QL 19/2/7) proposed to revise the Manchu name for the bird called *to gi*. This word was given an entry in Kangxi's Manchu *Mirror*. The dictionary did not say so, but *to gi* was a loan from the Chinese *tuoji*, literally "camel chicken." It is not clear exactly what bird was meant in the entry in Kangxi's *Mirror*. Yet whatever the case, the name *to gi* was no longer acceptable during Qianlong's language reform. The memorial read,

> An investigation revealed that there is no Manchu word for *tuoji* 駝雞. In the *Mirror of the Manchu Language*, a transcription of the Chinese characters is given. Its body is very large. With that in consideration, I have established the name *temen coko* 'camel chicken' and written it on this yellow slip that I have pasted onto the attached picture. Please advise whether it is fitting or not.
>
> 駝雞 *be baicaci, manju gisun akū manju gisun-i buleku bithede nikan hergen de toodame arahabi, beye umesi amba ofi, acara be tuwame temen coko seme toktobufi, afahari arafi nirugan de latubuha, acanara acanarakū babe dergici jorime tucibureo.*

Qianlong's rescript read "noted" (*saha*).[64] The mention of a picture (*nirugan*) in this memorial is interesting, as is the date. It is tempting to conclude that the memorial is related to the *Catalog of Birds*. Perhaps the camel chicken, whatever it was, was a candidate for inclusion in the album. Yet if so, it did not make the cut, because no bird with this name is included. The cassowary is included, but under a different name, and it does not have colorful tailfeathers in the picture featured there, which is how *to gi* is described in the 1708 *Mirror*. A picture was

---

64. Manchu palace memorial file copy, 03-0172-0810-001, FHA.

apparently submitted with this memorial, however, although not with the copy currently accessible in the archive.

The name proposed in the memorial, *temen coko*, is similar to many other new Manchu coinages in that it mimics the semantic structure of the Chinese expression, whose two characters mean precisely "camel chicken." The memorialist seems to suggest that he chose the word "camel" to describe this bird because of its size, but that is hardly convincing. Many things are big; "camel" was in all likelihood chosen in analogy with *tuo*. Perhaps the proposed name sounded a little too direct, however. In the revised *Mirror*, the word does not appear as *temen coko* but as *temege coko*, where *temege* is a neologism.[65]

New Words Discussed at the Grand Council (2): An Insect

Our second example of the names of animals discussed at court concerns locusts and their offspring.[66] The document is of a similar character as the one on the camel chicken, but because it lacks a rescript, it is possible that it is not a memorial but instead some other kind of communication submitted to superiors within the Grand Council. The note is undated. On the basis of its placement in the archive, however, it has been dated to the spring of 1754 (QL 19/1).

The reason for this note was probably the translation of *Qiju zhu* for 1752 into Manchu. That year, Qianlong made a statement on the origin of locusts to the court. The emperor voiced the common opinion that "if fish and shrimp place small eggs by riverbanks and in low-lying damp areas, which then enter onto land, they will all change into locusts (*huang*)" 水濱沮洳魚鰕微卵，入土即皆變蝗 after first passing through the stage of immature locust (*nan* 蝻).[67] The difficulty of translating this statement into Manchu was probably what prompted the note, which read as follows:

---

65. *Yuzhi zengding Qingwen jian*, vol. 233, 30:16b (220, upper panel).
66. The following paragraphs largely copy the corresponding passages in Söderblom Saarela, "On the Manchu Names," 53–55.
67. *Gaozong*, vol. 14 (part 6), 415:5b (428). Entry dated 1752/6/27 (QL 17/5/*bingzi*).

Upon examination: *huangcong* 蝗蟲 'locust'. A kind of grasshopper (*sebsehe*) with a body that is slightly bigger. Taking the ordinary pronunciation of *sebsehe*, I suggest we call the locust *sabsaha*. As for *nanzi* 蝻子 'immature locust', I suggest that we combine the two words [lit. characters] *use* 'seed, insect egg' and *honika* 'fish fry' to form *unika*. Please advise as to whether it is appropriate.

baicaci, 蝗蟲 sebsehe-i duwali, beye majige ambakan, sebsehe-i an mudan be gaime, 蝗蟲 be sabsaha obuki, 蝻子 be, use honika juwe hergen be šošofi, unika obuki, acanara acanarakū babe dergici jorime tacibureo.[68]

The note thus proposed two new names. The first, *sabsaha*, was intended to differentiate locusts from grasshoppers by replacing the vowel *e*, which was sometimes identified as "feminine" (and occurred in words such as *hehe* 'woman' and *eme* 'mother'), with *a*, which was "masculine" (and occurred in words such as *haha* 'man' and *ama* 'father').[69] Metaphorically, the two vowels could represent other distinctions, such as big versus small, or robust versus delicate. Thus the slightly larger body of the locust relative to the grasshopper probably inspired using the "male" vowel here. Whoever wrote this note does not seem to have considered using one of the several terms for various locusts and grasshoppers listed in Kangxi's *Mirror* and unambiguously redefining it as the generic name for "locust."

The other word, *unika*, was proposed as the word for immature locust. There was no existing Manchu word to work with. Yet the anonymous Manchu scholar did not create a calque on the corresponding Chinese expression (to late imperial users of Chinese, *nan* was a morpheme that could not be further semantically divided). Instead, the new word was formed in reference to current ideas about locusts. Echoing Qianlong's words that immature locusts were born from fish and shrimp eggs that ended up on land before hatching, new word *unika* is formed from two parts that reference fish and eggs. The theory that fish and locusts had the same origin but differed because of their environment evidently motivated the new word.

68. Manchu palace memorial or note file copy, 03-0172-0810-002, FHA.
69. Söderblom Saarela, "Manchu and the Study," 336.

Curiously, the neologism that was ultimately chosen for "locust" was not *sabsaha*, as proposed in the anonymous note. In the event, the new word chosen was *sebseheri*, which also was clearly coined on the basis of *sebsehe*. We have not seen a document explaining the reasoning behind this choice. We find it, however, in the Manchu translation of *Qiju zhu* for June 1752, which as mentioned might have been carried out in 1754 or later.[70] The suggestions raised in the anonymous note were thus only partially accepted by the Grand Council, much as in the case of the "camel chicken."

New Words Discussed at the Grand Council (3): Deer

Earlier in this chapter, we discussed how the editors of the pictorial *Catalog of Beasts* consulted Kangxi's *Mirror* but then diverged from it. This section focuses on the *Catalog* editors' discussion of the *Mirror* in reference to the names of different kinds of deer. It then follows the discussion as it outgrew the *Catalog* and came to involve the phrasing used in the imperial calendar.

The starting point is a document on the Manchu and Chinese names for several kinds of deer dating from 1761 (QL 26/7/8), thus around the time that the Manchu text in the *Catalog of Beasts* was being written. The document originated somewhere in the central government and was sent to the Mukden Imperial Household Department. It is accessible in the form of a transcript in a copy book made there, which has since been published.

This document represents a moment in the revision of the terms for different kinds of deer in both Manchu and Chinese that eventually involved both references to classical texts and contemporary translations, as well as observation of animals in the wild. Yet the moment the discussion passed from literary usage to observation, the discussion also shifted from Manchu to Chinese. Unlike in the example of a gourd

---

70. *Ilire tere be ejehe dangse*, QL 17/5/part two/*fulgiyan singgeri* [= *bingzi*], no pagination. The new words are, as expected, likewise used in the Manchu translation of the *Shilu* for this event, which postdate *Qiju zhu* (*Daicing gurun-i g'aodzung*, QL 17/5/part two/*fulgiyan singgeri* [= *bingzi*], 415:7a–8a).

that we discuss later in this chapter, the Manchu discussion remained focused on how to translate various Chinese terms, not on the study of the animal itself.

In chapter 6, we discussed some of the words referring to deer used in the *Ode to Mukden*. Evidently, the terminology for deer was still not settled twenty years later. Scholars and officials at court were for a long time not sure how to talk about animals that were endemic to Manchuria in Manchu:

"Upon investigation following an edict, [it was found that] in the *Mirror of the Manchu Language, she* 麝 'musk deer' is said to be 'called *miyahū*, also *mikcan*'. *shefu* 麝父 'musk buck' is another name for *miyahū*. Thus *miyahū* was made to mean simply 'musk deer' and *mikcan* changed to mean 'musk buck'. In the *Mirror of the Manchu Language*, it says that 'the Chinese call a deer with a long tail *mi*.' In the *Book of Calculations of Propitious Times* [i.e., the official calendar, which was issued in Manchu, Mongolian, and Chinese editions, the phrase] 'the [*mi* deer] sheds its antlers' ([*mi*] *jiao jie* [麋]角解) [in the month of the winter solstice, when *yin* starts to wane and *yang* to wax] is written as 'the long-tailed deer (*uncehen golmin buhū*) sheds its antlers'.[71] Therefore *arfu buhū* was proposed as the new word for [*mi* deer].[72] Now, upon investigation of *Mencius*, [we found that] the phrase 'rejoicing that he had his *mi* deer (*milu* 麋鹿), his fishes, and turtles' is rendered as 'rejoices at the existence of his small deer [or Sika?] (*suwa*) deer, fish, and turtles'.[73] With this in

---

71. The phrase [*mi*] *jiao jie* is puzzling. The version of the document that was entered into the copy book that we consulted has *she* 麝 for *mi* 麋, which is a mistake, as the calendar had the phrase *mi jie jiao*, not *she jie jiao*. We have emended it. See Miu, *Da Qing shixian shu jianshi*, 1040:685 for an explanation of the phrase.

72. As in the previous instance, the version of the document that we consulted has *she* 麝 for *mi* 麋. In the previous case, the error is obvious on textual grounds. Here, it is obvious because the musk deer (*she*) does not have a long tail, whereas Père David's deer does.

73. The translation of *Mencius* is adapted from Legge's *The Chinese Classics* (vol. 2, 128), which has "large deer." *Yuzhi fanyi Mengzi*, 189:1:4a (460, lower panel). A Manchu edition of the *Four Books*, which included *Mencius*, was published in 1677 (reprinted in 1691 and 1733). Subsequently, two editions were published under Qianlong. The first, dated 1741, is from before the language reform. The second, dated 1756 (according to Fuchs) or 1760 (according to Yeh), contains neologisms. We have only been able to consult the latter version, which accords with the wording in this

consideration, [we propose] *suwa* 'Sika deer' [as the word for] *mi*, and 'long-tailed deer' (*uncehen golmin buhū*) [as the word for] *zhu* 麈. After it has been authorized by the throne, please send it to the translation office of the Imperial Board of Astronomy and emend the usage so that it is uniform. We respectfully report on this matter." This was respectfully presented in a memorial on Qianlong 26/3/30 [1761/5/4]. The imperial rescript read "Noted."

hese be dahame baicaci, manju gisun-i buleku bithede 麝 *miyahū geli mikcan sembi sehebi,* 麝父 *miyahū-i encu gebu ofi tuttu* 麝 *be an-i miyahū obume mikcan be* 麝父 *de guribuhe. manju gisun-i buleku bithede uncehen golmin buhū be nikan mi sembi sehebi, erin forgon-i ton-i bithede* [麋] 角解 *be uncehen golmin buhū-i uihe sumbi seme arahabi, uttu ofi* [麋] *arfu buhū seme banjibufi ibebuhe. te baicaci mengdzi bithede* 樂其有麋鹿 魚鼈 *sehe gisun be terei suwa buhū, nimaha, aihūma bihede sebjelembi sehebi. bahaci* 麋 *be suwa obuki,* 麈 *uncehen golmin buhū obuki. acanara acanarakūbe dergici jorime tacibuha manggi abka be ginggulere yamun bithe ubaliyambure boode afabufi emu adali halame dosimbuki. erei jalin gingguleme ibebuhe seme abkai wehiyehe-i orin ningguci aniya ilan biyai gūsin de wesimbuhede, hese saha sehe.*[74]

The discussion in this document concerns three Chinese terms—*she*, *mi*, and *zhu*—and their Manchu translation.[75] These three words are all headings for animals in the *Catalog of Beasts*. To establish Manchu corresponding terms, the officials at court discussed Manchu terminology for deer entirely in reference to classical Chinese texts, established Chinese usage, and Manchu translations by the court. No reference was made to anything beyond these texts, such as the animals themselves.

---

document. We do not know if the earlier Qianlong version—or, indeed, the even earlier 1677 version—renders the phrase this way. See Fuchs, "Eine unbeachtete Mandju-Übersetzung"; Yeh, "Man-Han hebi," 14–15.

74. Zhao and Liaoning sheng dang'an guan, *Heitu dang: Qianlong*, vol. 6, 104; Zhang Hong, Cheng, and Tong, "Qianlong chao 'Qinding Xin Qingyu' (qi)," 30–31.

75. Regarding *she*, the musk deer, the authors note that Manchu uses a pair of synonyms, which they then match to two Chinese synonyms (*shefu* was already an established synonym for *she*). See Li Shizhen, *Bencao gangmu*, vol. 774, 51:51b (491, upper panel).

Among the most interesting words for deer discussed in this document is *mi*. The authors of the report say that it is given as a Chinese translation of "long-tailed deer" in the 1708 *Mirror*. We note in chapter 2 that the entry there, whose headword is a cumbersome Manchu noun phrase, could very well have been formed on the basis of the Chinese *mi*, of which the Manchu name would then be a kind of gloss. The defining characteristic of this deer was that its tail could be made into a fly whisk. By contrast, the 1684 *Shengjing tongzhi* said that the *malu* 馬鹿 'elk' (lit. horse deer) could swat away flies with its tail,[76] which thus ought to have been somewhat long. It is possible that *mi* was used as a synonym for elk. The Manchu word commonly used in reference to elk, *ayan buhū* (described as large and yellow; *ayan* was used in the names of birds characterized by their large size as well), was translated as *milu* 麋鹿 by Shen Qiliang in 1683, which suggests that *milu* and *malu* were synonyms for some Chinese speakers.[77]

The word *zhu* was an ancient term with obscure significance, but it was thought to be a kind of *mi*.[78] The *zhu* was described, and depicted in an illustration, in *Gujin tushu jicheng* alongside *lu* 鹿 'deer' and *mi*. The *lu* and *zhu* look similar, both with antlers and a tail, but the *lu* is spotted. By contrast, *mi* is spotted but has no antlers.[79] It was carried over from *Gujin tushu jicheng* to the *Catalog of Beasts*, which is why it shows up in this document. The *Catalog* followed this document in translating *mi* as *suwa*.[80] The word *zhu* was notably used in the

---

76. *Shengjing tongzhi*, 21:20a.

77. See *Han-i araha manju gisun*, 1708, 19:*gurgu-i hacin*:44a, s.v. *ayan*; for *ayan gaha*, a big crow, see 19:*gasha-i hacin*:32b; for *ayan coko*, a neologism for a bird with a thick and rather big body, see *Yuzhi zengding*, vol. 233, supplement, 4:83a (446, lower panel). Hayata and Teramura, *Daishin zensho*, vol. 1, 90 (line 0610b2).

78. On the meaning of *zhu*, cf. *Taiping yulan*, vol. 901, 906:16a (133, upper panel); Shi, *Jijiu pian*, 55a. The translation of the title of *Jijiu pian* in the bibliography follows Foster ("Study of the *Cang Jie Pian*," 34). See also Li Shizhen, *Bencao gangmu*, vol. 774, 51:32b (481, lower panel). On *zhu* as a kind of *mi*, see Duan, *Shuowen jiezi zhu*, vol. 1, 822 (lower panel).

79. Chen Menglei and Jiang, *Qinding Gujin tushu jicheng*, vol. 521, 9b (bottom panel).

80. Chuang, Shoupu *Manwen*, 197. The word *suwa* that the document proposed be used to translate *mi* had been defined as a small-bodied and light pink deer in Kangxi's *Mirror* (*Han-i araha manju gisun*, 1708, 19:*gurgu-i hacin*:44a), and it was arguably a poor

compound *zhu wei* 麈尾 '*zhu* tail', referring to a fly whisk associated with the medieval aristocracy long in disuse by the Qing period.[81] The legend accompanying the painting of *zhu* in the *Catalog of Beasts*—representing an imaginary animal, according to Wang Zuwang 王祖望 and his colleagues—is based on related lore but does not quote any sources.[82] In the Manchu legend, *uncehen golmin buhū* 'long-tailed deer' was used to translate *zhu* in reference to this imaginary animal. This usage further suggests that this Manchu name already had an unclear or legendary referent when it was entered into the original Manchu *Mirror*, given that the editor and translators of the *Catalog of Beasts* could not easily have used it in this sense if it had had a well-established, real-world referent.[83]

The document further mentions that at an earlier point, *arfu buhū* had been suggested as a possible translation for *mi*. As is clear from the document, this word was not adopted. It is not used in the *Catalog of Beasts* or in the later, revised *Mirror*. The reason was perhaps the association of *mi* and *zhu* that the proposed Manchu translation would

fit. The anonymous authors of the report made this recommendation upon consideration of the Manchu translation of *Mencius*. In the Manchu translation from 1677, interestingly, *milu* had been translated as *mafuta buhū* 'buck deer' (Lasari, *Inenggi giyangnaha*, 13:7b), which is also paired with *milu* in Sangge's Manchu-Chinese dictionary from 1700 (Sangge, *Man-Han leishu*, 26:24b). The 1708 *Mirror* quoted the passage from *Mencius* in the entry for *mafuta* (*Han-i araha manju gisun*, 1708, 19:*gurgu-i hacin*:44a). Curiously, in Shen Qiliang's compendium of phrases from the *Four Books*, the word was translated as *jolo buhū* 'doe deer'—the exact opposite (Shen, *Sishu yaolan*, *mengdz*:18a).

As for *suwa*, Shen Qiliang translated it much like the *Mirror*, as "patterned deer" (*hualu* 花鹿) and "generic term for small deer" (*xiao lu zong ming* 小鹿総名) (Hayata and Teramura, *Daishin zensho*, vol. 1, 115 [0725a3]). This Manchu word was thus in use in reference to one or several common animals. Wang Zuwang and colleagues note that the description of the animal called *mi* and *suwa* in the *Catalog of Beasts* suggests a reference to Père David's deer, but that the image shows something else: A slender, spotted deer with no antlers, not unlike a Sika hind and clearly related to the corresponding illustration in *Gujin tushu jicheng* (Yu and Zhang, *Qinggong shoupu*, 408). The legend in the *Catalog* used the Chinese word *mi* in the sense it had in ancient texts, not in the sense of the binome *milu* 'elk' of the contemporary vernacular. The Manchu translation *suwa* had little to do with it, but was a better fit than the Chinese word for the image in the *Catalog*.

81. Zhang Qingwen, "Zhuwei de qiyuan."
82. Yu and Zhang, *Qinggong shoupu*, 408.
83. Chuang, *Shoupu Manwen*, vol. 1, 199.

imply. The neologism *arfu* is formed from *arfukū*, used in the 1708 *Mirror* in the sense of a fly whisk made from the long tail of a deer or mule.[84] Li Yanji translated it as a "fly broom" (*cangying zhouzi* 蒼蠅 箒子) and "fly whisk" (*yingfuzi* 蠅拂子).[85] Later, in the revised *Mirror*, however, the court replaced it with the resurrected *zhuwei* and its associations of royalty and leisure.[86]

The immediate purpose of the document ought to have been the writing of the *Catalog of Beasts*. The contents ought to have already been picked out from *Gujin tushu jicheng*, and at this stage what was needed were Manchu translations of the relevant terms. Yet the document ended with an instruction to send the finalized terms to the Board of Astronomy, which suggests that the discussion had already outgrown its initial purpose. Indeed, it was in relation to the imperial calendar, established by the Board of Astronomy, that the discussion of *mi* and *zhu* continued.

The discrepancy in the use of *mi* in antiquity and in the Qing period surfaced soon again. Deer and antlers were a familiar and symbolically laden topic to the Qing emperors, with deer being an important game animal and antlers decorating the imperial throne. Around the time of the report quoted in the memorial, Qianlong appears to have taken a particular interest in them. A little more than half a year after the writing of the memorial that included translations to be used in the *Catalog of Beasts* (early 1762, QL 27), Qianlong authored "Lujiao ji" 鹿角記 (Note on the antlers of deer). This note has been studied by Shi-yee Liu. Qianlong remarked that the animals that he knew as *mi* did not shed their antlers in winter, which is what the calendar claimed based on a passage in a classical text. What *mi* referred to here is not entirely obvious. It could have been elk, which as we suggested were called *milu* by some Chinese speakers, or a small deer akin to the *mi* (Ma. *suwa*) depicted in the *Catalog of Beasts* (Qianlong later, sometime between 1772 and 1783, indeed wrote that *mi* was a small deer).[87]

---

84. See *Han-i araha manju gisun*, 1708, 16:*tetun baitalan-i hacin*:3b.
85. Li Yanji, *Qingwen huishu*, 1:13a (9, lower panel).
86. *Yuzhi zengding Qingwen jian*, 1782, vol. 233, 25:13a (45, lower panel).
87. Gaozong, *Yuzhi shi si ji*, vol. 1308, 57:15b (269, upper panel). This collection contains poetry written "between the year after our emperor completed a sexagenary cycle [thus 61 *sui*, 1771] and three years after he reached the age rare since antiquity

Qianlong was perplexed: "How could the *mi* of antiquity be different from the *mi* of today?," the emperor wrote. "This is something we have no way of finding out" 豈古之麋非今之麋乎？是不可得而知矣.[88]

A few years later, at the time of the winter solstice of 1767 (QL 32), Qianlong suddenly remembered that he had heard that in the southern hunting park near Beijing, they were raising so-called *zhu* (*suowei zhu zhe* 所謂麈者).[89] Qianlong said that this animal in the vernacular (*su* 俗) was called *changwei mi* 長尾麋 'long-tailed *mi*', a name that echoes the Manchu name for *mi* in the 1708 *Mirror* and the translation of *zhu* in the *Catalog of Beasts*.[90] Perhaps Qianlong had the Manchu name in mind when he referred to the vernacular. The emperor dispatched Ufu 五福 (d. 1774), one of his guards of the antechamber, to the park to investigate whether the "long-tailed *mi*" were shedding their antlers at that time.[91] It turned out that they were, and Ufu brought a set back to the city to show Qianlong. In all likelihood, these "long-tailed *mi*" or *zhu* deer were Père David's deer (*milu* 麋鹿 in modern Chinese) that famously survived in the imperial hunting park even though they had become rare elsewhere. In light of the new evidence, Qianlong ordered the phrase "the *mi* deer sheds its antlers" in the official calendar changed to "the *zhu* deer sheds its antlers" (*zhujiao jie* 麈角解).[92]

This edict was issued in Chinese to the Grand Secretariat and sent in Chinese to Mukden. It did not mention whether the Manchu version of the calendar also needed revision. The report on Manchu translations to be used in the *Catalog of Beasts* said that the calendar rendered

---

[thus 73 *sui*, 1783]" 皇上週甲之逾年至古稀之三歲所製也 (Gaozong, *Yuzhi shi si ji*, vol. 1307, *zouzhe*:2a [2, upper panel]).

88. Liu Shi-yee, "Containing the West," 58, fig. 1.

89. Liu Shi-yee, "Containing the West," 59, fig. 2.

90. This is even more so in the version of the edict that was recorded in the *hetu* archive copybooks in Mukden. That version has *changwei lu* 長尾鹿 'long-tailed deer' rather than 'long-tailed *mi*'. We should perhaps not make too much of this, however. The version in the *hetu* archive is garbled. Perhaps the scribes who handled this otherwise largely Manchu archive were confused by the Chinese text? See Zhao and Liaoning sheng dang'an guan, *Heitu dang: Qianlong*, vol. 7, 7.

91. On guards of the antechamber, see Brunnert and Hagelstrom, *Present Day Political Organization*, 27 (item 100).

92. FHA, *Qianlong chao shangyu*, vol. 5, 236 (item 674). Relatedly, Gaozong, *Yuzhi shi san ji*, vol. 1306, 82:1b (599, upper panel).

mi as "long-tailed deer" in Manchu. This was still the case as of 1769 (QL 34), a year after the change went into effect in the Chinese version of the calendar.[93] The variant *buhekū* for *buhū* is seen in the calendar for 1869, however, but we have been unable to find out when or why it was introduced (we have not encountered the word anywhere else).[94] The issue deserves closer scrutiny.

One kind of deer that was not discussed in the document was *kandahan* 'moose'. This animal, however, does not appear in the *Catalog of Beasts* either (*kandatu* [< *kanda* 'dewlap' + Mongolian suffix -*tu*] was included, but it served to translate the name of a legendary Chinese bovine).[95] Perhaps the reason is the same in both cases: The *kandahan* did not have an established Chinese name. The ascription of the name "camel deer" to *kandahan* in *Shengjing tongzhi* of 1684 had not caught on (Li Yanji translated the *Mirror*'s entry but did not give a Chinese name, whereas Juntu did not mention it among the deer).[96] As we saw, Kangxi still referred to it only as *kandahan*. Qianlong did take an interest in this animal, however, owing to its association to archery thumb rings, on which Kangxi had already remarked. In 1752 (QL 17), almost a decade earlier than the document on the Manchu translations of Chinese deer names, Qianlong wrote a poem in which he theasserted the superiority of moose thumb rings of the north over those made from southern ivory in stronger terms than his grandfather had done. Likewise unlike Kangxi, Qianlong explicitly connected the Manchurian name of *kandahan* with a deer mentioned in *Erya*. This deer was said to be very powerful. *Erya* called it *jian* 麢, which in the normative pronunciation of the court would have been *kjɛn*.[97] Qianlong

---

93. *Daicing gurun, shiyi yue*:1a. According to Lu Dingpu (*Lenglu yihua*, supplement, 24), it began in Qianlong *wuzi* (1768, QL 33). In the Kangxi period (we consulted a calendar from 1680), *mi* was translated as *mafuta*, which accorded with the then current translation of *Mencius* (*Daicing gurun-i elhe taifin-i, shiyi yue*:1a).

94. *Daicing gurun-i yooningga dasan, shiyi yue*:1a.

95. Chuang, Shoupu *Manwen*, 469.

96. Li Yanji, *Qingwen huishi*, 2:37a (47, upper panel); Juntu, *Yi xue san guan Qingwen jian*, 53b (144, upper panel).

97. *Jian* is a variant of *jian* 麢 (see Hao, *Erya yishu*, vol. 2, 898). The commentary to *Erya* said that *jian* 麢 was homophonous with *jian* 堅, which the official transcription manual (*Qinding Qing-Han duiyin zishi*, 51a) transcribes as *giyan* in Manchu (i.e., not the *kan-* of *kandahan*).

said that *kjɛn* "was similar [to *kandahan*] both in pronunciation and meaning" 音義均近焉.⁹⁸ Qianlong appeared to be making a connection between an ancient Chinese word and the Manchurian—Mongolic and now Manchu—word for the moose (*kjɛn* presumably appearing similar to the syllable *kan-* in *kandahan*). The word *kandahan* was finally made Chinese when the Manchu terminology for deer was further revised in the *Expanded and Emended Mirror of the Manchu Language*.⁹⁹

New Words Discussed at the Grand Council (4): A Berry

Our fourth example is the most interesting. In addition to that of birds, insects, and deer, the Manchu vocabulary relating to plants was discussed at the Grand Council in the 1760s. We focus on the discussion in 1765 surrounding what appears to have been a kind of herbaceous raspberry. It concerns the natural-historical description and Manchu name of the fruit called *cao lizhi* 草荔支 'grass lychee'. In Qianlong's Manchu-Chinese *Mirror*, this word came to be used to translate *ilhamuke*. This so-called flower water was defined in the 1708 *Mirror* and had been discussed by Chinese travelers to Manchuria. In 1735, Mingdo translated the word as *di shen* 地椹 'ground mulberry fruit'.¹⁰⁰ No note was taken of this vernacular rendering at the Qianlong court, where they opted for a different Chinese translation.

---

98. Gaozong, *Yuzhi shi er ji*, vol. 1303, 36:17a–18a (664, lower panel; 665, upper panel); transcribed and punctuated in Wu Weiping, "Qianlong huangdi yu yu banzhi," 108, table 2; Shih, "Cong *Huoji dang* zhong"; see also Ruan, *Erya zhu shu*, chap. 10, 84 (middle panel).

99. Another kind of deer not discussed in the document sent to the Mukden Imperial Household Department was the muntjac (*jun*), which we encountered as discussed in the 1736 edition of *Shengjing tongzhi* and as *sirgan* (i.e., *sirga*) in Shen Qiliang's Manchu dictionary. This deer evidently had an established name in Manchu, but in the *Catalog of Beasts* that name occurred only in order to translate *zhang* 麞 'Chinese water deer', which was given as a synonym for muntjac in the *Catalog*. The word *jun* was rendered with the neologism *sirgatu* (< *sirga* + *-tu*), so that the lexical variation in Chinese was duplicated in Manchu.

100. Mingdo, *Yin Han Qingwen jian*, 18:224b.

The process by which the Chinese "grass lychee" and the Manchu "flower water" came to be equated is not clear in all its details, but a key episode of this process is revealed by a report submitted to the Grand Council on June 2, 1765 (QL 30/4/14, according to the FHA catalog) by Hoiling 惠齡 (1743–1804)—who had worked as a translator in various capacities at court and in the central government for nine years at this point—and Horonggo 和隆武 (d. 1782).

The report was prompted by an edict transmitted to them by a message from the Grand Council (a "court letter"). The edict ordered the pair to investigate the word *cao lizhi*. Qianlong claimed it occurred in his poetry and in the "catalog of fruits" (*guopu* 果譜), a section of *Guang Qunfang pu*. It appears that the edict added the additional instruction that this word should be translated into Manchu, because that is what Hoiling and Horonggo then did, but if so, this part of the edict was not quoted in the report. The Grand Council transmitted the edict to Hoiling and Horonggo with additional instructions. The two officials researched the word, but did not come up with very much. It appears from their report that Qianlong was mistaken in thinking that the *cao lizhi* was mentioned in *Guang Qunfang pu*. Hoiling's and Horonggo's report reads in full:

Submitting: Hoiling, concurrently serving in the [Manchu-Chinese Translation] Office, and Horonggo respectfully submit a letter to the Grand Secretary.[101] A [court] letter from the tenth of this month said:

"An edict was issued that said: '*cao lizhi* 草荔支 [sic] occurs in my poetry and in the *guopu* 果譜 'catalog of fruits'.[102] Investigate and then report.' We respectfully act on this edict. Our office will send all the instances of *cao lizhi* 草荔支 that are found in the imperial poems to President Yu [Minzhong 于敏中 (1714–1779)] for investigation. Besides, [the poems] should be sent from your office to the Manchu-Chinese Translation Office for investigation in the *guopu* 果譜 'catalog of fruits'. After it has been satisfactorily investigated [by the Manchu-Chinese

---

101. The identification of the original's *bithei boo* with the Manchu-Chinese Translation Office follows Zhao Zhiqiang ("Lun Qingdai de Nei Fanshu Fang," 22).
102. Not *zhi* 枝.

Translation Office], look through the result and send it together with the original books."

Acting according to the edict, the name *cao lizhi* 草荔支 was then respectfully entered into the *Mirror of the Manchu Language*. Furthermore, we repeatedly searched for the name of the fruit called *cao lizhi* 草荔支 in the *cao mu dian* 草木典 'canon of herbs and trees' in the book *Gujin tushu jicheng* 古今圖書集成 and in the *guopu* 果譜 'catalog of fruits' included in the book *Guang Qunfang pu* 廣群芳譜, but did not find the name *cao lizhi* 草荔支. Thus we respectfully encase the two books *Gujin tushu jicheng* 古今圖書集成 and *Guang Qunfang pu* 廣群芳譜 and forward them according to the [court] letter. In the meantime, we will go on and, as usual, search repeatedly in other books. After we have found it, we will send it to the Grand Secretaries for consideration. We respectfully submit [this report] on this matter. Fourth month, fourteenth day.

aliburengge, bithei boode yabure hoiling, horonggo sei gingguleme aliha da de alibume jasiha, ere biyai juwan de yabure coohai nashūn baci afabuha jasigan de, hese, 草荔支 *mini irgebun, jai* 果譜 *dolo gemu bi, baicafi wesimbu sehebe gingguleme dahafi, ejen-i irgebun de bisire* 草荔支 *be ubaci ioi aliha amban de alafi baicabuha ci tulgiyen, suweni baci ubaliyambure boode afabufi* 果譜 *dolo baicakini baicafi bahara be tuwame da bithe be unggi seme afabume jasiha be, uthai hese be dahame* 草荔支 *sere gebu be manju gisun-i buleku bithede gingguleme dosimbume arahabi, jai* 草荔支 *sere tubihe-i gebu be* 古今圖書集成 *bithei dorgi* 草木典並廣群芳譜 *bithei dorgi* 果譜 *de fuhašame baicaci umai* 草荔支 *sere gebu be baicame bahakū, uttu ofi,* 古今圖書集成廣群芳譜 *sere juwe hacin-i bithe be gingguleme hiyabsalafi, afabume jasiha songkoi alibuha, ere sidende, be an-i gūwa bithede narhūšame baicaki, baicame baha manggi, aliha da de tuwabume alibuki, erei jalin gingguleme alibuha. duin biyai juwan duin.*[103]

The "grass lychee" did occur in Qianlong's poetry composed between 1748 and 1759 (QL 13–24). The official in charge of editing his poems of this period was none other than Yu Minzhong, the official who was to investigate the use of the term *cao lizhi* in the emperor's

---

103. Report kept among the Manchu palace memorial reference copies, 03-0181-2137-031, FHA.

poetry. Yu was Chinese (from Jiangsu) but had studied Manchu upon his assignment to the Hanlin Academy in 1737.[104] At that time, the emperor was staying in the Rehe summer resort built by his grandfather and now substantially expanded. While there, Qianlong received fresh lychee dispatched from Fujian and wrote three short poems about the fruit. Presumably on the same day or within a few days, he then compared the Fujian lychee with the so-called grass lychee that he claimed was a plant growing only beyond the Great Wall in the Khingan and Ula regions. According to Qianlong, it was the Kangxi emperor who ordered the plant transplanted to Rehe, much like what we saw Kangxi do in the case of *yengge*. Qianlong's poem, which we fully annotate and translate here, is not just about grass lychee but presents a daring juxtaposition between the Manchurian berry and the famous southern delicacy:

Sweet and full taste by the mouthful, with a powerful fragrance to the nose	滿口甘腴撲鼻芳
The Flower Goddess, with graceful understanding, bestows the cloud-like juice	女夷雅解遺雲漿
I lyricize over things that are created, and thus can also name them	品題造物能名物
Yet cannot tell among the two brothers [i.e. the two fruits] which one is superior than another	比似元方與季方
In earlier catalogs, they compare the flesh and shell [of lychee] to crimson snow	漫譜瓤平如絳雪
And boast about its tiny pits the size of mother cloves [the dried fruit, not flower bud]	繞稱核小尚丁香[105]

---

104. Fang Chao-ying, "Yü Min-Chung," 942.
105. See Cai, *Lizhi pu*, 3a (156, top panel).

> In my Studio of Exalted Spices, this 崇椒嘉植煙霞外[106]
> auspicious plant [i.e., the grass lychee]
> grows beyond smoky clouds
> With roasted beef and lamb that await 那更羶薌待蔡襄[107]
> a modern-day Cai Xiang!

This poem helps shed light on the context for the bilingual report on the grass lychee. Both documents show that from the perspective of the Qianlong court—the emperor himself and the Manchu and Chinese scholars at his command—the project of naming things through lexicography (*mingwu* 名物 'naming things') was intimately connected with the study of things that were created (*zaowu* 造物 'created things'). Furthermore, the earlier Chinese genres of encyclopedias and catalogs (*pu*) were very much on Qianlong's mind as precedents for his own undertaking, and Qianlong saw late Kangxi-era books such as *Gujin tushu jicheng* and *Guang Qunfang pu* as more immediate examples of works in these genres. Qianlong taunts earlier authors such as Cai Xiang, who wrote a catalog, for not being able to partake in the emperor's enjoyments of roast beef and lamb, and for their ignorance of delicacies from "beyond the border" that were as good as the coveted southern fruits.

It is somewhat comical, then, that the emperor was so sure that the grass lychee must have been entered into the earlier Qing court compilations, only to get back the gingerly report that in fact the berry was not found there. However, because the emperor had by now written a poem about it, a new entry had to be created and archived in lexicographical form. Because Hoiling's and Horonggo's report mentioned that *cao lizhi* should be "entered into the *Mirror*," it would seem that, first, work on the new *Mirror* had begun at this time (the compilation is mentioned as under way on June 20, 1765 [QL 30/5/*dingchou*], a few weeks after the writing of this report), and, second, that the two officials'

---

106. Chongjiao (Exalted spices) is the name of a studio built on top of the artificial hill in the imperial resort (modern-day Beihai Park). See Tojin, *Da Qing huidian shili*, 662:14a.

107. Gaozong, *Yuzhi shi er ji*, vol. 1304, 49:27a–b (64, lower panel). Cai Xiang (1012–1067) was the Song scholar-official who authored the *Lizhi pu* 荔枝譜 (Catalog of lychees).

research in Chinese scholarly literature was intended to produce a description of the plant for entry into the revised dictionary.[108] This is why we see the pair of grass lychee and *ilhamuke*, a frontier herbaceous berry that had not in fact been commented on by the late Kangxi courtiers, now entering the enlarged Manchu dictionary of the 1770s.

The bureaucratic reverberations of the lexicographic revisions of the kind discussed here could reach quite far. Judging by the available evidence, the court at times continued to investigate a plant even after a pair of Manchu-Chinese names had been established. One last example concerns a gourd, which was discussed at court a few months earlier than the grass lychee exchange.

Natural-Historical Investigation of a Gourd

The discussion regarding the gourd can be gleaned from a letter (*jasigan*) sent on the "first of the second month" (of QL 30, according to the FHA catalog), or February 20, 1765, but as with the grass lychee, it is evidently only one part of a longer paper trail. The telegraphic, almost colloquial letter was sent by the Manchu-Chinese Translation Office to the "gentlemen" (*looye*) at the Grand Council.[109] The Trans-

---

108. *Gaozong Chun huangdi shilu*, vol. 18, 100, lower panel (*juan* 73, page number not visible on the centerfold in the reprint).

109. The Manchu text has *yabure coohai nashūn ba-i looye sa*, with *looye* < Ch. *laoye* 老爺 'master', marked with a plural suffix, and *yabure coohai nashūn* might be the Manchu equivalent of *junji xingzou* 軍機行走. Brunnert and Hagelstrom translate this expression as "Probationary Grand Councilor" and define it as a designation for newly appointed Grand Councilors (*Present Day Political Organization*, 42 [item 129A]). However, a perusal of the tables of official appointments reveals that no new Grand Councilors were appointed around this time. Brunnert and Hagelstrom's translation reflects the situation in the early twentieth century, which might have been different from that in the mid-eighteenth. That said, official records note that Yu Minzhong held the position of *junji xingzou* at this time ("Renming quanwei ziliao ku" database). Other Chinese sources say that there was a position called *junji chu xingzou* 軍機處行走, which would correspond word for word to *yabure coohai nashūn ba*. That is to say, there is uncertainty as to whether *ba* 'place [of work]' refers to a title (*junji chu xingzou*) or to an office (the office of *junji xingzou*). By extension, it is not clear whether the "gentlemen" to which the letter is addressed are the Grand Councilors themselves, or the secretarial staff of one of the Grand Councilors, perhaps

lation Office had received a request transmitted through the postal relay stations, which inquired about "the word *guwalase* for *kegua* 客瓜" (*guwalase sere* 客瓜 *emu gisun*).[110] The author of the document wanted to know what the plant looked like and "what kind of herb or tree" (*ya hacin-i orho moo de banjire*) it grew from. The staff of the Translation Office "searched repeatedly in *Gujin tushu jicheng* but did not find anything at all" (*uthai gu jin tu šu ji ceng bithede dahime baicaci umai akū*). They had been told to investigate further, and would do so. Meanwhile they asked the "gentlemen" to let the Grand Secretary (which could refer to one of several individuals) know about their intentions.[111]

That it arrived through the postal relay system suggests that even though the request was not an edict, it originated with the emperor's retinue. Qianlong was on one of his southern tours at this time, passing through Shandong when this report arrived in Beijing.[112]

Interestingly, the Grand Council copy of the letter—the only version available to us—has a note slip pasted onto it with the word "seedling" (Ma. *fursun*; Ch. *yangzi* 秧子) written on it in Manchu and Chinese. The Manchu word was not lemmatized with this meaning until in the expanded *Mirror*, so it might have been that the slip refers to the establishment of its new Chinese translation.[113] Alternatively, it might be that a seedling was sent along with the letter. Plants were evidently sent in conjunction with similar discussions. Around the same time as the seedless green grape was given the Manchu name

---

Yu Minzhong. The discussion of the relevant Chinese and Manchu terms in Zhao Zhiqiang (*Qingdai zhongyang*, 324–27) is inconclusive. We thank Masato Hasegawa and Tsai Ming-che for help in reading this document.

110. Transmitted on 1/29 (1765/2/18), FHA.

111. Letter, 03-0181-2123-031, FHA.

112. *Gaozong Chun huangdi shilu*, vol. 18, ch. 727 (14, lower panel; original page number not visible in reprint).

113. *Yuzhi zengding*, vol. 232, 21:36a (713, upper panel). In the original *Mirror*, the word *fursun*'s only acceptation was "sawdust" (*Han-i araha manju gisun*, 1708, 19:*moo-i hacin*:21a). This is the only instance of the word listed in Alt'aiŏ yŏn'guso ("*Ŏje Ch'ŏngmun'gam*" *saegin*, 2:260).

kišimiši, for instance, Qianlong ordered the plant "uprooted and transported for plantation in the imperial garden" 命取根移植禁苑.[114]

Whatever the case, the request under discussion here did not concern the translation of the words *guwalase* or *kegua* 客瓜. The Manchu word was already lemmatized and described clearly in the original *Mirror*. It was translated into Chinese as *kegua* in two commercially published Chinese *Mirror* translations published before this document was written.[115] The translation was thus well established and not controversial. Perhaps the fact that it concerned a scholarly matter involving Manchu-language material was enough for it to fall within the purview of the Translation Office. The request, however, was for more information about the plant itself, not the word.

Conclusion

In his article on Qianlong's prodigiously voluminous poetry collection, Yan Zinan 顏子楠 remarked on the "routine production" aspect of court culture during the second half of the eighteenth century. As we have shown in this chapter, the same *routinization* of neologism-creation is similarly characteristic of Qianlong's overall approach to language reform. As long as a word is logged as in need of "investigation" (Ch. *cha* 查; Ma. *baicambi*), court officials mobilized the available reference works from the Kangxi reign to come up with a satisfactory candidate for a new Manchu name for a plant or animal. Thanks to this routinized production of neologisms, we have been able to partially reconstruct the backstage philological work, as it were, that buttressed the bilingual annotations of court-produced pictorial albums (themselves a continuation of earlier *Catalogs* of flora), as well as the increasingly standardized official parlance seen in Manchu governmental

---

114. *Qinding Huangyu Xiyu tuzhi*, 251:43:12a (820, top panel). Qianlong specifies that this happened "the year before last" (*qiannian* 前年), which would correspond to 1761.

115. Mingdo, *Yin Han Qingwen jian*, 18:214b; Juntu, *Yi xue san guan Qingwen jian*, 25a (130, upper panel).

documents churned out by the "Qing palace machine."[116] Such new developments substantially altered the Manchu lexicon's vocabulary for plants and animals, which had earlier been codified in the Kangxi-era monolingual Manchu *Mirror*.

With this context in mind, we now turn to the bilingual *Expanded and Emended Mirror of Manchu Language*, which was completed in 1772. This text not only included many word pairs used in the bilingual albums, but also summarized and consolidated the creation of Manchu neologisms at court that had been going on since the 1740s. The Qianlong emperor would come to reframe Manchu-language learning from a different angle than his grandfather seventy years prior. We examine the ways in which Qianlong's ideological commitment to the ideal of the "unification of writing" (Ma. *hergen be emu ombime*; Ch. *tongwen* 同文) reconfigured the *Mirror*'s nature as an encyclopedic text, in particular its chapters on flora and fauna.

---

116. Yan, "Routine Production"; Siebert, Chen, and Ko, introduction to *Making the Qing Palace Machine Work*, 24–27. We borrow the phrase "palace machine" to describe the Grand Council's role in generating neologisms even though this term was coined by the editors of this volume to primarily investigate the Imperial Household Department and other inner court institutions.

CHAPTER EIGHT

Plants and Animals in the Expanded and Emended Mirror of the Manchu Language

The coining of new words at court, discussed in the previous chapter, went too quickly for the bureaucracy in the provinces to keep up with. Work on a new edition of the *Mirror of the Manchu Language* began in 1750 but was not finished until more than twenty years later.[1] This book is the focus of this chapter. The bureaucracy awaited its publication with anticipation. In 1765, the Grand Councillors advised that the expanded edition, once compiled, be circulated within the government apparatus in order to reduce mistakes and inconsistencies in Manchu usage.[2] The resulting *Expanded and Emended Mirror of the Manchu Language*, a Manchu-Chinese edition of the *Mirror*, was finished in 1772 (the date of the imperial preface, QL 36/12/24).[3] With the publication of this work, the process of word creation reached its apex. As we will see, the proliferation of neologisms revealed certain deep-seated assumptions about both languages and the nature of scholarly investigation into words and things.

1. Huang, "*Yuzhi zengding Qingwen jian*," 45.
2. Huang, "*Yuzhi zengding Qingwen jian*," 30.
3. Imanishi, "*Shinbunkan*," 134.

This chapter begins with a close analysis of Qianlong's political and intellectual agenda in the bilingual *Mirror*, concerning in particular the meanings of *tongwen* (Ma. *hergen be emu ombime*), an ancient ideal of imperial rule under which institutions, infrastructure, and written language were unified across the realm and a subject that has been extensively covered by scholars of Qing imperial rulership. We then give a general overview of the *Expanded and Emended Mirror of the Manchu Language*'s sections on plants and animals in comparison with the 1708 *Mirror*, in terms of the arrangement of chapters and sections. Going over the substantive changes made to each chapter, we try to deduce the rationales behind such changes with regard to the principles laid out by the emperor in his preface. Using detailed case studies, we show how the *Expanded and Emended Mirror* represents an engagement in lexicography and philology that is likewise seen elsewhere in contemporary Chinese sources. The chapter closes with the plants and animals that were allocated to the supplement (*bubian* 補編) to the new *Mirror*. Overall, we argue that the heavily revised sections on plants and animals reflected Qianlong's effort for Manchu to become a universal language with a lexicon on par with—and sometimes exceeding—the richness of the Chinese lexicon. However, contrary to the emperor's assertion in the preface, actual lexicographical decisions made by the new *Mirror* rather reflected the constructed and messy nature of "everyday language," imperfectly captured by textual references.

The Politics of *tongwen* in the Imperial Prefaces to the New Mirrors

The problems involved in using the script of a dominant language to transcribe other tongues preoccupied Qianlong repeatedly in the 1770s, as reflected in the prefaces to the new *Mirrors*. In the preface to the expanded Manchu-Chinese edition of 1772, the emperor proceeded to demonstrate that plenty of mistakes happen when sounds do not perfectly match, and yet worse is the excess of meaning when the text, once transcribed in Chinese characters, invests the original sound with connotations that were not originally there.[4] Hence a great many ridiculous mistakes had resulted, according to the emperor, from a too-heavy dependence on Chinese (Ch. *Hanwen* 漢文; Ma. *nikan bithe*) to write histories involving non-Chinese-speaking people, rendering the latter's names, customs, and institutions impenetrable and confusing. In Qianlong's mind, this is how past rulers of a different ethnicity became "hindered and entangled" in the maze of the "Chinese documents" (Ma. *nikan bithe de gocimbume hūsibufi*; Ch. *wei Hanwen suo qianche* 為漢文所牽掣).[5]

This remark sets the Qianlong-era Manchu *Mirrors* apart from the Kangxi-era precedents in that the *Mirrors* not only sought to help the Manchus gain a better command of their own language, but also to tame and exert control over the unruly Chinese script and the documents written in it. In other words, the Manchu *Mirrors* under Qianlong point to a loftier goal of revealing the "essence of sound and rhymes" (Ma. *jilgan mudan-i fulehe*; Ch. *sheng lü zhi yuan* 聲律之元) of any language.[6] In the same preface, Qianlong sought to negate the primacy of Chinese by pointing to the arbitrariness of language sounds through the word "Heaven" as understood not only in Manchu (*abka*) and Chinese (*tian*), but also Mongolian (*tngri*), Tibetan (*namkha*), and "Muslim" (Chagatay) language (*asman*). By entering into a kind of multilingual philological detail not seen in earlier *Mirrors*—one that arguably even appears out of place in this work that after all was only bilingual—Qianlong took aim at a brand of Chinese exegesis that only dealt with foreign words in Chinese-character transcription. Qianlong contrasted old transcriptions with the new, reasoned method of phonetic transcription used in the new *Mirror*. The concern with the accurate representation of sound, present in the revised Manchu-Mongol *Mirror* from 1743, was even more pronounced here.

Whereas Qianlong in 1772 argued for equality on the basis of the arbitrary nature of speech sounds—and by extension, of their written

---

4. The preface to the *Expanded and Emended Mirror of the Manchu Language* is found at the beginning of the trilingual *Mirror* in the *Siku quanshu* edition, which is the most accessible (*Yuzhi Manzhu-, Menggu-, Hanzi*, 12–20).

5. *Yuzhi Manzhu-, Menggu-, Hanzi*, vol. 234, *xu*:9b (16, top panel).

6. *Yuzhi Manzhu-, Menggu-, Hanzi*, vol. 234, *xu*:10a (16, bottom panel).

representation—the emperor in the preface to the Manchu-Mongol-Chinese *Mirror* of 1780 turned to the universal principles of the world on which human institutions were built, language included. Yet even when approached from this angle, the representatives of the Chinese language—unnamed scholars—frustrated Qianlong by refusing to yield to his vision of linguistic parity. Having described how he studied Mongolian, Tibetan, and Uighur to understand the situation of his subjects from near and far, Qianlong declared that "from this I learned that even though the languages of the world are of ten thousand [different] kinds, the principles of the world are the same" (Ma. *tereci abkai fejergi-i gisun, tumen hacin gojime, abkai fejergi-i giyan emu be sahabi*; Ch. 因悟天下之語萬殊，天下之理則一). Everywhere in the world, heaven is above one's head and earth below one's feet, just as everywhere in the world, people think that what is right is right and what is wrong is wrong. Rulers and parents are universally respected, and so on. Even so, "when it comes to the sounds of language and the script of books, the Chinese (Ma. *nikasa*; Mo. *kitad*) vainly consider their own to be correct and everyone else's to be wrong" (Ma. *tenteke nikasa gisun mudan bithei hergen-i dorgide balai murime beyeingge be uruseme, weringge be wakašara, adalingge wesihuleme, encungge be fusihūšarangge*; Ch. 彼其於語言文字中，謬存我是彼非). The Chinese text of the preface curiously omitted that it was the Chinese, and not people in general, who acted this way.[7]

The complex message of the two imperial prefaces is telling in two regards. First, it shows Qianlong's ascription to a "Learning of the Way" style understanding of the relationship between language and reality.[8] The history of non-Chinese regimes relying on the metaphysical Confucian philosophy of the Song period for legitimacy was a long one, going back to the Mongol support for Neo-Confucianism, or the Learning of the Way, in the fourteenth century.[9] It is easy to see why a cosmology centered on a universal, hierarchical order beyond any

---

7. *Yuzhi Manzhu-, Menggu-, Hanzi, xu*:10b–13 (7, upper panel; 8, lower panel). We do not transcribe the Mongolian version.

8. We follow Elman, Duncan, and Ooms ("Rethinking 'Confucianism,'" 530) in using Learning of the Way rather than Neo-Confucianism.

9. Elman, *Cultural History of Civil Examinations*, 33.

cultural and ethnic distinctions would appeal to non-Chinese rulers. If the world order affirmed and upheld by Confucianism was not specifically Chinese, then Manchus could occupy the throne. Similarly, if the articulation of that order in the canonical books were somehow independent of their Chinese linguistic form, those books could be translated into another language without losing any of their canonical status. The Qing court thus published Manchu versions of the Confucian classics.[10]

Qianlong's understanding of language sheds light on the nature of the Manchu-language reform. The reform was entirely focused on words, not on grammar or style. It proceeded on the assumption that the principles of the world could be itemized as words. Here, the Qianlong emperor's reasoning reads as Gong Dingzi's and the Kangxi emperor's description of the relationship between the world and language albeit extended to an explicitly multilingual context. For Gong, the written—or "refined," actually (*wen*)—characters were without question the characters of the Chinese writing system, and these individual characters, as words, represented the principles of the world. When Gong's words (or some very similar statement) were translated into Manchu for inclusion in the preface to Kangxi's *Mirror*, the fit was somewhat awkward. Qianlong's statement, by contrast, did not privilege the Chinese language. Touting the addition of "more than five thousand newly established Manchu words" (*ice toktobuha manju gisun sunja minggan*) in the dictionary, Qianlong saw no equivalent in the Chinese tradition for the unprecedented—and artificial—expansion of the Manchu lexicon.

Second, and important for our purposes here, the goal of *tongwen* (Ma. *hergen be emu ombime*) involves not only the establishment of equivalences between the Manchu and Chinese lexicon, but also revising the Sinocentric order of words and therefore also that of things. Qianlong's criticism of a Sinocentric interpretation of terms in other languages launched him into laying out a new vision for Manchu lexicography that built on—but also departed from—his grandfather's monolingual Manchu *Mirror*. Most significant, Qianlong was clearly dissatisfied with the references to the Chinese literary tradition in the

---

10. Kornicki, *Languages, Scripts, and Chinese Texts*, 210–12.

Kangxi *Mirror*. In his preface, Qianlong ordered that the new *Mirror* should prioritize the "everyday language" (Ma. *inenggidari baitalara an-i gisun*; Ch. *riyong changyan* 日用常言) included in the original *Mirror* over its obscure classical references. Yet of particular interest here is how plants and animals again received a somewhat special treatment, for the emperor did allow that "whatever is of assistance for the investigation of categories—such as the names of ancient offices, clothing, hats, vessels, utensils, beasts, birds, flowers, and fruits—has been made into a separate supplement that has been placed at the end of the book" (*julgei hafan-i gebu, etuku, mahala, tetun, baitalan, gurgu, gasha, ilha, tubihe-i jergi hacin-i acabume kimcire de tusa bisirengge be, encu niyeceme araha banjibun obufi, bithei wajime de sindaha*; Ch. 若古官名冠服器用鳥獸花果等，有裨參考者，別為補編，系之卷末). He was referring to the new *Mirror*'s supplement (*bubian*).[11] Its purpose clearly went beyond language learning. It was intended to provide references for that which did not fall into "everyday language," but instead constituted proper objects for research (Ma. *acabume kimcimbi* 'to investigate and bring together'). Plants and animals, in their various manifestations as birds, beasts, flowers, and fruits, constituted one such venue for research on the same footing as the investigation into the ancient political order, historical institutions, and material culture.

By focusing on the chapters on plants and animals in Qianlong's bilingual *Mirror*, we test how the lofty imperial vision for *tongwen* was in fact implemented in the dictionary. Qianlong's acknowledgment of the world's linguistic diversity goes hand in hand with a firm commitment to a shared underlying ontology of the world. His project of language reform was predicated on the notion that languages differed primarily in terms of sound and written representation, not necessarily in the structure of their lexicon, which represented a common set of referents. There does not seem to have been anything similar to the recognition of the difficulty of finding matching terms (*ming* 名) in two languages, which theorists of Buddhist translation expressed centuries earlier.[12] Qianlong appears to have recognized that not all Manchu

---

11. *Yuzhi Manzhu-, Menggu-, Hanzi, xu*:15a–17b (19). The translation is adapted from Fraser, *Tanggu Meyen*, 173.

12. Harbsmeier, "Very Notions of Language," 149.

words had Chinese equivalents, as our discussion of the Manchu-Chinese *Mirror* demonstrates in the section that follows. That does not appear to have bothered him. If a word found in Chinese did not have a Manchu equivalent, however, a new Manchu word would be coined. Thus, to some extent the reform can be seen as one step in a longer process to align the Manchu lexicon with that of Chinese and ipso facto to the principles of the world. Qianlong's criticism of Chinese scholars' weddedness to their own language notwithstanding, Chinese vocabulary was often the starting point for new Manchu coinages. In many cases, the bilingual new *Mirror* ensured that at least in the "main chapters" (*zhengbian* 正編), the richness of the Manchu lexicon remained greater than that of Chinese, which the editors achieved by frequently rendering multiple Manchu words with only one Chinese word. Elsewhere, however, new Chinese words were also created in the process of bringing more of the world's variety of plants and animals into both languages.

Preparatory Work for the Expanded *Mirror* at the Qianlong Court

Work on the Manchu-Chinese *Mirror* began in 1750, the imperial preface was penned in 1772 (late in QL 36), and the book was probably printed that year. Huang Wei-Jen 黃韋仁 has elucidated the new dictionary's compilation history, showing the Qianlong emperor's direct participation and confirming that Fuheng 傅恆 (1721–1770) was probably the high-ranking official most involved in the work. Qianlong was reading drafts of the work by 1767.[13] Furthermore, Chunhua 春花 has studied a Manchu-Chinese vocabulary that appears to have been produced in preparation for the new dictionary, with which it shares a similar structure and page layout. Stickers pasted onto this book include imperial corrections. It gives some insight into the difficulties involved in establishing Chinese translations for Manchu flora and fauna, and thus provides a case inverse to those described in the Grand Council documents discussed in chapter 7, where the issue in most

---

13. Huang, "*Yuzhi zengding Qingwen jian*," 45–55.

instances was to coin new Manchu words in reference to existing Chinese terms.

The difficulties of establishing Chinese equivalents for Manchu words can be inferred from the fact that some of the Manchu words included in the preparatory vocabulary have Chinese translations that differ from what we see in the finished *Expanded and Emended Mirror of the Manchu Language*. One of these words is *niomošon* 'Siberian salmon'. The preparatory vocabulary translated this word as *eyu* 鮇魚, where *e* represents an obscure character that is not precisely defined in Chinese reference works, including the court's own *Kangxi zidian*.[14] As for *niomošon*, it was an established word in Manchu. Tulišen had left it without Chinese translation. Qianlong's lexicographers similarly did not know how to render it into Chinese. Unlike Tulišen, however, they could not avoid it. They had to provide a Chinese name, and thus chose this rare Chinese character that served as an empty signifier in this context. Yet their translation was evidently ultimately considered inappropriate. One might speculate that the reason was that it went contrary to Qianlong's stated preference for "everyday language" over obscure phrasing and classical references.[15] Whatever the case, the finished Manchu-Chinese *Mirror* did not follow the preparatory vocabulary, but instead the Manchu version of the *Ode to Mukden*, where *niomošon* was used to translate *xilin* 'narrow-scaled'. In the new *Mirror*, the word was rendered into Chinese as *xilinbai* 細鱗白 'narrow-scaled white'. Unlike the obscure classical name, this word gave a Chinese-speaking reader some idea of what the fish called *niomošon* would look like.[16]

The word for Siberian salmon was not an isolated example. Several birds for which Manchu-Chinese equivalences existed received new names in the *Expanded and Emended Mirror*, which the aristocrat

---

14. *Yuding Kangxi zidian*, vol. 231, 35:3a (512, top panel): "the name of a fish" (*yu ming* 魚名).

15. *Yuzhi Manzhu-, Menggu-, Hanzi*, vol. 234, *xu*:14b (18, bottom panel). Remember that this is the preface to the Manchu-Chinese *Mirror*; the *Siku quanshu* editors included it in the Manchu-Mongol-Chinese *Mirror*.

16. *Yuzhi zengding Qingwen jian*, 1782, vol. 233, 32:32a (293, upper panel). Chunhua, "Gugong cang zhenben," 79.

lexicographer Ihing 宜興 (1747–1809) noted.[17] Sometimes the new names were as vernacular as the ones they replaced, but sometimes less so, and on occasion even classical.[18] The history of Chinese names for the bird called *ihan mušu* 'bull quail' is an example. Two "bull quails" were entered into the *Catalog of Birds*, one northern and one southern.

17. Consider the cases of *alhacan niyehe* 'mottled duck' (*Han-i araha manju gisun*, 1708, 19:*gasha-i hacin*:31a; *Yuzhi zengding Qingwen jian*, 1782, vol. 233, 30:22a [223, upper panel]; Ihing, *Qingwen buhui*, 1:1:18a). Cf. Li Yanji (*Qingwen huishu*, 1:18b [12, upper panel]), who described *alhacan niyehe* as "a wild duck with white mixed in among its feathers" 毛有白雜的野鴨. These examples could be further multiplied. Consider *ija cecike* 'gadfly sparrow' (a finch?), whose most current vernacular Chinese name appears to have been "Minister Stringing-things-together" (*chuanhe lang* 串合郎) or "Minister Stringing-flowers-together" (*chuanhua lang* 串花郎), which became *zizi hei* 仔仔黑 'the little black one' (alt. 'the little chirping one') by the *Catalog of Birds* and then the *Expanded and Emended Mirror*. See Hayata and Teramura, *Daishin zensho*, vol. 2, 226; Li Yanji, *Qingwen huishu*, 1:38b (22, upper panel). Chuang, Niaopu Manwen, vol. 12, 183; Ihing, *Qingwen buhui*, 1:38a; *Yuzhi zengding Qingwen jian*, 1782, vol. 233, 30:49a (236, lower panel).

18. Consider the jay (Ma. *isha*), which was originally *shan heshang niao* 山和尚鳥 'mountain monk bird' (Hayata and Teramura, *Daishin zensho*, vol. 1, 28; Sangge, *Man-Han leishu*, 26:17b; Li Yanji, *Qingwen huishu*, 1:41b [23, lower panel]), which in turn was listed in *Shengjing tongzhi* (21:18a). Yet, as we saw in chapter 6, *isha* became *songya* 'pine crow' (a more elegant name) in the *Catalog of Birds*, which was carried over into the new *Mirror*.

Curiously, however, the vernacular vocabulary for small birds was perpetuated elsewhere. Consider the *soncoho cecike* 'pigtail bird' (*Yuzhi zengding Qingwen jian*, 1782, vol. 233, 30:44b–45a [234, upper and lower panels]), given the Chinese translation *san heshang* 三和尚 'monk three' in the new *Mirror*. The character *san* here in all likelihood represents a Mandarin dialect in which the Middle Chinese retroflex sibilant initial *sr-* (*sheng* 生 initial) had merged with the dental sibilant initial *s-* (*xin* 心 initial) in syllables with first-division rhymes (*yideng yun* 一等韻). This merger is a feature of the Jiao-Liao dialects of northern vernacular Chinese, which were spoken in southern Manchuria when the Manchus moved into the area in the seventeenth century. Early Manchu transcriptions of Chinese reflect this merger, so that characters that would be pronounced *shan* in the modern Beijing standard are written identically to characters pronounced *san* (Sukita, "Manshūji hyōki no Kango," 106–7, 118). Thus the Chinese name *san heshang* probably reflects a variant pronunciation of *shan heshang*. The name was removed from the jay only to be attached to the "pigtail bird" in nonstandard orthography. Ihing implied that *shan heshang* and *san heshang* were two forms of the same name when he wrote that the former had been changed to "pine crow" and "*soncoho cecike* has separately been established as *san heshang*" 另定 *soncoho cecike* 曰三和尚. Ihing added that the "old [language] also had this word"

The legend that accompanied the picture of the southern bull quail quoted ancient Chinese authorities that identified the bird with *ru* 鴽, which was said to be a quail or a small bird resembling a quail. The legend also quoted a more vernacular-sounding name: *huang'an* 黃鵪 'yellow quail', which is reconcilable with Li Yanji's description of the bird as yellow ("small yellow quail" [*xiao huang anchun* 小黃鵪鶉]). In Qianlong's *Expanded and Emended Mirror of the Manchu Language*, the word *ihan mušu* was not translated using this vernacular term, however, but as *ru*, the classical term.[19] The new *Mirror* thus standardized not only Manchu names, but Chinese names as well, evidencing a tension between Qianlong's directive to use vernacular language and the established scholarly practice of associating new things with ancient precedents, especially when the new words proved difficult to pin down.

The example of the Siberian salmon shows that the Manchu-Chinese *Mirror* was a book with a history almost as long as that of Qianlong's language reform. The process did not really end with the publication of the dictionary in 1772, as the bibliographical research of Imanishi Shunjū 今西春秋 (1907–1979), Sŏng Paek-in 成百仁 (1933–2018), Li Xiongfei 李雄飛, and Huang Wei-Jen has shown.[20] Original palace prints and later reprints of the dictionary exist, some of which are found bound with new words in separately labeled sections (Ma.

(*jiu yi you* 舊亦有), which we take to refer to the Manchu name *soncoho cecike*. See Ihing, *Qingwen buhui*, 1:41a.

We should note that the description of *soncoho cecike* is not easy to reconcile with its identification with the hoopoe in Norman, *Comprehensive Manchu–English Dictionary*, 329; Corff et al., *Auf kaiserlichen Befehl*, vol. 2, 918 (item 4198.1). Today, the current Chinese name for the hoopoe is *daisheng* 戴勝, which was called *indahūn cecike* 'dog bird' in Manchu. See *Han-i araha manju gisun*, 1708, 19:*cecike-i hacin*:33b; Ihing, *Qingwen buhui*, 1:40b.

19. For the word's early history in Manchu and Chinese, see *Han-i araha manju gisun*, 1708, 19:*cecike-i hacin*:24a; Li Yanji, *Qingwen huishu*, 1:34b (20, upper panel); Mingdo, *Yin Han*, 19:222b (*niu an* 牛鵪 'bull quail', apparently a calque on the Manchu). See Yanagisawa, "Kōkyū Hakubutsuin zō Man-Kan," 36; Chuang, *Niaopu Manwen*, vol. 2, 183. Moreover, Yanagisawa noted that a related bird album adds the remark that the "southern" quail refers to *runiao* 鴽鳥. Lai Yuchih suspects this is a forgery based on a consultation of the court album ("Qinggong, Ouzhou").

20. Imanishi, "Zōtei Shinbunkan no ihan ni tsuite"; Sŏng, "Ŏje chŭngjŏng Ch'ŏngmun'gam"; Li Xiongfei, "Yuzhi zengding Qingwen jian"; Huang, "Yuzhi zengding Qingwen jian," 38–44.

sirame tosimbuha ice manju gisun; Ch. 續入新清語 and Ma. *jai mudan* idem; Ch. *er ci* 二次 idem), and some with a few new words added on existing pages.[21] It is probable that these later copies were not printed at court. In addition, several fair manuscripts were produced for Qianlong's *Siku quanshu* and its abridgment.[22]

The existence of several editions, including copies visibly struck from worn—and, it follows, much used—blocks, suggests a wide circulation, which is confirmed by the many copies still found in libraries around the world. In addition, the lemmata were subsequently rearranged and included in Ihing's dictionary *Manchu Collected, Supplemented* (*Manju gisun be niyeceme isabuha bithe* | *Qingwen buhui* 清文補彙) from 1786, a continuation of Li Yanji's adaption of the original *Mirror* that included new words from various sources, Qianlong's Manchu-Chinese *Mirror* among them. Ihing's dictionary appeared in a revised edition in 1802 and ensured an even greater circulation for the new Manchu words.

Rearranged Sections in the Revised *Mirror*

The new *Mirror* kept the basic structure of the original but added new segments (Ma. *meyen*) and rearranged existing sections to make the semantic arrangement more thorough. For example, one of the segments under "Vegetables and dishes" was now headed by the word *suwaliyasun* 'seasoning' (< *suwaliyambi* 'to mix'), which was not listed in Kangxi's *Mirror*. The word here almost worked like a rubric. The segment assembled spices, aromatics, and products used for seasoning, including words such as *suwanda* 'garlic' (< Ch. *suan* 蒜 + Ma. *da* 'root') that in the Kangxi *Mirror* had been listed among the edible plants of the section rather than with other aromatics.

Furthermore, words were grouped more consistently within segments as well. Thus *guwalase*, a *hengke* 'gourd' that in Kangxi's *Mirror* had been listed at some remove from the other gourds in the section "Vegetables and dishes" and that was discussed at court in 1765 was

21. See, e.g., *Yuzhi zengding Qingwen jian*, 1772.
22. *Yuzhi zengding Qingwen jian*, 1778, 1782.

PLANTS AND ANIMALS IN *ZENGDING QINGWEN JIAN*   285

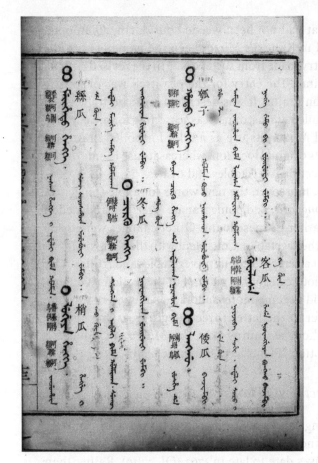

FIG. 8.1. A page from the *Expanded and Emended Mirror of Manchu Language* (*Zengding Qingwen jian*) discussing new kinds of gourd (Beijing: Wuying Dian, 1771, 27:13b). Palace edition (on microfilm), New York Public Library.

placed with the other gourds in the revised dictionary. Two new gourds were also added here: *hoto hengke*, lit. skull gourd, translated as *huzi* 瓠子 'squash', and *langgū*, translated as *wogua* 倭瓜 'Japanese gourd', referring to the pumpkin (see fig. 8.1).[23]

Similarly, in an apparent effort to keep entries focused, some were split up and placed into different sections. For example, *giyangdu* 'cowpea' (perhaps from early Mandarin *kjaŋtəw'* 豇豆, in any case a

23. *Yuzhi zengding Qingwen jian*, 1782, vol. 233, 27:18a (129, upper panel), 27:30a–33a (135, upper panel; 136, lower panel); Fèvre and Métailié, *Dictionnaire Ricci des plantes*, 467.

Chinese loanword that had not been weeded out during Qianlong's language reform) had in Kangxi's *Mirror* been the name of both the pea and a steamed pastry made from such peas.[24] In the revised edition, it was listed a third time. The entry for the pea itself was originally found in the section on grains. It was described and said to be eaten mixed in with other boiled grains, or used as seasoning for meat in side dishes. In the revised *Mirror*, its use as a filler for boiled grains remained described in its original location, whereas its use in side dishes was entered into the section "Vegetables and dishes."[25]

Among the fruits, moreover, the new words for apple (*pingguri*), Chinese crabapple (*yonggari*), and pomegranate (*useri*), introduced in early 1748, replaced earlier Chinese loans. Other changes were made in a similar fashion as the vegetables. Consistent with the Kangxi *Mirror*, the new *Mirror* did not attempt to combine vegetables and fruits under a new and more capacious category of plants, but the sections "Grasses, Trees, and Flowers" did become more elaborate as they went through revision. Here we refer to the latter three categories as "plants" for simplicity's sake without assuming that the *Mirror* introduced a broader taxonomic notion "Plants" (or, similarly, "Animals").

The New Mirror's Section on Plants

A glimpse into the Qing court's revision process for plant terminology can be found in a Grand Council document that the archivists at the First Historical Archives date to late in 1762 (QL 27/10). Rather than finding an appropriate Manchu name for any particular species, the document discussed the anatomical structure of plants:

24. Transcription from Pulleyblank, *Lexicon of Reconstructed Pronunciation*. The word might have been borrowed into Manchu from a vernacular Chinese dialect where the velar stop initial *k-* (*jian* 見 initial) had not palatalized in syllables with second-division rhymes (*erdeng yun* 二等韻). The homophonous character *jiang* 江 was most often written as *giyang* in early Manchu transcriptions of Chinese (Sukita, "Manshūji hyōki no Kango," 91–93; *Han-i araha manju gisun*, 1708, 18:*efen-i hacin*:18b, *bele jeku-i hacin*:44a).

25. *Yuzhi zengding Qingwen jian*, 1782, vol. 233, 27:20b (130, upper panel), 28:38b (168, lower panel).

Upon examination, [we found that] in the *Mirror of the Manchu Language,* huaxin 花心 'center of a flower' is called *jilha.* Given that *huaxu* 花鬚 'stamen' are the long tassels in a flower, we propose to call the stamen *jilhari* on the basis of the sound of *jilha.* Could the throne please advice as to whether it is appropriate?

baicaci, manju gisun i buleku bithede, 花心 *be jilha sembi,* 花鬚 *ilha de banjiha narhūn suihe, be dahame, uthai jilha-i mudan be gaime* 花鬚 *be jilhari obuki, acanara acanarakū babe dergici jorime tacibureo.*[26]

It is noteworthy that, first, Kangxi's Manchu *Mirror,* a nominally monolingual dictionary, was here referenced to find Manchu translations of Chinese words that were not lemmatized, indexed, or otherwise listed. In other words, court scholars in the 1760s were intentionally using the Kangxi *Mirror* as a reference work to clarify corresponding Chinese terms. Second, Kangxi's *Mirror* did indeed contain an entry for *jilha* (but of course not one for *huaxin*), which was defined as "the innermost part in the center of a flower" (*yaya ilha-i dulimbai niyaman be, jilha sembi*). Adjacent headwords included *fiyentehe* 'petal' and *bongko* 'bud' but no word for stamen.[27] Curiously, however, neither *jilhari* nor any other word for stamen was entered into the ensuing bilingual revision of the *Mirror,* this document notwithstanding.[28] Perhaps the partial semantic overlap of "the innermost part in the center of a flower" and "stamen" made the court refrain from introducing the new word into the dictionary? Whatever the case, this instance suggests that the routinized creation of Manchu neologisms in the 1760s, most of which were carried out by clerks at the Grand Council, received further scrutiny and screening during the compilation process of the new *Mirror.*

Furthermore, although the new *Mirror* kept the basic categories "Grasses" (*orho*), "Trees" (*moo*), and "Flowers" (*ilha*), the order by which words and phrases appeared in the chapters and sections

---

26. Manchu palace memorial file copies, 03-0180-1975-015, FHA.
27. *Han-i araha manju gisun,* 1708, 19:*ilha-i hacin*:23a.
28. If present, we would expect the word to be listed at *Yuzhi zengding Qingwen jian,* 1782, vol. 233, 29:66a–67b (211).

headed by these words was substantially altered. For instance, whereas the Kangxi *Mirror* starts the chapter "Grasses" with very homely words such as *niyanciha* 'green growths of grass' and *hakda* 'grass remains after burning,' the new *Mirror* starts the section with the auspicious fungi (*lingzhi* 靈芝), *orhoda* 'ginseng' (lit. chief of the grasses), and the fragrant Manchurian herbs we discussed in some length in chapter 7. The section "Trees" starts with *jakdan* 'pine' and a host of other evergreen trees, instead of *cuse moo* 'bamboo' and *u tong moo* 'parasol tree' (with Chinese classical references), as had been the case in the original *Mirror*. The section "Flowers" pushed back the discussion of lotus and peony, both subjects of a rich tradition of Chinese writings, and started with the added entry on frost-resistant blossoms such as *nenden ilha* 'plum' (lit. first flower) and its variations (e.g., *suwayan nenden ilha* 'yellow plum').[29] The reordering of words was clearly motivated by the meanings attributed to these real-world plants as reflected through a Qing imperial sensibility that prioritized supposedly Manchu nature.

Although each of the three plant sections saw substantive additions in terms of new plant names and morphological terms (about forty-five in "Grasses," some fifty in "Trees," and more than one hundred in "Flowers"), as far as we can tell the sources of those additions were by no means homogenous. Notably, despite the emperor's clear dislike of quoting Chinese classics in a Manchu dictionary, many added entries in the plant sections were nevertheless covert quotations from Chinese sources. Consider, for example, the new entries on *jijiri orho* and *mijiri orho*:

> *jijiri orho*. Name of grass. Its leaves can be woven into sitting mats; can also make shoes.
>
> *jijiri orho, orho-i gebu, erei abdaha be derhi jodombi, sabu araci inu ombi*.[30]

---

29. *Yuzhi zengding Qingwen jian*, 1782, vol. 233, 29:3b (179, bottom panel), 29:21a (188, bottom panel), 29:47a (201, bottom panel).

30. *Yuzhi zengding Qingwen jian*, 1782, vol. 233, 29:7b–8a (181, bottom panel; 182, top panel).

In fact, this line was most likely paraphrased from the *Kangxi zidian* entry for the character *bao* 苞, which in turn quotes two ancient sources:

> The *Shuowen* says: *bao* is a kind of grass, and the people of Nanyang uses it to make coarsely woven shoes.... The commentary to the *Book of Han* says that it is nowadays used to make sitting mats.
> 《說文》：草也，南陽以爲麤履....《漢書》註卽今所用作席者。[31]

Next is the entry for *mijiri orho,* which the Qianlong *Mirror* paired with the Chinese name *micao* 靡草:

> *mijiri orho.* The leaves resemble those of shepherd's purse. Grows in spring, then dries up and turn yellow in the fourth month.
> *mijiri orho, abdaha niyajiba-i abdaha-i adali niyengniyeri banjifi duin biyade sorome olhombi.*[32]

This quote, again, borrows a crucial detail about *micao* from from the "Monthly ordinances":

> In this month [i.e., the fourth month] they collect and store up the various medicinal herbs. Delicate herbs [*micao*] (now) die; it is the harvest time (even) of the wheat.
> 是月也，聚畜百藥。靡草死，麥秋至。[33]

In other words, despite the emperor's severe prohibition against reliance on Chinese classics, they nevertheless remained in the new *Mirror* as an undercurrent of knowledge. The addition of classical quotations such as those just cited, which appeared in the new *Mirror*

---

31. *Yuding Kangxi zidian*, vol. 231, 25:14b (8, lower panel).
32. *Yuzhi zengding Qingwen jian*, 1782, vol. 233, 29:8a (182, top panel).
33. English translation from Legge, *Sacred Books of China*, 271; see also Couvreur, *Li ki*, vol. 1, 357; Ruan, *Liji zhengyi*, 137, middle panel (chap. 15).

without explicit attribution, could have been facilitated by the fact that *Guang Qunfang pu* and other late Kangxi-era projects such as *Kangxi zidian* already sorted through old literature and made it very easy to browse.

Strong evidence also suggests that the Qianlong *Mirror* not only added frontier plants discussed in the late Kangxi-era *Guang Qunfang pu*, but also relied on the latter text for crucial details. See, for instance, this vivid description of jasmine (Ma. *moli ilha*; Ch. *moli* 茉莉), which was another addition in the new *Mirror*:

> Its stem is delicate, its branches many; its leaves is like that of the tea tree; its flowers are white with small buds; it blooms at twilight, emitting pure fragrance.
>
> *cikten uyan, gargan labdu, abdaha cai i abdaha i gese, ilha šanyan gubsu ajige, yamjishūn ilambi, bolgo wangga.*[34]

Although the text does not say so, it follows very closely Wang Xiangjin's description of jasmine in *Guang Qunfang pu*, which is in part borrowed from earlier pharmacopoeia but unique in many details:

> Its stem is delicate, its branches many; its leaves resemble tea tree and yet larger, green-colored and round-shaped with a pointed tip. In summer and autumn small white flowers bloom invariably at twilight. Its fragrance is pure, gentle, and elegant with a winning flavor.
>
> 弱莖繁枝，葉如茶而大，綠色團尖。夏秋開小白花，花皆暮開，其香清婉柔淑，風味殊勝。[35]

Elsewhere, the new *Mirror* added a whole cluster of new words that were varieties of orchid listed in *Guang Qunfang pu*, with the Manchu descriptions closely following the Chinese text.[36]

---

34. *Yuzhi zengding Qingwen jian*, 1782, vol. 233, 30:58a (207, upper panel).

35. For Wang's original description, see Wang Hao, *Yuding Peiwen zhai Guang Qunfang pu*, vol. 846, 43:12a (510, lower panel).

36. On the cluster of names for rose campein (*jian chunluo* 剪春羅), see *Yuzhi zengding Qingwen jian*, 1782, vol. 233, 29:60a–b (208, upper panel); on *jian chunluo*,

Last are a substantial number of new entries that cannot be traced to Chinese classical references or other court publications in Chinese. For instance, a series of trees commonly used for fine furniture wood were added to the section "Trees" in the new *Mirror* that were discussed in neither the 1708 *Mirror* nor *Guang Qunfang pu*. A "speckled *nan* wood," for instance, bears the following description:

> Also *nan* wood. As the years pass, on the side exposed to the sun, the veins of this tree will generate whirl-like patterns.
>
> alhangga anahūn moo, uthai anahūn moo inu aniya goidafi šun goire ergi moo-i jun foron-i gese ilha banjinambi.[37]

The *Expanded and Emended Mirror* rendered this word as *douban nan* 豆瓣楠 'bean-lobe *nan*' in Chinese, a term briefly mentioned in late Ming literati treatises on furniture and woodwork. None of these treatises, however, explained how the whirling pattern was generated.[38] That knowledge—of the exposure to the sun generating the pattern—probably originated in a vernacular discourse in Chinese. Here, we thus see an example of vernacular discourse being brought to bear on a new Manchu phrase, one that likewise probably came from "everyday language" with no clear classical point of reference.

A more subtle pattern of vernacular knowledge finding its way into the new *Mirror* concerned the marked predilection for fragrances (Ma. wa). The 1708 Manchu *Mirror* included the word *okjiha* 'sweet flag', a common water plant whose furry, dried spadix (Ma. *ibaga hiyabun*; Ch. *changpu* 菖蒲) could be useful in making flammable tinder, along with mentions of an ostensibly identical plant in the *Poetry Classic*. The new *Mirror*, however, added an entry meaning "root of sweet flag" (*okjihada*) immediately following *okjiha*, and attributed the Chinese name *cangzhu* 蒼朮 'rhizome of *Atractylodes lancea* or *Atractylodes chinensis*'

---

see also Fèvre and Métailié, *Dictionnaire Ricci des plantes*, 224–25); on the cluster of names for orchid, see *Yuzhi zengding Qingwen jian*, 1782, vol. 233, 29:48a–49a (202).

37. *Yuzhi zengding Qingwen jian*, 1782, vol. 233, 29:25a (190, lower panel).

38. On using *douban nan* for inkstone cases, see, for instance, Gao Lian, "Zunsheng bajian," 684.

to the former.[39] The Chinese *bencao* pharmacopoeia tradition clearly distinguishes *changpu* and *cangzhu* as two very different plants. The confusion of the latter to be the root of the former may have to do with the fact that dried and shredded calamus root and *cangzhu* root were frequently combined to make fragrant medicines in folk remedies. The editors of the new *Mirror* might not have consulted the pharmacopoeia closely, and indeed were explicitly told not to rely too much on written Chinese sources. Yet they could not so easily be prevented from relying on the vernacular knowledge around them. As a result, the new *Mirror* crafted a neologism (*okjihada*) that in turn was drawn from a closely entangled practical use of the two very different herbs in everyday life.

The New Mirror's Section on Birds

In the chapter "Birds large and small," the number of names in the "Large birds" section had more than doubled and the number in "Small birds" almost tripled. Several names from the original *Mirror* had been changed, replacing Chinese loanwords such as *funghūwang* (< Ch. *fenghuang* 鳳凰) with Manchu neologisms such as *garudai* 'phoenix'. This word was derived from an existing *gerudei*, which the Kangxi *Mirror* gave as a synonym of the phoenix. The revised *Mirror* retained *gerudei*, but now with the restricted meaning of "female phoenix" (*garudai-i emile*).

Many of the new kinds of birds were lifted from the *Catalog of Birds*, and in several instances names follow the order of the *Catalog*. The introduction of birds from the *Catalog* meant that descriptions of avian fauna from far beyond Manchuria and northern China could now be read about in a Manchu reference work. For example, *alhari coko* 'mottled chicken' (Ch. *shanhua ji* 山花雞 'mountain mottled chicken') is introduced to the *Mirror*'s readers as follows:

---

39. *Yuzhi zengding Qingwen jian*, 1782, vol. 233, 29:11b (183, lower panel). On common names and Latin names for those plants, see Fèvre and Métailié, *Dictionnaire Ricci des plantes*, 47, 40.

A kind of pheasant found in Fujian. Its head is black, the feathers on its cheeks protrude out from the head. On its back, the feathers grow in a mix of red and white so that it looks spotted. The wings and tail are black with yellow stripes.

ulhūma-i duwali, fugiyan-i bade tucimbi, uju sahaliyan, šakšaha-i funggaha uju be duleme banjiha, huru-i funggaha fulgiyan šanyan suwaliyaganjame banjiha bime mersen bi, asha uncehen sahaliyan, bederi suwayan.[40]

The Qing court "consumed delicacies from every region of the empire," including pheasant,[41] of which several kinds were described in the new *Mirror*. As we see here, it was done in some detail. Curiously, although the description accords with that in the legend for the picture of this Fujianese pheasant in the *Catalog of Birds*, the phrasing is different.[42] It was either revised for inclusion in the dictionary, or goes back to an earlier, unknown document in the *Catalog*'s paper trail.[43]

Some divergences from the *Catalog of Birds* are evident.[44] For example, Tsai Ming-che notes that *nukyak gasha* 'Malabar pied hornbill', the Manchu name for which was probably derived from a South or Southeast Asian language, had its name changed to *senggelengge gasha* 'combed bird'. The Chinese name was *nukeyake* 弩克呀克 and thus a transcription from the same underlying non-Qing word. The *Mirror*'s editors might have thought that the Manchu's similarity to the Chinese in this case made the word run afoul of the principles of Qianlong's language reform, hence the change.[45]

---

40. *Yuzhi zengding Qingwen jian*, 1782, vol. 233, 30:17a–b (220, lower panel).
41. Schlesinger, *World Trimmed with Fur*, 22.
42. Chuang, *Niaopu Manwen*, vol. 3, 273–75.
43. Other entries for birds in the *Catalog of Birds* differ in the phrasing as well.
44. Cf., e.g., the *Catalog*'s *ala gasha* 'hill bird' as the translation of *yuanniao* 原鳥, a name for the pheasant, and the revised *Mirror*'s revision of the word into *ala ulhūma* 'hill pheasant' (Chuang, *Niaopu Manwen*, vol. 2, 215; *Yuzhi zengding Qingwen jian*, 1782, vol. 233, 30:20a [222, upper panel]).
45. Tsai, "Qing qianqi Manzhou," 242–43. Tsai proposes that the origin of the name is Thai.

More important, at least some of the new entries in the new *Mirror* were not lifted from the *Catalog of Birds* at all. Yet it seems that these words too were neologisms, which shows that the creation of Manchu bird vocabulary extended beyond the already very great number of words introduced in the *Catalog*. One such word was *heturhen*, which with the fairly vague definition of "a type of [crow] falcon[46] [or] hawk, it is good at catching things like rabbits" (*nacin giyahūn-i duwali gūlmahūn-i jergi jaka be forire mangga*) appears as a desk creation.[47] It was inserted right after *itulhen*, defined as "a type of gyrfalcon used to catch rabbits" (*šonggo nacin-i duwali gūlmahūn forire de baitalambi*). This word was listed already in the original *Mirror* and had the established Chinese translation *tugu* 兔鶻 'rabbit falcon', which was used also in the *Ode to Mukden*.[48] The Chinese translation of the new word *heturhen* was *lan hushou* 攔虎獸 '*lan* tiger beast (?)', where *lan* clearly represented the same underlying morpheme variously seen as *lan* 蘭 or *lan* 藍 in names for larks. The name *lan hushou* was a vernacular term attested in *Shengjing tongzhi* but not used in the *Catalog of Birds*. The gazetteer presented a taxonomy of birds of prey quite different from what emerges in the entries in the revised *Mirror*.[49] The Manchu word *heturhen* did thus not come from the *Catalog of Birds* but was probably a neologism all the same.[50]

---

46. Li Yanji (*Qingwen huishu*, 1:16a [36, lower panel]) translated *nacin* as *yagu* 鴉鶻 'crow falcon'.

47. The word *forimbi* (> *forire*), which also means "to beat, to strike" is used here as a technical term for when birds of prey pierce their quarry with their talons (Corff et al., *Auf kaiserlichen Befehl*, vol. 2, 924 [item 4229.3]). The Chinese word *da* 打 idem was also used in this sense: Li Yanji, *Qingwen huishu*, 1:36b (21, top panel), s.v. *itulhen*.

48. Li Yanji, *Qingwen huishu*, 1:36b (21, top panel).

49. In the 1684 *Shengjing tongzhi*, *tugu* was listed as a name for the larger kind of falcon (*gu*). The smaller kind was called *yagu* 鴉鶻 'crow falcon', which in the eighteenth century was the established translation of *nacin*, used in the definition of *heturhen* in the revised *Mirror*. The gazetteer listed *lanhu shou*—the Chinese translation of *heturhen*—as the name for one of the several kinds of sparrow hawk (*yao* 鷂) present in the Liao region. See *Shengjing tongzhi*, 21:16b–17a; Chen Menglei and Jiang, *Qinding Gujin tushu jicheng*, vol. 76, 13b (top panel).

50. It is phonetically suspiciously similar to *itulhen*, the headword of the preceding entry. The first syllable might have been taken from the word *hetumbi* 'to transverse, to cross', which could be used to describe birds flying across the sky (diving to catch rabbits?). See Hu, *Xin Man-Han da cidian*, 404. That the physical features of the bird

The New Mirror's Section on Beasts

The names of kinds of animals in the chapter "Beasts" had almost doubled in the *Expanded and Emended Mirror*. Like that on birds, it added more mythical creatures recorded in Chinese sources and revised the names of ones already there. The received Chinese exegesis of these animals took the form of a commentary on their representation in the Chinese script. Thus the *Catalog of Beasts*, following *Gujin tushu jicheng*, which in turn was quoting older sources, presented *qilin* 麒麟 'unicorn'—an animal of great importance in ancient Chinese writings on nature, and one whose appearance was an auspicious omen—as a binome, where *qi* referred to the male and *lin* to the female.[51] This explanation probably has nothing to do with the actual history of the word, however, and the individual characters are not attested as having been used in this sense in ancient China. "Although the interference of real monosyllabic etymons cannot be ruled out," writes Juha Janhunen, "there is no overruling evidence against *qilin* . . . being a single etymologically indivisible bisyllabic entity."[52]

The Manchu editors of the *Catalog of Beasts* were nevertheless compelled by the conventions of that book to translate the Chinese gloss. They said of the unicorn that "male individuals are called *sabitun*, female individuals *sabintu*" (*haha ningge be sabitun sembi, hehe ningge be sabintu sembi*).[53] Thus when the editors of the *Expanded and Emended Mirror* removed the Chinese loan *kilin* (< early Mandarin *kʰi'lin'* or a similar vernacular dialect in which the palatalization *k-* > *q-* in this context had not yet happened),[54] they replaced it with two

---

are not described in the *Mirror* suggests that we are dealing with a paper tiger, as it were. The issue would be resolved if we had access to a court document of the type discussed in chapter 6, but so far we have not found one.

51. Chen Menglei and Jiang, *Qinding Gujin tushu jicheng*, vol. 519, 40b (lower panel). Lu Ji, *Mao Shi cao mu niao shou chong yu shu*, vol. 70, chap. 2, part 2:1b (123, lower panel). See also Carr, "Linguistic Study," 52–53.

52. Janhunen, "Unicorn, Mammoth, Whale," 195.

53. Chuang, Shoupu *Manwen*, vol. 1, 145.

54. Early Qing transcriptions of *qi* 麒 reflect a palatalized pronunciation (*ci*): Sukita, "Manshūji hyōki no Kango," appendix, 6. It is tempting to conjecture that the word entered Manchu or Jurchen early on, before the palatalization had taken place, but we have not researched the issue.

entries: one for *sabitun*, translated as *qi*, and one for *sabintu*, translated as *lin*.⁵⁵ The first, however, used the old definition of *qilin*, so perhaps the reader was supposed to infer that it meant both the male unicorn and unicorn in general. Scholars have noted that both words are created from *sabi* '(auspicious) sign, omen'.⁵⁶

The terminology for deer was confused in both Manchu and Chinese, as we saw in earlier chapters. Now the two were aligned.⁵⁷ The name *mi* deer, which had caused so much confusion, was simply left out of the new *Mirror*.⁵⁸ Furthermore, the moose, *kandahan*, was not associated with a name with Chinese morphology, such as *tuolu* 'camel deer', which the 1684 *Shengjing tongzhi* had used. Instead, the transcription *kan-da-han* that Qianlong had used in his poem was used as a Chinese translation.⁵⁹

One large animal known from the empire's northern periphery that had been discussed by Kangxi in his *Kangxi ji xia gewu bian* and orally at court found its way into the new *Mirror*: the frozen mammoth referred to as a rodent (*shu*). No reference was made to Kangxi's discussion of this creature, however, which notably specified that it could still be found in the present in frozen form and described it similarly to a mammoth. The entry in the *Expanded and Emended Mirror* was thus probably not based on Kangxi's writings or discussions, but on *Gujin tushu jicheng*, which in turn quoted the *locus classicus* in *Shenyi*

---

55. *Yuzhi zengding Qingwen jian*, 1782, vol. 233, 31:4b (245, upper panel).

56. Janhunen, "Unicorn, Mammoth, Whale," 196; Chuang, *Shoupu Manwen*, vol. 1, 38.

57. Unsurprisingly, *buhū* was *lu* 鹿 'deer' and *ayan buhū* was *malu* 馬鹿 'elk'. Curiously, however, *suwa buhū* 'Sika deer'—also the "generic name for small deer" (*xiaolu zongming* 小鹿總名) according to Shen Qiliang (Hayata and Teramura, *Daishin zensho*, vol. 1, 115 [line 0725a3]) —was not translated as *mi*, as proposed in the document sent to the Mukden Imperial Household Department that we studied in chapter 7, but as *meihua lu* 梅花鹿 'plum blossom deer', which is the current name for the Sika in Chinese. Shen similarly gave *hualu* 'blossom deer' as a translation for *suwa buho* (i.e., *buhū*). By contrast, Li Yanji (*Qingwen huishu*, 5:17a [101, upper panel]) did not translate the word.

58. Because *suwa*, which the *Catalog of Beasts* had used to translate *mi*, was in the new *Mirror* given a more accurate translation that reflected current usage in both Manchu and Chinese, it was no longer available as a translation for *mi*.

59. *Yuzhi zengding Qingwen jian*, 1782, vol. 233, 31:12b–13a (249, upper and lower panels).

jing.⁶⁰ The new *Mirror* called the creature *juhen* (< *juhe* 'ice') *singgeri* 'iced rodent' and defined it as follows:

> This rodent nests in the ground beneath thick ice in the north. Its flesh weighs a thousand catties. It is edible. Its fur is several feet long. If made into a cushion, it can shut out the evening chill.
>
> *ere singgeri amargi ba-i jiramin juhe-i fejile boihon-i dolo tomombi, yali minggan ginggen ujen, jeci ombi, funiyehe ududu jušuru golmin, sektefun araci, šahūrun ya be gidambi.*⁶¹

In this truncated form, the entry is still compatible with the description of a mammoth or some other frozen beast. Yet the entry seems to suggest that the creature is alive beneath the ice. We show in the conclusion that it was of interest to later readers who used the *Expanded and Emended Mirror* as a source for the natural environment of the Qing empire.

The entries on simians that the new *Mirror* lifted from the *Catalog of Beasts*, like the description of the unicorn, drew on Chinese sources, and in most cases, the simians cannot be easily identified with known animals. Some readers might certainly have found the abbreviated Manchu descriptions that the revised *Mirror* gave of these creatures more accessible than earlier descriptions in Chinese. Yet the most interesting thing about the new monkey entries is their rearrangement as they passed from the *Catalog of Beasts* to the new reference work. Some entries were placed among the "Beasts" in the main body of the dictionary, others in the supplement. We have been unable to discern the logic behind this division with certainty, but have found ground to speculate. Furthermore, the arrangement of those simians that made it into the main body of the dictionary appears reasoned as well.

Readers of the original *Mirror* had already been made acquainted with *monio* 'monkey; macaque' and *yuwan* 'gibbon', a Chinese loanword. In the *Expanded and Emended Mirror*, the latter was replaced by an existing—possibly dialectal—variant word, *bonio*, repurposed as the

60. Chen Menglei and Jiang, *Qinding Gujin tushu jicheng*, vol. 521, 52b (top panel).
61. *Yuzhi zengding Qingwen jian*, 1782, vol. 233, 31:22b (254, upper panel).

new name for the gibbon.[62] In addition, several more simians were added, all of which came from the *Catalog of Beasts*, which in turn based its descriptions on *Gujin tushu jicheng*.[63] Furthermore, additional

---

62. The word *monio* is attested already in Jurchen language of the late fifteenth century (Kane, *Sino-Jurchen Vocabulary*, 218 [item 424]). See also Vovin, "Some Thoughts on the Origins," 120. Both *monio* and *bonio* are translated as *yuanhou* 猿猴 'simian' in Sangge (*Man-Han leishu*, 26:36a). The form *bonio* was used to refer to the ninth of the twelve earthly branches (Ch. *shen* 申) in the sexagenary cycle of timekeeping. See *Han-i araha manju gisun*, 1708, 19:*gurgu-i hacin*:46b; see also Hayata and Teramura, *Daishin zensho*, vol. 2, 80 (item 0601a3).

63. The dictionary acquainted its readers with *sirsing*, which is clearly formed on the basis of Ch. *xingxing* 猩猩. This latter word is listed in *Erya* along with four other creatures that Robert van Gulik (1910–1967) identified as primates. The canonical text says simply that it is "small and likes to cry" (*xiao er hao ti* 小而好啼), but other ancient sources give it a human face and the ability to speak. In the new *Mirror*, it is said that it "lives in the state of Jiaozhi [交阯], looks similar to a monkey, walks like a human, and can speak. Its blood is used as a dye" (*giyao jy gurun de banjimbi, monio de dursuki, niyalmai adali yabumbi, gisureme bahanambi, erei senggi fulgiyan icere de baitalambi*) (*Yuzhi zengding Qingwen jian*, 1782, vol. 233, 31:16a [251, upper panel]). These details are given in the *Catalog of Beasts*, albeit with a different phrasing (Chuang, *Shoupu Manwen*, vol. 1, 287–89). The information can all be found in the sources cited in the section on *xingxing* in *Gujin tushu jicheng* (Chen Menglei and Jiang, *Qinding Gujin tushu jicheng*, vol. 522, 22b–23a). Jiaozhi, the location of the *xingxing*, corresponds to northern Vietnam (Gulik, *Gibbon in China*, 26), and indeed, in a story in a collection probably dating from 1661, a Chinese traveler to Vietnam encounters simians with these characteristics, which suggests the currency of the description in the Qing period (McMahon, *Causality and Containment*, 140; relatedly, Tina Lu, *Accidental Incest*, chap. 4). The illustration of *xingxing* in *Gujin tushu jicheng* shows a creature with a human physiology and hair (no fur), whereas the *Catalog of Beasts* clearly depicts a simian. Wang Zuwang and his colleagues go so far as to identify it with the orangutan, called *xingxing* in modern Chinese (Yu and Zhang, *Qinggong shoupu*, 409).

In addition to *xingxing*, the editors entered four other simians from the *Catalog* into the chapter on beasts in the main body of the new *Mirror*. They were *sahaldai monio* (Ch. *guoran* 果然), a macaque named for its black (*sahaliyan*) body (Hauer, *Handwörterbuch der Mandschusprache*, 2007, 396); *sobonio* (Ch. *rong* 狨) 'yellow gibbon' (< *sohon* + *bonio*; Hauer, 422); and *ukeci* (Ch. *penghou* 彭猴), which looks like a dog without a tail (*Yuzhi zengding Qingwen jian*, 1782, vol. 233, 31:16a–b [251, top panel]). The depictions of these creatures in the *Catalog of Beasts* do not correspond to any animals known, according to contemporary specialists (Yu and Zhang, *Qinggong shoupu*, 409).

simians listed toward the end of the *Catalog of Beasts* as fantastical creatures from ancient Chinese literature ended up in the new *Mirror*'s supplement, along with three more simians of *Gujin tushu jicheng* provenance. We have been unable to establish with certainty why four new simians were added to the main body of the new *Mirror* and the rest consigned to the supplement. One possibility is that simians that Li Shizhen, quoted in *Gujin tushu jicheng*, described confidently as varieties of ape were placed in the main body of the *Mirror*, whereas creatures of uncertain status were placed in the supplement.[64] There is, then, some reason to believe that a consultation of the most authoritative scholar, as quoted in the great imperial encyclopedia, influenced the selection of simians for the main body of the revised *Mirror*.[65]

64. Chen Menglei and Jiang, *Qinding Gujin tushu jicheng*, vol. 522, ch. 87 (18b, bottom panel; 23b, bottom panel), ch. 88 (21b, bottom panel).

65. The *sirsing*, which the new *Mirror*'s editors placed first among the simians, is a special case. *Xingxing*, the Chinese translation—and origin—of *sirsing* was, unlike *guoran*, *rong*, and *penghou*, mentioned in the headword of another entry as well. In the section "Silk and cloth" (Ma. *suje boso-i hacin*; Ch. *bubo lei* 布帛類), the dictionary listed something it called *fulgiyan nunggasun* 'red felt', which in Chinese was *xingxing zhan* 猩猩氈 'xingxing felt'. The definition of this word read "felt dyed to a red like the scarlet blood of the *sirsing*" (*nunggasun be sirsing ni senggi-i fularjame fulgiyan obume icehengge be, fulgiyan nunggasun sembi*; *Yuzhi zengding Qingwen jian*, 1782, vol. 232, 23:27b [777, lower panel]). In Qing usage, the word "*xingxing* felt" referred to a cloth that was a common element in the material culture of upper-class life in north China (repeatedly mentioned in *Story of the Stone*: Cao, "Honglou meng 'xingxing' kaobian"). Despite the reiteration of the idea that monkey blood can dye cloth in the *Expanded and Emended Mirror*, "*xingxing* felt" was not made in this way. In fact, it was not even necessarily red. Wu Siyu 吳思雨 and Wang Le 王樂 believe the cloth was imported from the Netherlands ("Qianlong shiqi," 98). Most likely, the name *xingxing* felt was similar to the *zhu* tail discussed in chapter 7. A famous Song-dynasty painting depicts imperial figures sitting on cushions made from the fur of large simians (Wang Cheng-hua, "Ting qin tu de zhengzhi yihan," 84). Probably, making reference to simians in the name of an imported cloth used at court in the eighteenth century was a way to endow it with antiquity and refinement, as in the case of the fly whisk made from a deer's tail hair. Perhaps the prominence of the cloth named after the *xingxing*—in Chinese only!—in daily life contributed to the elevation of *sirsing* to the head of the list of simians in the dictionary.

The New Mirror's Sections on Fish and Bugs

New words were added to the sections on river fish, sea fish, and bugs as well. Yet it appears that the editors spent less time on these sections than on the sections on birds and beasts, perhaps because no bilingual pictorial albums had been created in these categories. New river fish included words such as *erin nimaha* 'time fish', obviously coined on the basis of the Chinese *shiyu* 鰣魚 'reeves shad', where the character *shi* contains the phonetic element *shi* 時 'time'. It is possible that the entry was based on *Gujin tushu jicheng*.[66]

Among the bugs, Chinese loans were removed and neologisms to translate rare words in the *Poetry Classic* added.[67] A few down-to-earth creatures such as *usi umiyaha* (Ch. *huichong* 蛔蟲), an intestinal worm, were added here as well.[68] Judging by the paucity and simplicty of changes, the editors appear to have spent comparatively little time on these sections.

Changes in the New *Mirror* on the Level of the Entry

Two changes on the level of the entry are immediately obvious in the new *Mirror*. First, true to the Qianlong emperor's claim in the preface, references to Chinese classics and other works were indeed systematically eliminated from the description of plants and animals. This starts from the very definition of "Grasses" (*orho*), "Trees" (*moo*), and "Flowers" (*ilha*) as general categories. Second, Manchu dialect terms or synonyms that were subsumed under the same lemma in the Kangxi *Mirror* were made into entries in their own right and placed after the original entry in Qianlong's expanded edition. Many of these new entries have no Chinese translation or Manchu definition. Instead, the note "Chinese name same as above" (*Hanming tong shang* 漢名同上)

---

66. Cf. *Yuzhi zengding Qingwen jian*, 1782, vol. 233, 32:39a–b (296, lower panel); Chen Menglei and Jiang, *Qinding Gujin tushu jicheng*, vol. 526, 32b (middle panel).
67. *Yuzhi zengding Qingwen jian*, 1782, vol. 233, 32:63a–b (308, lower panel).
68. *Yuzhi zengding Qingwen jian*, 1782, vol. 233, 32:71a (312, lower panel).

and a corresponding phrase in Manchu let the reader know that the word was a synonym.

An example—one of many—where a Manchu synonym was created but no new Chinese translation supplied is among the sea fish. Here, the word *ica* had in the original *Mirror* carried the Manchu-script Chinese translation *miantiao yu* 'noodle fish'. The 1684 *Shengjing tongzhi* presented this Chinese name as a "vernacular" (*su* 俗) synonym for *yinyu* 銀魚 'silvery fish'.[69] The word was retained in the *Expanded and Emended Mirror* but gained a synonym: *honokta*, which was a word with a history. The Kangxi *Mirror* mentioned *honokta* among the river fish as a synonym for *honggoco*, defined as "white fish mixed in with ice and frozen in winter" (*tuweri juhe suwaliyame gecehe šanggiyan nisiha*).[70] In the revised *Mirror*, the word *honggoco*—which after all referred to a method of preserving fish rather than to a specific kind of fish—was moved to the section "Cooked grains and meats."[71] Its synonyms were removed from that entry, and *honokta* was remade into an alternative name for "noodle fish," but without a new Chinese name.[72]

69. The "silvery fish" was described under the headword *kuaican yu* 鱠殘魚: *Shengjing tongzhi*, 21:22b. See also Li Shizhen, *Bencao gangmu*, vol. 774, 44:24b (292, lower panel).

70. *Han-i araha manju gisun*, 1708, 20:*birai mimaha-i hacin*:34a. Li Yanji identified the fish named here as *yinyu* 'silvery fish' (*Qingwen huishu*, 3:8b [61, lower panel]), which in the original *Mirror* corresponded to Ma. *menggun nisiha* 'silvery fish' (*Han-i araha manju gisun*, 1708, 20:*birai mimaha-i hacin*:33b). The revised *Mirror*, however, called it *huang guyu* 黃䱛魚 'yellow silver xenocypris' (*Yuzhi zengding Qingwen jian*, 1782, vol. 233, 32:38b [296, upper panel]; Ihing, *Qingwen buhui*, 3:6a). Note that the spelling of the Manchu word reflects the replacement of *šanggiyan* 'white' with *šayan* idem. The word *huanggu yu* was listed in the 1684 *Shengjing tongzhi*, where it was said that it "looked like white fish but without a raised head and tail" 似白魚而頭尾不昂. The name referred not to its color, but to the yellow oil that was extracted and used for lamp fuel (*Shengjing tongzhi*, 21:22b).

71. *Yuzhi zengding Qingwen jian*, 1782, vol. 233, 27:10b–11a (125, both panels).

72. In the Kangxi *Mirror*, the kind of "white fish" that was mixed with ice in winter had a second synonym: *mungku*. This word disappeared altogether in the *Expanded and Emended Mirror*. Ihing wrote that *mungku* was changed to *mūnggu* 'bird's nest'. This word was not in the original *Mirror*. It is best treated as an independent new addition in the *Expanded and Emended Mirror*, rather than as a transformation in both form and meaning of *mungku*. See Ihing, *Qingwen buhui*, 3:6a, 6:28b.

In some cases, however, words that had been synonyms in the original *Mirror* were made into separate entries with distinct definitions. In these cases, the synonymity of the words was no longer obvious, with the result that the number of creatures described in the dictionary appeared to increase. For example, *ferehe singgeri* 'bat' (lit. aged [?] mouse) had the synonym *ashangga singgeri* 'winged mouse' in the original *Mirror*. In Qianlong's expansion, *ferehe singgeri* was translated as *bianfu* 蝙蝠 'bat'. As for *ashangga singgeri*, it was made into a separate entry, still defined in Manchu as a synonym of *ferehe singgeri*, but given a different Chinese translation, *fei lei* 飛鸓 'flying squirrel'.[73] Ihing followed the Manchu and said that this latter word also meant bat (*bianfu*). In his understanding, the two Manchu synonyms had simply been given separate Chinese names.[74] Yet a reader unaware of the history of the language would get the impression that the words had no relationship and referred to different creatures.

The case of *anggir niyehe* and *lama niyehe*, two ducks, is a little more complicated. The original *Mirror* only lemmatized the former, and gave it the definition "Among the ducks, this is the biggest. Yellow color. It is called 'lama duck'" (*niyehe-i dolo ere amba, boco suwayan ofi, lama niyehe sembi*). In the Manchu-Chinese *Mirror*, *lama niyehe* was lemmatized, following the *Catalog of Birds*, placed before *anggir niyehe*, and given the Chinese translation *tu yuanyang* 土鴛鴦 'rustic mandarin duck' and the Manchu definition "similar to the mandarin duck, yellowish color" (*ijifun niyehe de dursuki boco suwayakan*).[75] The entry that followed was *anggir niyehe*, now translated as *huang ya* 黃鴨 'yellow duck' but retaining the definition from 1708 sans the addendum regarding *lama niyehe*.

Curiously, the mandarin duck was listed as *ijifun niyehe* in both the original and expanded *Mirrors*. The original said that "the Chinese

---

73. *Han-i araha manju gisun*, 1708, 19:*gurgu-i hacin*:48b; *Yuzhi zengding Qingwen jian*, 1782, vol. 233, 31:21b (253, lower panel). For a Japanese translation, see Imanishi, Tamura, and Satō, *Gotai Shinbunkan yakukai*, vol. 1, 916 (item 16073).

74. Ihing, *Qingwen buhui*, 1:15a, 8:25a. For another example, see Ihing, *Qingwen buhui*, 3:40a, s.v. *bunjiha*. Moreover, there is some uncertainty as to the reference of the Chinese word *fei shu* 飛鼠 'flying rat' (or 'bat'?) and related words in *Shanhai jing* (Mathieu, *Étude sur la mythologie*, vol. 2, vol. 1, 174n7).

75. Chuang, *Niaopu Manwen*, vol. 4, 195.

call it *yuanyang* 'Mandarin duck'" (*nikan yuwan yang sembi*) and the expansion added the Chinese characters *yuanyang* 鴛鴦.[76] Ihing noted that "in earlier usage it was interchangeable with 'lama duck' [*lama ya*; Ihing here gives a Chinese version of the name not seen in the *Mirror*], but now they have been separated" 舊話與喇嘛鴨通用，今分定.[77] Thus the kinds of duck described increased by a certain sleight of hand.

The entries that in the original *Mirror* had only Chinese translations (in Manchu script) in lieu of definitions were in some cases given Manchu definitions in Qianlong's revised edition, but not in all. For example, *tama* 'sole', a saltwater fish, received the definition "shaped similarly to the sole of a foot" (*banin sabu fatan-i gese banjihabi*), a paraphrase of its Chinese name *xiedi yu* 'shoe-sole fish'.[78] Yet not all of the entries were complemented with new information. Thus for *sotki* and *uyu*, we still only read that they "live in the sea."[79] They seemed destined to remain as hapax legomena of Qing court lexicography.

The process by which these entries received new information does not appear to have been entirely straightforward. In Kangxi's *Mirror*, *koojiha* was translated into Chinese as *latun* (Ma. *la tun*) and not further defined. The word there followed immediately after *kosha* 'globefish', which had a definition, but which was similarly also glossed in Chinese as *hetun* (Ma. *ho tun*). In the revised *Mirror*, *kosha* was given the Chinese name *hetun* 河豚, lit. river piglet, where the character *tun* 'piglet' was an alternative form of *tun* 魨 'globefish'. The word *hetun* was commonly written using the character for "piglet," so the editors were not wrong in using it here.[80] Li Yanji, however, had given the form *hetun* 河魨 'river globefish'.[81] Indeed, it seems as if whoever wrote up the entry in the new *Mirror* might have actually thought it referred to a pig of some sort; the revised *Mirror* reads that the *kosha* is a "yellow

---

76. *Han-i araha manju gisun*, 1708, 19:*gasha-i hacin*:30b, 31a; *Yuzhi zengding Qingwen jian*, 1782, vol. 223, 30:22b (223, upper panel). The order of the entries changed between the two editions: *ijifun niyehe* was moved to the top of the section.

77. Ihing, *Qingwen buhui*, 1:14a.

78. *Yuzhi zengding Qingwen jian*, 1782, vol. 233, 32:50b (302, upper panel).

79. *Yuzhi zengding Qingwen jian*, 1782, vol. 233, 32:46b–47a (300).

80. See, e.g., Zhang Zilie, *Zhengzi tong*, 1995, vol. 235, *hai ji zhong* 亥集中, *yu bu* 魚部:4a (770, lower panel).

81. Li Yanji, *Qingwen huishu*, 2:57a (57, upper panel).

pig" (*suwayan judura*) whereas the original 1708 *Mirror* said it had "yellow stripes" (*suwayan juduran*).⁸²

The following word, *koojiha*, which originally had no definition, was given the Chinese-character translation *latun* 臘豚 'cured piglet' (Li Yanji had translated it as *latun* 臘魨 'cured globefish') and a new definition: "Globefish eaten after the winter solstice are called *koojiha*" (*tuweri ten-i amala jetere kosha be, koojiha sembi.*).⁸³ Thus *la* 'cured' is here to be read as a reference to *layue* 臘月, the twelfth month in the calendar. We have not seen the name *latun* anywhere else, so are unable to determine whether it is a product of the editors' fancy, or an example of vernacular knowledge that, as we showed, evidently made its way into the new entries on plants.⁸⁴ Furthermore, some words that had been clearly defined in the original *Mirror* received revised definitions in Qianlong's expanded edition.⁸⁵

---

82. Both the digitized blockprint copy held at the Berlin State Library (they hold several copies; only one of them was digitized in full at the time of writing) and the Wenyuan Ge *Siku quanshu* manuscript copy have this change. See *Yuzhi zengding Qingwen jian*, 1782, vol. 233:32, *haiyu lei* 海魚類:38b; 32:49a (301, lower panel). For the original, see *Han-i araha manju gisun*, 1708, 20:*mederi nimaha-i hacin*:37b.

83. Li Yanji, *Qingwen huishu*, 2:57a (57, upper panel); *Yuzhi zengding Qingwen jian*, 1782, vol. 233, 32:49a (301, lower panel).

84. Another example is found in the definition of *huhucu*, the bellflower that had a definition consisting of only two Chinese translations in the 1708 *Mirror*, one of which might have been a word already borrowed into Manchu usage. In the revised *Mirror*, such a definition was unacceptable. The bellflower thus received the new definition "like the great willow herb, with broad leaves and a lot of juice. It is blanched and then eaten" (*honggocon sogi-i adali bime, adbaha onco, šugi labdu hethefi jembi*). See *Yuzhi zengding Qingwen jian*, 1782, vol. 233, 27:29a (134, lower panel); Fèvre and Métailié, *Dictionnaire Ricci des plantes*, 281, s.v. *liǔ yè cài*.

85. One example is the word pair *buka*—which the 1708 *Mirror* had presented as a word for male sheep and goats—and *kūca*, a word for intact male sheep. These words were discussed at court in 1760 and then used in the description of a sheep in the *Catalog of Beasts*. In the *Expanded and Emended Mirror*, the definitions of the words received some alterations. Now, *buka* was translated into Chinese as *gong yang* 公羊 'male sheep' and accordingly defined in Manchu (*haha honin* 'male sheep'). The word was no longer said to refer to both male sheep and male goats. This meaning was taken over by *kūca*, now translated as *gong shanyang* 公山羊 'male goat' and similarly defined in Manchu (*haha niman* 'male goat'). There was no longer any word for intact male individuals of either type of animal. See *Yuzhi zengding Qingwen jian*, 1782, vol. 233, 31:29b (257, lower panel); Ihing, *Qingwen buhui*, 3:38a.

So far, we have surveyed the newly added entries in the main chapters of the 1772 bilingual *Mirror* and changes made to existing entries. Many of the changes were clearly made on the basis of Chinese encyclopedic works, pictorial albums, and the discussions on neologisms previously carried out at court. Contrary to what Qianlong claimed, the main chapters of his revised Manchu dictionary did not only draw from "everyday language" (*an-i gisun*), but in fact also owed its richness to serious engagement with documentations of flora and fauna in ancient and modern Chinese sources. In the following section, we move on to examine the supplement of the 1772 bilingual *Mirror*, which in Qianlong's design should serve as a kind of reservoir for plants and animals that did not appear in "everyday language," but were placed solely as reference points for future research.

The Supplement to the Manchu-Chinese *Mirror*

The natural historical sections in the supplement to the 1772 *Mirror* are centered on the notion of the "marvelous" or "strange" (Ch. *yi* 異 or *yiyang* 異樣; Ma. *encu-i hacin*). They begin with four subsections on ninety-seven kinds of "marvelous fruits" (*yiyang guopin* 異樣果品), followed by fifty-six kinds of trees (only eighteen of which were marked as being "marvelous," more on which later) and 114 kinds of "marvelous flowers." Like the main chapters, the supplement includes Manchu definitions for each word.

Oliver Corff, in his study of the *Pentaglot* that follows the same word list as the 1772 bilingual *Mirror*, noted that "virtually the complete list of flower names from *Guangqun fangpu*, fascicle 53, 'huapu' 32 . . . were copied" into the supplement's section on "strange flowers" (*Hier wurde die nahezu vollständige Liste von Blumennamen aus dem Guangqun Fangpu, Fasz. 53, Huapu 32 . . . übernommen*).[86] It is perhaps not surprising at this point that the "marvelous" plants were all Manchu neologisms coined to pair with the entries added to Wang Xiangjun's *Qunfang pu* by Wang Hao and other Qing court scholars in 1708. The same also applies to the sections on marvelous fruits and

---

86. Corff et al., *Auf kaiserlichen Befehl*, vol. 1, xxxvi.

trees, which largely derived from the *Guang Qunfang pu* but with the order of entries adjusted and the gloss paraphrased from the Chinese. An example of this kind involves a giant fruit described in *Shenyi jing*, an ancient Chinese source from which Kangxi drew connections with the mammoth:

> *endurin soro* 'divine jujube'. This fruit is yellow-colored, its flower red. It blooms once every three hundred years, and bears fruit once every nine hundred years. If one cuts the fruit open with an iron knife, it tastes sour; if one cuts it open with a stalk of reed, its taste becomes pungent. People who eat it can become immortals.
>
> *endurin soro, ere tubihe boco suwayan ilha fulgiyan, ilan tanggū aniya emgeri ilha ilambi, uyun tanggū aniya emgeri tubihe banjimbi, ere tubihe be selei huwesi-i hūwalaci amtan jušuhun ulhū-i jisuci amtan hargikan ombi, jeke urse endurin oci ombi.*[87]

The Chinese name for this Manchu "divine jujube," *ruheshu shi* 如何樹實 (lit. fruit of Whatever Tree), reveals the source of this passage from an entry recorded in *Guang Qunfang pu*, which in turn simply cited *Shenyi jing*.[88] A closer comparison indicates that the Manchu passage omitted many details included in the Chinese passage, including the height of the tree, texture of its trunk, and size of the leaves. It also omits the crucial details of the fruit being "with a pit, and shaped like a jujube" 有核形如棗子, with a size of five feet in length and the same in circumference—quite a monstrous size. All in all, the editorial approach toward creating this Manchu word was not all that different from some lemmata in the main section that were reverse-engineered from Chinese sources.

Corff furthermore made the interesting remark that even for those lemmata carried over from *Guang Qunfang pu*, the rendering of the Chinese names is nevertheless not always identical between the two works. Differences between character variants aside, the lemmata in

---

87. *Yuzhi zengding Qingwen jian*, 1782, vol. 233, supplement, 3:9b–10a (382, bottom panel; 383, top panel).

88. Wang Hao, *Yuding Peiwen*, vol. 846, 67:27b–28a (786, bottom panel; 787, top panel).

the *Mirror* more frequently take the form of noun phrases headed by *hua* 花 'flower' than is the case in Kangxi's great flower catalog. Although the irregularities suggested to Corff that the list of flowers might have been based on an earlier treatise, we have found no evidence so far that an intermediate source existed besides *Guang Qunfang pu* itself.[89] Instead, the effort to adjust the *Chinese* terminology in the bilingual *Mirror* is significant in and of itself. By making the Manchu nomenclature uniform by giving all the noun phrases "flower" (*ilha*) as their head, and likewise adding *hua* 'flower' whenever the earlier Chinese name was too vague or figurative, the editors of the bilingual *Mirror* were trying to impose order as they saw it existing in the real world unto their dictionary. Again, the case of the monstrous divine jujube illustrates the silent labor that went into this reordering of the world by way of moving lemmata around. The *endurin soro*, or divine jujube, that as we have seen came from an ancient Chinese source, was the first entry followed by five additional "jujubes" (*soro*) gathered from different places in *Guang Qunfang pu* and given Manchu names.[90] Thus it is important for us to see here that the bilingual *Mirror* not only leaned on Kangxi-era encyclopedic works in Chinese as textual sources, but also sought to further a kind of natural historical agenda that used Manchu neologisms toward greater consistency and clarity in representing the order of things.

One last observation about plants in the supplement of the 1772 *Mirror* concerns precisely the frontier flora "beyond the border" (Ma. *jasei tulergi*), or beyond the Great Wall, which we discussed in chapter 5. The "Golden lotus flower," which acquired iconic status following Kangxi's repeated visit of Mount Wutai, appears in the supplement as *aisin su ilha*. Its Manchu description is largely derived from the corresponding entry in *Guang Qunfang pu*, which as discussed was based

---

89. Corff et al., *Auf kaiserlichen Befehl*, vol. 1, xxxvi.

90. *Yuzhi zengding Qingwen jian*, 1782, vol. 233, supplement, 3:10a–b (383, top panel). They are *sorotu* (Ch. *tianzao* 天棗), *tumsoro* (< *tumen* + *soro* [Ch. *wan sui zao* 萬歲棗]), *mimsoro* (< *minggan* + *soro* [Ch, *qian sui zao* 千歲棗]), *yesoro* (Ch. *nuozao* 糯棗), and *bosoro* (Ch. *bosi zao* 波斯棗 'Persian jujube'). They came from Wang Hao, *Yuding Peiwen*, vol. 846, 58:18a–19a (622, top and bottom panels).

on court scholars' writings from the imperial tours.[91] It is worth asking, then, why the 1772 *Mirror* put this flower, which had received such glorifying depictions at court, in the supplement along with fantastical plants from older Chinese sources rather than in the main section of the dictionary. Similarly, the 1772 *Mirror* seems to have disregarded the Kangxi emperor's speculation in *Kangxi ji xia gewu bian* regarding the possible common origin of the Manchu word *umpu* 'haw' and the Chinese term *wenpu*. On the contrary, it created a clumsy Manchu neologism, *wemburi*, to match the Chinese term in the supplement.[92] In both instances, we see that the dictionary's editors were clearly beholden to the specific demands by the Qianlong emperor; that is, to separate "common usage" from bookish references. Unlike the courtiers who accompanied Kangxi for tours and then edited *Guang Qunfang pu*, the Qianlong-period editors did not pay much attention to the existence, significance, and identification of frontier flora, but merely used *Guang Qunfang pu* as a textual source to draw on and reorganize what was already identified and codified in these court compilations. A similar approach can be detected in the entries on strange animals in the supplement in the 1772 *Mirror*.

Animals covered in the supplement included large and small birds, beasts, but no fish or bugs. The sources used for these sections were the pictorial *Catalog of Birds*, *Catalog of Beasts*, and the encyclopedia *Gujin tushu jicheng*. Particularly numerous among the birds, but occurring among the beasts too, are different names (Ma. *encu gebu*) for creatures already listed in the main body of the dictionary. These alternative names reproduce the lexical variety of Chinese, which had accrued in a written tradition stretching back millennia and covering vast stretches of space. Words that some Chinese source at some point mentioned as regional terms for a certain bird are thus lemmatized in the supplement. For example, when *Gujin tushu jicheng* says of the stork that "west of the pass, they call it *guanque* 'foremost sparrow'" 自關而西謂之「冠雀」, the supplement says "the people of the west call

---

91. *Yuzhi zengding Qingwen jian*, 1782, vol. 233, supplement, 3:35a–b (395, bottom panel).

92. *Yuzhi zengding Qingwen jian*, 1782, vol. 233, supplement, 3:13a (384, bottom panel).

the stork *durujun* [< *durun* 'model' + *weijun* 'stork']" (*wargi ba-i niyalma weijun be durujun sembi*).⁹³

Similarly, legendary creatures mentioned in Chinese texts are listed in the supplement. The many strange beasts described in *Shanhai jing* are there, as in the *Catalog of Beasts*, as are the "strange beasts" (*yi shou* 異獸) listed in *Gujin tushu jicheng* but not included in the *Catalog*.⁹⁴ Thus the reader of the dictionary's supplement encounters creatures such as the *fulgiyentu* (< *fulgiyembi* 'to blow with the mouth'; Ch. *xiushi* 嗅石 'stone smeller'), which has a "shape somewhat similar to a unicorn, eats no raw herbs, and drinks no muddy water." If this creature "smells a stone, it will know if it contains gold or jade. If the stone is split, the gold and jade will appear" (*banin sabitun de adalikan, banjire orho be jeterakū, duranggi muke be omirakū, wehe de wangkiyaci, aisin gu bisire be sambi, fulgiyeci, wehe fakcafi aisin gu tucinjimbi*).⁹⁵

The supplement furthermore lists foreign creatures that *Gujin tushu jicheng* had lifted from Verbiest, who in turn relied on European Renaissance natural histories. The *salmandara* 'salamander' (Ch. *salamandala* 撒梓漫大梓) from "Germany" (*zen ma ni ya*) is there, as are the *hiyena* 'hyena' (Ch. *yiyena* 意夜納) from "Libya" (*lii wei ya*), the *g'amuliyang* 'chameleon' (Ch. *jiamoliang* 加默良 [< *kamoliang*]) from Judea (*zu de ya*), all as in the *Catalog of Beasts*.⁹⁶ The *onasio* 'giraffe' (Ch. *enaxiyue* 惡那西約 < Lat. *orasius*) from the *Catalog* is now called *onasu*, however; perhaps the first Manchu rendering was too close to the Chinese—transcribed *o na si yo* in the *Expanded and Emended Mirror*—for Qianlong's scholars to be comfortable with it (see fig. 8.2).⁹⁷

---

93. Chen Menglei and Jiang, *Qinding Gujin tushu jicheng*, vol. 517, 17b, top panel; Hauer, *Handwörterbuch der Mandschusprache*, 1952, vol. 1, 224; *Yuzhi zengding Qingwen jian*, 1782, vol. 233, supplement, 4:7a (408, lower panel).

94. See *Yuzhi zengding Qingwen jian*, 1782, vol. 233, supplement, 4:72a (441, top panel) onward; Chen Menglei and Jiang, *Qinding Gujin tushu jicheng*, vol. 525, 18a (middle panel) onward.

95. *Yuzhi zengding Qingwen jian*, 1782, vol. 233, supplement, 4:72a–b (441, top panel).

96. *Yuzhi zengding Qingwen jian*, 1782, vol. 233, supplement, 4:69b (439, lower panel), 4:71a (440, lower panel); Chuang, *Shoupu Manwen*, vol. 2, 619–27.

97. Walravens, "Konrad Gessner," 97; Chuang, *Shoupu Manwen*, vol. 2, 633; *Yuzhi zengding Qingwen jian*, 1782, vol. 233, supplement, 4:71b (440, lower panel). As noted, that was the case with the *nukeyake* 'combed bird.'

FIG. 8.2. A page from the *Expanded and Emended Mirror of Manchu Language* (*Zengding Qingwen jian*)'s supplement discussing giraffe (Beijing: Wuying Dian, 1771, supplement, 4:53b). Palace edition (on microfilm), New York Public Library.

The supplement finally includes the simians from the *Catalog of Beasts* that were left out of the main body of the dictionary, as discussed. Yet the *Mirror* inflates their number further. In the *Catalog of Beasts*, *falintu monio* 'binding macaque' (< *falimbi* 'to bind, to tie a knot') was a synonym for "macaque."[98] In the *Expanded and Emended Mirror*, however, it was defined as "a macaque expert at climbing" (*moo de tafara mangga monio be, falintu monio sembi*), not as "another name" (*emu gebu*) for the same monkey.[99]

98. Hauer, *Handwörterbuch der Mandschusprache*, 1952, vol. 2, 273; Chuang, Shoupu *Manwen*, vol. 1, 271.

99. *Yuzhi zengding Qingwen jian*, 1782, vol. 233, supplement, 4:49b (429, lower panel).

Similarly, *hoilantu* was used in the *Catalog of Beasts* in reference to a kind of *jue* 玃, a large simian mentioned in ancient texts that was rendered as *elintu* (< *elinture* 'watching from afar').[100] The legend that accompanied the image in the *Catalog* quoted *Erya*, which used the term *juefu* 玃父, which might be treated as a synonym for *jue* or taken to mean 'male *jue*'. Either way, it was treated in the *Catalog* as a *jue* or a subset of *jue*, and both were described as good at looking or seeing (Ch. *gu* 顧; Ma. *tuwambi*).[101] The Manchu version of the legend translated *juefu* as *hoilantu* (< *hoilacambi* 'look to both sides').[102] In the new *Mirror*, it was described as "bigger than a macaque, has jet black fur, is expert at looking from afar" (*monio ci amba, funiyehe sahahūkan, elintume tuwara mangga*). The following entry was *hoilantu*, defined as a "type of macaque, expert at looking to both sides" (*monio-i duwali, hoilacame tuwara mangga*).[103] Thus two words that had been synonyms in Chinese canonical sources and in the Manchu text to the *Catalog of Beasts* were made into two kinds of simians, one characterized by its long-distance vision, one by its peripheral vision. The new *Mirror* was splitting terms and creating a level of biological diversity in Manchu that rivaled that of Chinese.

Conclusion

Did the chapters on plants and animals in the *Expanded and Emended Mirror* follow and accomplish the Qianlong emperor's lofty goals set up in his preface? Our answer, at the end of this meandering tour through both the main and supplemental chapters of this bilingual

---

100. Hauer, *Handwörterbuch der Mandschusprache*, 1952, vol. 1, 244. There is no reason to infer that -*tu*, a common suffix for neologisms (see following) would be derived from *mahūntu*, another simian neologism, *pace* Hauer.

101. Chuang, Shoupu Manwen, vol. 1, 283.

102. For the derivation of the word, see Hauer, *Handwörterbuch der Mandschusprache*, 1952, vol. 2, 449. For the identification, based on the *Catalog of Beasts* and thus of a Qing-era understanding of this ancient name, see Yu and Zhang, *Qinggong shoupu*, 409, s.v. *jue* 玃.

103. *Yuzhi zengding Qingwen jian*, 1782, vol. 233, supplement, 4:50a–b (430, upper panel).

dictionary, is yes and no. On the one hand, it is important to reckon with the significant additions, alterations, and distinctions in the bilingual *Mirror* that made it a very different kind of book from the 1708 *Mirror*. On the other hand, however, many of Qianlong's ambitious demands could not be met without numerous compromises, byways, and inconsistencies. Unlike its predecessor, the bilingual *Mirror* tried to showcase the richness of the Manchu lexicon as on par with that of Chinese, and preferably going beyond it, coining a large number of neologisms in the process. Likewise departing from previous practice, it purged explicit citations of Chinese sources, only to reintroduce covert translations of Chinese-language court texts back into explanations of Manchu neologisms. In another departure from the 1708 *Mirror*, the bilingual Manchu-Chinese *Mirror* made a clear distinction between "common usage" and bookish references, but could not always maintain and police the ontological distinction implicit in the division of common language and "anomalous/marvelous" (*yi*) kinds of plants and animals. Even if these boundaries proved shaky, it is nevertheless important to register their importance to the changing political and cultural agenda set forth at the Qianlong court.

In the conclusion that follows this chapter, we build on the close reading of entries provided here to show that the 1772 Manchu *Mirror* itself in fact inspired more than one plausible reading as a standard reference for the Manchu language. The encyclopedic dictionary posed a problem even for contemporary bibliographers who were also working at the Qianlong court. At the same time, however, readers in and beyond Qing China continued to refer to it for natural historical investigations.

CONCLUSION

Reception of the Manchu Mirrors in Qing China and Beyond

My dear Wolfgang!
Thanks for your message from the 12th. Your essay will be a nice and useful contribution! Note 8 is in order. Somehow an error has slipped into note 14: the word is *yolo* (not *iolo*) and it has two acceptations: 1) a big sea eagle 2) a doe. That means you can delete the note. The [*Wuti Qingwen jian*] 五體清文鑑 gives "big eagle" for *yolo* and "eagle" for *damin* (*ô-washi* alternatively *washi* in jap. translation). It is always a bit tricky with the animal names. I'm not able to investigate the basis of H[auer]'s [translation of the word as] bearded vulture at the moment. It's best if you simply omit note 14.

Mein lieber Wolfgang! | Habe Dank für Deine Zeilen vom 12. Deine Arbeit wird ein schöner und nützlicher Beitrag werden! Anm. 8 ist in Ordnung. Bei Anm. 14 ist irgendwie ein Fehler unterlaufen: das Wort heisst yolo (nicht iolo) und hat zwei Bedeutungen: 1) ein grosser Seeadler 2) Hirschkuh. Damit entfällt die Anm. Der [Wuti qingwenjian] 五體清文鑑 *gibt für yolo «Grosser Adler» und für damin «Adler» (in jap. Übersetzung ô-washi bzw. washi). Mit den Tiernamen ist es so eine Sache. Worauf H[auer]'s Lämmergeier beruht, kann ich jetzt nicht nachprüfen. Am besten ist, Du lässt Anm. 14 einfach fort.*[1]

---

1. Walter Fuchs to Wolfgang Seuberlich, February 14, 1975, in Walravens and Gimm, *Wei jiao zi ai*, 200.

These words are excerpted from a letter that Walter Fuchs (1902–1979) sent to Wolfgang Seuberlich (1906–1985) in 1975. Fuchs and Seuberlich were both Sinologists who had spent much of their lives in China. Fuchs was one of the pioneering Manjurists of the Republican period, living first in Shenyang and then in Beijing. By the time he wrote this letter to Seuberlich, Fuchs had retired from the University of Cologne.[2]

Seuberlich had received a Russian secondary and tertiary education in Harbin and then worked for a time in the Manchukuo civil service before writing a PhD dissertation in Berlin on the administrative history of Manchuria. When he received this letter from Fuchs, Seuberlich had retired from a career at the Staatsbibliothek in West Berlin.[3]

The letter—Fuchs's commentary on an article draft by Seuberlich—is suggestive of the problematics of the Manchu vocabulary relating to the natural world and, at the same time, of the role of Qing court-sponsored lexicography for twentieth-century Manjuristics. The core of Seuberlich's footnote 14 is apparently the semantics of the Manchu word *yolo*, which Fuchs has searched for in the multilingual capstone of Qing court lexicography—the Qing *Pentaglot*—and in Fuchs's former teacher Erich Hauer's *Handwörterbuch der Mandschusprache* (published posthumously in the 1950s).[4] The *Pentaglot* was much used by twentieth-century Manjurists after it appeared in a searchable edition in 1966.[5]

Fuchs cites the definition "big eagle" from the *Pentaglot*, which contrasts markedly with the second acceptation "doe." He was certainly right that dealing with the Manchu names of animals is tricky; he would have made his point even more strongly had he cited a second instance in which *yolo* occurs in the *Pentaglot*, where it means "Tibetan dog" (Ch. *Zanggou* 藏狗, referring to the Tibetan mastiff).[6] Hauer,

---

2. Walravens and Gimm, *Wei jiao zi ai*, 15–19.

3. Krempien, "Wolfgang Seuberlich zum 75."

4. Hauer, *Handwörterbuch der Mandschusprache*, 1952; *Handwörterbuch der Mandschusprache*, 2007.

5. Imanishi, Tamura, and Satō, *Gotai Shinbunkan yakukai*.

6. Corff et al., *Auf kaiserlichen Befehl*, vol. 2, 942 (item 4303.3). Remember that Parrenin had used the word in reference to a dog with a "big and long muzzle, a similar-looking tail, big ears, and drooping cheeks."

Fuchs's other source, was similarly aware of the difficulties involving the animals as well as plants. Indeed, zoological and botanical names were among the terms for which Hauer stated that a translation into Chinese would have to be included in his Manchu-German dictionary.[7] For Hauer, the identification in Manchu was often uncertain or imprecise, and only a Chinese translation, lifted from Qing lexica, would anchor the Manchu word somewhat for the reader. In this regard, Hauer was working with the same constraints as the lexicographers at the Kangxi court, as we saw in chapter 2. Hauer, moreover, was aware of the presence of "new words artificially forged by imperial committees" (*von kaiserlichen Kommissionen künstlich gebildeten neuen Wörtern*) in works like the *Pentaglot*.[8] Nevertheless, he remained—like many other Manchu lexicographers since—reliant largely on imperial dictionaries for both words and their meanings.[9]

Fuchs, for his part, apparently thought of Qing court lexicography as simple reference works. In 1952, when Hauer's dictionary was not yet in print, Fuchs had written to Denis Sinor (1916–2011) that Haneda Tōru's 羽田亨 (1882–1955) *Man-Wa jiten* 滿和辭典 (Manchu-Japanese dictionary) from 1937 was not as good as Zhi-kuan's 志寬 (fl. 1867) and Pei-kuan's 培寬 (fl. 1873) *Qingwen zonghui* 清文總彙 (Comprehensive collection of Qing writing) from 1897 because the latter included the supplement to Qianlong's Manchu-Chinese *Mirror*—full of new coinages, many of which known only from the dictionary—whereas the former did not.[10] The examples of Fuchs and Hauer cited here are indicative of the relatively uncritical approach taken toward Qing court lexicography within twentieth-century Manjuristics, notwithstanding the particular difficulties that the treatment of flora and fauna posed in these works.

7. Hauer, "Ein Thesaurus der Mandschusprache," 640.
8. Hauer, "Ein Thesaurus der Mandschusprache," 630.
9. Interestingly, Hartmut Walravens criticized Hauer's dictionary precisely on the grounds that it did not include the "terms from natural history" found in the *Pentaglot*. See Rudolph and Walravens, "Comprehensive Bibliography of Manchu Studies," 296 (item 439).
10. Dates of Zhi-kuan and Pei-kuan from Zhi-kuan and Pei-kuan, *Duiyin jizi*, *xu*:3b–4a. Walter Fuchs to Denis Sinor, June 21, 1952, in Walravens and Gimm, *Wei jiao zi ai*, 46.

In the half-century since Fuchs and Seuberlich discussed the words for eagles and vultures in the *Pentaglot*, the field of Qing history has moved in several directions, including toward intellectual history, the history of science, and environmental history. The study of the Manchus involved a turn to the Qing court as a focus for social, cultural, and intellectual history, enabled by the opening of the imperial archives and collections of objects and paintings. This book has thus taken a different approach to the Manchu *Mirrors*, bringing the plurilinguality of the empire to bear on the history of knowledge in early modern China.

In the preceding chapters, we have presented two main arguments. The first is that the original and expanded Manchu *Mirrors*—completed in the 1700s and 1770s respectively—reveal how the political significance of the Manchu language in Qing China goes far beyond the expression and maintenance of Manchuness as a constructed ethnic identity and the institutional privileges attached to this identity. Instead, the Manchu *Mirrors* also reflected and contributed to the reordering of both Chinese and Manchu textual studies, especially with regards to the question of how to account for the growing knowledge about flora and fauna living within and beyond the Qing empire. This process was first epitomized in the series of encyclopedic texts commissioned at the late Kangxi court in the 1700s—the monolingual Manchu *Mirror* among them—and culminating in the compilation of *Gujin tushu jicheng* in 1725. In this period, the Manchu language served as a capacious receptacle into which old and new sources were blended and rebalanced, generating a new ideal of *gewu* (the "investigation of things") articulated through an open-ended lexicographical encyclopedism, as we showed in parts 1 and 2.

The second argument is that the role of Manchu lexicography did not stop with integrating a vital stream of information about Inner Asian nature into the Chinese encyclopedic tradition. In part 3, we described the substantial, if hidden, effort behind the coining of more and more Manchu neologisms during the Qianlong reign between the late 1740s and the 1770s. In some cases, such as the pictorial catalogs of birds and beasts, the Manchu lexicon contributed to the entry of exotic plants and animals into the larger purview of the Qing imperial order; in other cases, Manchu neologisms were coined to achieve a sort of artificial parity with the Chinese lexicon, whose great variety had accumulated since ancient times. The result of this effort gave us the capacious and heterogeneous collection of words found in the bilingual *Expanded and Emended Mirror of the Manchu Language*, completed in 1772. In this process, the Manchu lexicon became radically stretched and transformed, so much so, in fact, that the flora and fauna chapters in the new *Mirror* came to little resemble those in its predecessor from only several decades earlier. Despite the seemingly clean façade now erected to separate the Manchu lexicon from words in Chinese and other languages, in reality, as we showed, a hard boundary between Manchu versus Chinese natural knowledge did not exist in the actual work of creating new lexicographical equivalences.

Our main arguments having been presented, the question remains as to why the Manchu *Mirrors* acquired the narrow image of being books only about language study and nothing else. In the rest of this conclusion, we briefly consider the bibliographical arrangement of the *Mirrors*, as well as other related court texts in Chinese, in the *Siku quanshu* project that began in the late 1770s and concluded in the 1780s. We then turn from the primarily Chinese *Siku* project to two examples from the *Mirror*'s reception history, taken from early nineteenth-century France and Japan, to point to other possibilities of reading and using the Manchu *Mirrors* as sources for natural-historical inquiry.

Siku quanshu and the Denouement of Court Scholarship

In 1773, only a year after the completion of the bilingual *Mirror*, another office (*guan*) opened at the behest of the Qianlong emperor. Unlike other offices that had been assembled at court, the *Siku quanshu* office was put in charge not to compile one new book, but to review, evaluate, and reproduce (or destroy) tens of thousands that were presented to the throne.[11]

---

11. Huang, Siku quanshu *zuanxiu yanjiu*, 16–39 (on the collection of books), 243–60 (on the question of "empty-case books" [*kong han shuji* 空函书籍] commissioned by Qianlong but not completed until later).

Obscured in the conventional narrative of *Siku quanshu* as a confrontation between the (Manchu) Qing state and Chinese scholars is the fact that the *Siku* office also duly reviewed, copied, and incorporated many books that had been commissioned and produced at the Qing court. The 1772 *Expanded and Emended Mirror of Manchu Language* was one of them, along with the trilingual *Mirror* completed in 1780. The *Mirrors* were shelved under the category of "Classics" (*jing*)—"Philology" (*xiaoxue* 小學)—"Character books" (*zishu* 字書).[12] In the *Summaries* (*tiyao* 提要) compiled by the *Siku* office, the 1772 *Mirror* received lavish praise with virtually no criticism, a privilege reserved for court compilations. Although this fawning gesture is in itself not surprising, the *Siku* editors treaded carefully in order to explain the work's significance vis-à-vis its precedents. Following the recently sanctified historical narrative about the Manchus' origin, the editors offered a brief history of script and writing in the Jurchen Jin dynasty (which ruled North China and parts of Inner Asia in the twelfth and early thirteenth centuries), reiterating the glorious moments of the creation and standardization of the Manchu script. Interestingly for our purposes here, the *Summaries* noted the considerable body of "new Manchu" words (*xin Qing yu* 新清語) that were added in the 1772 *Mirror*. To contextualize this change, the editors explained that Chinese characters, too, tended to proliferate over time, from around nine thousand in *Shuowen jiezi* to more than twenty-six thousand in *Guangyun* 廣韻 (The expanded rhymes, 1008 CE), writing that

---

12. Ji, *Qinding Siku quanshu zongmu*, vol. 1, 41:54a–b (858, upper panel). This discussion is not included in the shorter summary (*tiyao* 提要) found in the *Siku* manuscript copy of the *Expanded and Emended Mirror*. See *Yuzhi zengding Qingwen jian*, vol. 232, *tiyao*:1a–2b (1, lower panel–2, upper panel). On the differences between summaries preceding each book and the later revised *General Catalog* (*zongmu* 總目), see Huang, *Siku quanshu zuanxiu yanjiu*, 336–51. The summary pointed out that the *Mirrors* listed words both in their (native) script and in (Chinese) phonetic transcription, and thus could plausibly be put under the category of "glossing" or "exegesis" (*xungu* 訓詁) as well. However, the practice of using phonetic transcription in a different script (e.g., Manchu for Chinese) that was in addition phonographic at a subsyllabic level (in the words of the editors, it "paired sounds to form characters" [*pian yin wei zi* 駢音為字]) went beyond the convention in books of glosses or exegesis. Hence the *Mirrors* were kept in the "character books" section.

This is because the names and things proliferate by the day, and in consequence, the written records have to incorporate them. The institutions of the sages [which include writing] are just like that: Based in that-which-by-itself-is-so (*ziran*) in the tendency [of things], they bring about that-which-ought-to-be-so in their actuality.
是由名物日繁，記載遂不能不備。聖人制作，亦因乎勢之自然，為事之當然而已。[13]

The considerable expansion of the Manchu lexicon, according to the *Siku* editors, follows a spontaneous (*ziran*) propensity of expanding the human knowledge horizon, a phenomenon that also happened to other languages, including Chinese. Here we see a tongue-in-cheek acknowledgment of the need for lexicographical expansion, driven by the sheer scale of modern novelties that breached the bounds of ancient sources.[14]

Yet the *Siku* editors were also obliged to recognize the Manchu *Mirrors* as primarily a book about characters, forcing it awkwardly under the study of (Chinese Confucian) classics (*jing*). As pointed out in chapter 8, however, this framing of the *Mirrors* as a derivative work of classical philology sat awkwardly with the Qianlong emperor's explicit instruction to purge the main sections of the *Mirror* of any Chinese classical references and to highlight everyday usage (Ch. *riyong*

---

13. Ji, *Qinding Siku quanshu zongmu*, vol. 1, 41:54a (858, upper panel).

14. Another court text that epitomizes this sense of pride for the empire's proliferating flora and fauna can be found in the *Huangchao tongzhi* 皇朝通志 (Comprehensive treatises of our august dynasty), one installment in a series of institutional histories completed at the court in the early 1780s. It follows the conventions set forth by Zheng Qiao's 鄭樵 (1104–1162) *Tongzhi* 通志 (Comprehensive treatises) from 1161, but also made conspicuous adaptations. Notably, the chapter on animals and plants—the "Outline of the myriad bugs, herbs, and trees" (*kunchong caomu lüe* 昆蟲草木略)—is populated by frontier flora and fauna that were "not produced in the central state but in distant lands, not found in previous ages but appearing today" 至於中國所無，而產於遐方，前代所無，而出於今日. The chapter drew its sources from *Guang Qunfang pu*, *Kangxi ji xia gewu bian*, Qianlong's poems, and various gazetteers including the *Shengjing tongzhi*. See the relevant "editorial principles" (*fanli* 凡例) passage in *Qinding Huangchao tongzhi*, vol. 644, *fanli*:8b–9a (8, lower panel; 9, upper panel); for the chapters on animals and plants, see vol. 645, *juan* 125–26 (665–85).

changyan 日用常言; Ma. *an-i gisun*) instead. We are tempted to infer from this awkward placement that the everyday usage of language by a single people—provided that their use of this language is true to the tradition of that community—gains a sublime status on par with the Confucian classics of antiquity, whose truth the Manchu court and its Chinese officials acknowledged as universal.

This flattened—and, at times, deeply awkward—framing of the Manchu *Mirrors* is by no means singular in the *Siku* summaries. Other court-commissioned texts we have discussed in this book were likewise placed alongside very strange companions. For instance, the *Siku* had very little good things to say about the entire tradition of encyclopedic "classified writings," claiming that the *leishu*, by virtue of their inclusion of quotes from all four major bibliographic sections (classics, histories, masters, and anthologies), could find no home in any one of them.[15] A perhaps unsurprising consequence of this attitude was that the summary had nothing to say about the voluminous court-commissioned encyclopedias such as *Yuanjian leihan* or even *Gujin tushu jicheng*.

Last, the genre of catalogs (*pulu*) was considered such a trivial (*suosui* 瑣碎) branch of the "masters" literature that the *Siku* editors simply used the category as a placeholder to arrange "miscellaneous books that do not belong to a particular category" 諸雜書之無可繫屬者.[16] It is almost comical, therefore, to find the hundred-*juan*, court-commissioned *Guang Qunfang pu* reviewed in this corner of the bibliography, flanked by two one-*juan* private connoisseur treatises on mushrooms and fowl.

We do not intend to launch into a detailed discussion of the *Siku* summary. These observations are instead intended to reveal—and problematize—the acute tension between the court publications that we have discussed in this book, on the one hand, and the bibliographical order that *Siku quanshu* seeks to impose on the totality of the

---

15. Ji, *Qinding Siku quanshu zongmu*, vol. 3, 135:1a–b (845, lower panel).
16. Ji, *Qinding Siku quanshu zongmu*, vol. 3, 115:1a–b (484, lower panel). Huang Aiping (Siku quanshu *zuanxiu yanjiu*, 373) also notes this awkwardness of the category. See also Gandolfo, "To Collect and to Order," which takes for granted the coherence of the *Siku* classification scheme without questioning possible alternatives.

Chinese cultural heritage, on the other hand. In a sense, to conflate *Guang Qunfang pu* with other "catalogs" that did not measure up to its epistemic ambition in any meaningful way has the unintended consequence of eliminating the uniqueness of Qing court as a space known for making radical—if politically charged—departures from scholarly convention. Indeed, after *Siku quanshu*'s completion in the 1780s—and the passing of the then retired Qianlong emperor in 1799—no major new court "offices" were convened to reconfigure any existing subject of scholarship.[17] R. Kent Guy has noted how the *Siku* imperial library marked the professionalization of scholarship as a result of the "evidential learning" movement. Our contention here is that the same moment also beheld the denouement of court scholarship as a kind of vanguard for that same movement. In the early nineteenth century, the centers for contesting and upholding this High Qing legacy moved away from Beijing and into the provinces.[18]

This long process of forgetting and selective remembering has also resulted in the estrangement between branches of learning as set forth in the imperial library. By assigning the *Mirrors* to one section of the bibliographical summaries and books such as *Guang Qunfang pu* into another distant corner, *Siku quanshu* made it very difficult for readers to see the context that these books, after all, shared: the late Kangxi court and a partially overlapping discourse on plants drawn from previous Chinese records and freshly gleaned from the Qing imperial tours beyond the Great Wall. In this book, we have sought to deshelve those texts from their respective sections in the imperial library and highlight this plurilingual body of knowledge that is visible only if we read them together.

In a sense, we are late to this task. In the nineteenth century, scholars already read the Manchu *Mirrors* for their natural-historical contents. The relegation of the *Expanded and Emended Mirror of the Manchu Language* to the section of character books in Qianlong's *Siku* project notwithstanding, the work had some influence on natural history within the Qing empire, if not, as far as we have been able to

---

17. Huang, Siku quanshu *zuanxiu yanjiu*, 150–74.
18. See, e.g., Elman, *Classicism, Politics, and Kinship*; Miles, *Sea of Learning*.

ascertain, within the Chinese tradition. Gu Songjie 顧松潔 and Gao Xi 高晞 have recently shown that a manuscript of a Manchu-language anatomical work produced at the Kangxi court by Jesuit missionaries and local scholars has been revised by replacing Chinese loans with Qianlong-era neologisms, including for technology of European origin.[19] Furthermore, at the margins of the sinosphere, we discern the Qianlong bilingual *Mirror*'s impact in a Tibetan-language Mongolian materia medica from the first half of the nineteenth century. Its author, 'Jam-dpal-rdo-rje (1792–1855), was a Buddhist monk born into an aristocratic family in Inner Mongolia who studied Mongolian and Manchu in childhood and later trained in Tibet.[20] His richly illustrated materia medica in many cases contains the names of plants and animals in Mongolian, Chinese, and Manchu alongside the illustrations and Tibetan legends. The Manchu names include neologisms introduced during Qianlong's language reform, such as *ihasi* 'rhinoceros'. Furthermore, the once dialectal variants *monio* and *bonio* are used to distinguish the macaque and the gibbon, as in the *Expanded and Emended Mirror*.[21] Thus the Qianlong-era lexicon for the natural world entered Tibeto-Mongol learning as well. One could even say that 'Jam-dpal-rdo-rje's book provides illustrations for Manchu words that had previously only had textual descriptions.

It is elsewhere that we find an explicit discussion of the Manchu *Mirrors* as natural-historical scholarship in their own right, however. In French and Japanese scholars' citations of "classified writings" from the late Ming, which shared a background with the early eighteenth-century Qing encyclopedias, or *Guang Qunfang pu* itself, we see clearly how their reading practices differed from that encouraged by the *Siku* summaries.

---

19. The example they give is the word for the "striking clock" (Ch. *ziming zhong* 自鳴鐘), but by their own account, their list is not exhaustive (Gu and Gao, "Guanyu Manwen chaoben *Geti quanlu*," 145).

20. 'Jam-dpal-rdo-rje, *Mengyao zhengdian*, 1–2; see also Nappi, "Making 'Mongolian' Nature."

21. 'Jam-dpal-rdo-rje, *Mengyao zhengdian*, 327, 336–37; *Illustrated Tibeto-Mongolian Materia*, 234, 238.

Manchu Natural History and Early French Sinology

To our knowledge, the earliest discussion of Qianlong's language reform in the context of natural history is found in Jean-Pierre Abel Rémusat's "Observations sur l'état des sciences naturelles chez les peuples de l'Asie orientale" from 1828. In this essay, Rémusat situated the Manchu *Mirrors* within a tradition of "small encyclopedic collections" arranged by subject matter, that is, collections in the tradition of the ancient *Erya*. "In terms of the classification of forms of life," Rémusat wrote, these collections could "be considered together with the natural-historical treatises." Rémusat identified two ramifications of this tradition. One was constituted by "primers intended for the education of children," which contained "a few additional divisions for the vegetal kingdom." The other was "the Tartar dictionaries organized by subject matter," to wit the *Mirrors*. Rémusat described Qianlong's Manchu-Chinese edition in particular, writing that

> the Manchu terminology was later enlarged with all the new words that the emperors have introduced into the language, and those concerning natural beings were gathered in the third book of the *Mirror*'s supplement, where they form individual chapters. There is one chapter for fruits of extraordinary appearance, one for trees, one for exotic trees, one for flowers, another for rare flowers, two for birds large and small, one for quadrupeds, a second chapter for little-known quadrupeds, and a third for animals of the various categories that by virtue of some singularity are remarkable.
>
> *Les petits recueils encyclopédiques, où tous les mots de la langue sont disposés suivant la matière à laquelle ils se rapportent, peuvent, en ce qui concerne la classification des êtres organisés, être assimilés aux traités d'histoire naturelle.... Les recueils élémentaires destinés à l'instruction des enfans offrent quelques divisions de plus pour le règne végétal; du reste, ils présentent à peu près la même disposition.... On retrouve à peu près la même distribution dans les dictionnaires tartares par ordre de matières. La nomenclature mandchou ayant été postérieurement accrue de tous les mots nouveaux que les empereurs ont introduits dans la langue, on a réuni ceux qui avoient rapport aux êtres naturels dans le troisième livre du*

supplément au Miroir, où ils constituent des chapitres particuliers. Il y en a un pour les fruits de forme extraordinaire, un pour les arbres, un pour les arbres exotiques, un pour les fleurs, un autre pour les fleurs rares, deux pour les oiseaux grands et petits, un pour les quadrupèdes, un second pour les quadrupèdes peu connus, et un troisième pour des animaux de différentes classes, remarquables par quelque singularité.[22]

Rémusat, who held the first European professorship of Chinese—and on his own insistence, of Manchu—had a strong interest in natural history. His father was a prominent physician and Rémusat studied medicine, albeit reluctantly.[23] A fortuitous encounter with a missionary copy of an illustrated late Ming pharmacopoeia made Rémusat turn to the study of Chinese. Rémusat used Manchu translations and reference works—of which the *Mirrors* were the best known—to learn the language.[24] During the first lecture for his inaugural course on Chinese and Tartar-Manchu language and literature, Rémusat announced that he would cover Chinese botanical scholarship.[25]

Rémusat continued to research both the *Mirrors* and natural history. In 1817, he wrote about either the Manchu-Mongol *Mirror* of 1717 or Qianlong's *Expanded and Emended Mirror*—his account does not make it explicit which edition he was using—that he had "translated about half of it into French" (*dont j'ai traduit à peu près la moitié en français*).[26] He described it as "more of an encyclopedia than a Dictionary, at least in terms of our definition of that word" (*plutôt une encyclopédie qu'un Dictionnaire, au moins suivant l'idée que nous attachons à ce mot*).[27] Rémusat's view of the *Mirrors* as encyclopedias is reflected in how he used them in his research.

22. Rémusat, "Observations sur l'état des sciences," 154–55.
23. Obringer, "Jean-Pierre Abel-Rémusat."
24. Elliott, "Abel-Rémusat," 983–84 (not a discussion of the *Mirrors* in particular).
25. Métailié, "Aux sources d'une vocation," 124.
26. The reference to the "great Chinese-Manchu Dictionary of which Emperor Khanghi [i.e., Kangxi] commissioned a translation into Mongolian" (*le grand Dictionnaire chinois-mandchou que l'Empereur Khanghi a fait traduire en Mongol*) is a bit ambiguous because the original *Mirror* was in Manchu only.
27. Jean-Pierre Abel Rémusat to Schilling von Canstadt, undated (Walravens estimated 1817), in Walravens, *Jean Pierre Abel Rémusat*, 124.

Among Rémusat's papers are two notebooks with cards mounted on the pages (see fig. 9.1). The cards contain mostly names of plants in Chinese, Manchu, Japanese, and Mongolian, with transcriptions and references to sources, arranged in alphabetical order (each notebook starting from *a*).[28] The source for the Manchu words, and some of the Chinese, is the *Expanded and Emended Mirror*.[29] Mongolian words like *kil[a]yan-a* 'bramble-bush' (> Ma. *kilhana*) and *qonin suyiq-a* 'white wormwood (?)' Rémusat got from the Manchu-Mongol *Mirror*.[30] Japanese words (in *katakana*) and some of the Chinese words are from the section on plants in *Wakan sansai zue* 和漢三才圖會 (Illustrated dictionary of the three realms in Japanese and Chinese) from 1713, an adaptation of a late Ming encyclopedia.[31] Rémusat cross-referenced some of the Chinese, Manchu, and Mongolian words taken from the *Expanded and Emended Mirror* and the Mongolian *Mirror* with this Japanese encyclopedia.

One card, for example, carries the Manchu word *hoošari moo* 'paper mulberry', a neologism (< *hoošan* 'paper') introduced in the *Expanded and Emended Mirror*, which is referenced on the card along with the third folio of *juan* 84 in *Wakan sansai zue*.[32] The recto of that folio describes and depicts, using an illustration, the tree called *kaji* 楮 (Ch. *chu*), a Chinese character that is likewise used to translate *hoošari*

---

28. There are also words like *buduhu* 'loach', however (Rémusat, *Table alphabétique du Thsing-wen-kien*, card no. 265).

29. Walravens, *Jean Pierre Abel Rémusat*, 97. See also Monnet, "Abel-Rémusat," 105, note 105. Curiously, one of the notebooks carries the title "Table de Thsing-wen-kien pou," referring to the dictionary's supplement, but this is hardly appropriate because this notebook includes numerous Chinese and Japanese words from other sources. By the same token, the other notebook, which also includes Manchu and Mongolian words, is confusingly titled "Table alphabétique de l'Encyclopédie japonaise." The titles ought to have been added after the tables left Rémusat's hands and entered the Bibliothèque nationale de France collection, given that each table also has a number on the title card (different from the current call number).

30. Regarding *qonin suyiq-a*, see Lessing et al., *Mongolian-English Dictionary*, 964. Establishing the meaning of the word by chasing its Manchu translation *sumpa* quickly leads to difficulties, so we refrain from doing so here. Rémusat, *Table alphabétique*, card nos. 2966, 3205; Rozycki, *Mongol Elements in Manchu*, 139.

31. Marcon, *Knowledge of Nature*, 81.

32. Rémusat, *Table alphabétique*, card no. 149.

N° 3424. Table de Thsing- wen-Kien par A. Rémusat.	阿勒 A-pho-le E.J.88.10v.	Aboukka Ikha Ths.w.k.29,116v. add. y° 117	アツヾ Adzousa E.J.83.6.	加 Agabousa Ths.w.k.S.III,11v. E.J.91.38
阿芙蓉 A-fou-young E.J.103.19.+ [R.F. stamp]	阿片 A-pian E.J.103.19.+	Acouna Ths.w.k.29,11v. E.J.94 B.6.	アハホ Afabo Rec.Jap.IV.13.	アカ子 Agane E.J.96.25.
	阿薩鱄 A-sa-tan E.J.88.8.		アハキ Afaki E.J.83.7.	アギ Agi E.J.82.28
阿虞 A-iu E.J.87.28.	阿鏾 A-Hiei E.J.82.16.	Abouragiri E.J.83.9.	アハモリサメ Afamorisame E.J.94 B.29.	艾 Ai E.J.94 A.42.
		アフラカソウ Abourakasou E.J.103.5v.+		アイ Ai E.J.94 B.11.
阿迦嚧香 A-kia-lou-hiang E.J.82.14v.	阿駞 A-tteu E.J.88.9.		アハユキサメ Afayukisame E.J.94 B.29.	艾蒿 Ai-hao E.J.94 A.42.
阿鬼此 A-kieou-tseu E.J.91.37v.	アサツキ Asa-tsuki E.J.99.6v.	Aboutki ikha Ths.w.k.29,47. add. y° 118	アヘン Afen E.J.103.19.+	愛韭 Ai-kieou E.J.94 A.30v
	阿魏 A-wei E.J.82.28.	Adjé-mouyari Ths.w.k.28,37v. add. E.J.91.f°	アサタツヒカメ Asa tatsi kame E.J.87.19.	艾納 Ai-na E.J.91.22v.
アキウ A-kiu Rec.Jap.V.11. E.J.83.27v.	アハシカミ Abasigami E.J.93.10.	Adzeda-djoskha Ths.w.k.28,37v. E.J.84.4v.	アフラ シメチ Afoura simotei E.J.101.5.	艾納香 Ai-na-hiang P.m.XIV.70.
阿羅漢草 A-lo-han-thsao E.J.94 B.9.	Abdari Ths.w.k.29,24. E.J.87.29v.	Adzirgan-segi Ths.w.k.27,20. E.J.102.4v.	アフサリサウリ Afousari savri E.J.94 B.38.	艾子 Ai-sirou E.J.94 B.11.
阿面桃 A-mian-thao E.J.LXXXII.7.	アブイ Aboui E.J.94 A.33v.			艾子 Ai-tseu E.J.89.21.
	Aboukha Ths.w.k.27,14v. E.J.102,3.	Adjisi Ths.w.k.S.III.16. E.J.91.7.v.	アフチ Afoutsi E.J.83.11.	蘆藥 Ai-yo P.m.XXI.19v.

moo in the *Expanded and Emended Mirror*.[33] It ought to have been the Chinese translation of the word that allowed Rémusat to link the Manchu-language definition of the *Mirror* with the classical Chinese description of the Japanese encyclopedia.

The card catalog does not appear finished. Yet even in its present state, it evidences research on Asian plants in which the Qing court's Manchu reference works played an important part. Next, we turn to how the *Expanded and Emended Mirror* was similarly used in Japan.

The Manchu-Chinese *Mirror* and *Guang Qunfang pu* in Edo Japan

Around the same time that Rémusat was researching the Manchu-Mongol and Manchu-Chinese *Mirrors*, a group of scholars in Japan, led by Takahashi Kageyasu 高橋景保 (1785–1829), was likewise studying the *Expanded and Emended Mirror of the Manchu Language*. Kageyasu was three years Rémusat's senior and died three years before him; that is, the two were close contemporaries. Yet Rémusat, despite living through a politically very tumultuous time, was really an academic, not a government servant. Takahashi, by contrast, was a shogunal astronomer with some experience with Dutch books who was ordered to study Manchu when Japanese foreign relations suddenly demanded it in 1808.[34]

Takahashi and his team used the Manchu-Chinese *Mirror* for natural-historical research in addition to their politically motivated translations. Uehara Hisashi 上原久 (1908–1997), who like Fuchs and Seuberlich had worked in Manchukuo early in his career, studied

---

33. Terajima, *Wakan sansai zue*, 1193 (top panel); *Yuzhi zengding Qingwen jian*, vol. 233, 29:26b (191, top panel).

34. Söderblom Saarela, *Early Modern Travels*, 112–13.

FIG. 9.1 (*opposite*). A selection of Rémusat's natural-historical note cards with material from the Manchu *Mirrors* and Japanese sources. Courtesy of the Bibliothèque nationale de France (call numbers Japonais 391 and Japonais 392).

Takahashi's Manchu manuscripts.[35] Most of them were based on the *Expanded and Emended Mirror* and aimed to produce a Manchu dictionary for use in Japan. The manuscripts show that the *Mirror's* sections on plants and animals presented particular difficulties for Takahashi, but also that he was prepared—interested, probably—to solve the problems that they posed. Natural history thus featured prominently in early Japanese studies of Manchu, as it did in France. Like Rémusat, Takahashi cross-referenced the *Expanded and Emended Mirror* with other encyclopedias. In Takahashi's case, the encyclopedias included *Guang Qunfang pu*.

The manuscript *Yakugo shō* 譯語抄 (Excerpted translations), which listed Chinese translations of headwords from the Manchu-Chinese *Mirror*, includes Japanese glosses for many—but far from all—of the Manchu lemmata. Some lemmata in addition have Dutch translations in a reflection of the linguistic expertise of the members of Takahashi's team. For example, the grass lychee discussed in chapter 7 is listed here with the Japanese gloss *tsuru reiishi* ツルレイシ 'vine lychee' and the Dutch translation *balsam appel manneken* 'Momordica balsamina, male', referring to a fruit native to Africa.[36] We see that the exotic plants described in the Manchu-Chinese *Mirror* were not easy to identify in Edo. Indeed, the sections on herbs, trees, and flowers are unique among the "excerpted translations" in that many entries carried the note "unknown" (*fushō* 不詳).[37]

A later manuscript titled *Shinbunkan meibutsu goshō* 清文鑑名物語抄 (Content words excerpted from the *Expanded and Emended Mirror of the Manchu Language*), like the "excerpted translations," reproduces parts of the Manchu-Chinese *Mirror*. Yet whereas the latter only had simple Japanese or Dutch glosses, the *Shinbunkan meibutsu goshō* includes Chinese translations of the *Mirror*'s Manchu definitions and thus gives considerably more information.[38] Interestingly, this manuscript also includes marginal notes that evidence

---

35. Hirayama, "Ko Uehara Hisashi."
36. Takahashi, *Yakugo shō*, vol. 6, unpaginated. We thank Kjell Ericsson and Mario Cams for help with transcribing and interpreting the Japanese and Dutch glosses.
37. Uehara, *Takahashi Kageyasu*, 1071.
38. Uehara, *Takahashi Kageyasu*, 1074.

Takahashi's continued research on the Qing empire and its environment. He used the Qing statutes and an account of Qianlong's conquest of Xinjiang to better understand the Qing state, but the sections on flora and fauna show the greatest research effort.[39]

One of the books that Takahashi cited often was Li Yanji's dictionary, which translated the definitions of the original, monolingual *Mirror* into Chinese. Comparing Li Yanji's translations with those in the *Expanded and Emended Mirror* allowed Takahashi to reach the same insight as Ihing (discussed in chapter 8), whom Takahashi had not yet read at this point: In addition to introducing new Manchu words, Qianlong's Manchu-Chinese *Mirror* had actually changed many of the Chinese words for flora and fauna.[40]

In addition, Takahashi at one point used an unofficial Qing court chronicle that had reached Japan.[41] The occasion was the *Expanded and Emended Mirror*'s entry on the "iced rodent," the translation of the great beast "resting" in the permafrost of the north. As mentioned in chapter 5, the Chinese *locus classicus* called this creature a *xi* rodent, but Kangxi had referred to it as a *fen* rodent in both *Kangxi ji xia gewu bian* and his second lecture as recorded in the court chronicles. Takahashi translated the *Mirror*'s Manchu definition into Chinese and added a note in the upper margin. The note quoted the Kangxi emperor's description of the rodent as mammoth (very heavy, body and tusks like an elephant), which were reported in the court chronicle. Takahashi thus brought the Kangxi emperor's "investigation of things"—articulated in Chinese but with great plurilingual awareness—in direct dialogue with the Manchu *Mirrors*. Thereby some of the different strands of Qing natural-historical scholarship were brought together here, in the margins of a Japanese manuscript.[42]

---

39. Uehara, *Takahashi Kageyasu*, 1085–87. On Takahashi's source on Xinjiang, Perdue, *China Marches West*, 481–86.

40. Uehara, *Takahashi Kageyasu*, 1090–91.

41. On the chronicle used, see Biggerstaff, "Some Notes on the Tung-Hua Lu," 102–5.

42. Uehara, *Takahashi Kageyasu*, 1090. Takahashi noted the discrepancy between *xi shu* and *fen shu*, and that Li Yanji calls the zokor *fen shu* 糞鼠 'dung rodent' and *di shu* 地鼠 'ground rodent' and correctly suspected that there had been a mix-up.

Furthermore, Takahashi frequently cited *Guang Qunfang pu*, which provided him with a great deal of information on the plants of the *Mirror*.[43] He likewise relied on it heavily for the marginal notes written on a copy of the *Expanded and Emended Mirror*. Not only did Takahashi use *Guang Qunfang pu* to translate entries in the sections on flowers and strange fruits, but he also pointed out that the order of the entries in the section on "strange flowers" followed *Guang Qunfang pu*.[44] In the last product of Takahashi's Manchu studies, *Zōtei Manbun shūin* 増訂滿文輯韻 (Compiled Manchu rhymes, expanded and emended), unfinished in the year that he died, the results of years of work are presented in a concise and measured way.[45] Under *ilhamuke*, the "grass lychee" identified in an earlier manuscript as the African balsam apple, Takahashi gave a Chinese-language translation of the *Mirror*'s Manchu definition and noted the alternative Chinese translation that Mingdo had given in 1735.[46]

Thus, once again, Takahashi brought Kangxi-era court scholarship in Chinese and Yongzheng-era commercial lexicography into his work on Qianlong's Manchu-Chinese *Mirror*, an exercise we have attempted to follow in this book. His studies of the Manchu reference work show not only how difficult it was to make sense of its sections on flora and fauna, but also that the scholars in Edo found it worthwhile to invest the time necessary to do it. The natural world of the Qing empire commanded great interest, and in the early nineteenth century, the Manchu *Mirrors* were important sources for learning about it. Scholars in Takahashi's circle used the *Expanded and Emended Mirror* in their own work as well: It was used for natural-historical research (on the camel) by one of Takahashi's students.[47]

The preceding sections have shown that as the academic study of the Qing empire began to take shape abroad in the early nineteenth century, the Qianlong emperor's *Expanded and Emended Mirror of the*

---

43. Uehara, *Takahashi Kageyasu*, 1092.
44. Uehara, *Takahashi Kageyasu*, 1109.
45. Takahashi, *Zōtei Manbun shūin*, vol. 3, 3:33a; Söderblom Saarela, *Early Modern Travels*, 118. This graphologically arranged dictionary has no preface and does not extend through the entire Manchu alphasyllabary.
46. Takahashi, *Zōtei Manbun shūin*, vol. 3, 33a.
47. Ōno, "Matsumoto Tokizō," 53.

Manchu Language was an essential source. The languages of Qianlong's Manchu-Chinese *Mirror* were challenging in both France and Japan, even though classical Chinese had been used extensively as a local literary language in Japan for centuries. The Chinese language in the *Mirror*, notably the many terms for plants and animals, was not the classical Chinese with which Japanese scholars were familiar. As made clear in chapter 8, the Chinese language included many vernacular elements and words that had risen to the status of official terminology only on the publication of this dictionary. The Chinese words were thus largely unfamiliar to Takahashi and the scholars in Edo. The natural world that these words described certainly was as well, not to mention the Manchu language of the dictionary's definitions. To the French scholar Rémusat, the short Manchu definitions were probably easier to digest than many texts in Chinese, classical and vernacular alike. To both Takahashi and Rémusat, the *Mirror* was a treasure trove of information on Qing natural history, but their engagement with it also suggests some of the difficulties presented by this plurilingual scholarship, which included, in a sense, two new nomenclatures in both Manchu and Chinese.

Looping back to the exchange between Fuchs and Seuberlich a century after Rémusat and Takahashi, we see that scholars were still struggling to make sense of the words for plants and animals included in the Manchu *Mirrors*. By the mid-twentieth century, the focus had moved away from natural erudition and toward languages per se, however. The *Mirrors* had become reference works—character books, as the *Siku* editors have classified them—that were consulted for linguistic information but perhaps not always treated as complex scholarly products in their own right.

One final thought before the end of this book as we take the liberty to engage in a little speculation: Is it possible to trace a history of natural history in Qing China, despite the lack of a local equivalence to this discipline, an unmistakable etic category to our actors? In this book, we explore the rich depository of knowledge concerning plants and animals in the Manchu *Mirrors* and court-commissioned Chinese encyclopedic works that in turn influenced Manchu lexicography. We show how a plurilingual discourse about plants and animals in High Qing China, especially species that were found in the empire's frontiers, indeed grew tremendously over the eighteenth century. This knowledge and the discourse in which it was articulated, however, must be understood in conjunction with the salience of Manchu as the "dynastic language" (*guoyu* 國語) of an empire under minority rule. The story we have told here thus certainly resists any easy identification with the parallel history of natural history in Enlightenment Europe, where the dynamics of language were quite different. Yet in both cases, linguistic study and natural erudition became intertwined. To paraphrase Michel Foucault's musings, a Qing order of things must thus also be reckoned through the order of words.[48] What we have presented in this book can thus be seen as an experiment in what Federico Marcon has called "metaphorical comparison" in the history of knowledge, in which the historian maintains a critical reflection on our usage of present categories (such as natural history) but nevertheless deploys them to shed new light on the past.[49]

In a nutshell, this plurilingual body of plant and animal knowledge could—and, as we argue, ought to—change a thing or two about what we customarily take to be the stable emic categories of Chinese learning in the late imperial period. Throughout this book, we have shown that genres such as the encyclopedic "classified writings" (*leishu*) and "catalogs" (*pu*)—with or without pictures—could cross-pollinate easily with the scholarly study of words and characters. In archived correspondences between clerks, scholar-officials of both Banner and non-Banner origin, and the emperor, textual sources in both Chinese and Manchu were invoked to deliberate on the appropriate Manchu name—and, surprisingly, at times also the Chinese name—for a creature from somewhere in the vast empire that for one reason or other caught the attention of the court. What rendered all these developments possible, we argue, was the great expansion and regimentation of cultural activities in this space we call the imperial court. It was the growing involvement of the court—and one might say the state—in scholarly production that enabled the confluence of Chinese encyclopedism and Manchu

---

48. Foucault, *Order of Things*, 157.
49. Marcon, "Critical Promises of the History."

lexicography, resulting in a decisive shift away from the literati-centered tradition of natural erudition.

This dominant role of court scholarship in China's eighteenth century, to be sure, was accomplished by violent disciplining of personnel and censorship of speech, as amply discussed in previous literature. Yet we demonstrate, following arguments by scholars such as R. Kent Guy and Benjamin Elman, that the court as a space for cultural and ideological reproduction should not be taken as the sworn enemy to an imagined community of Chinese scholars who saw themselves as the main protagonists of evidential learning.[50] In this book, we use the Qing court as a new vantage point from which to reconsider the history of knowledge during China's contentious eighteenth century, bookended by the two Manchu *Mirrors*. The study of plants and animals was a key site in which a composite epistemic change took place—one that revisited, reorganized, and transformed many premises of China's intellectual tradition up to the late Ming. Beyond the exegesis of classical texts and the spirited debate over Confucian doxa, we suggest that the study of plants and animals, including their plurilingual nomenclatures, should be considered a crucial development in the intellectual history of the period. What Elman memorably called the shift "from philosophy to philology" could, we suggest, be more capaciously understood as "from philosophy to various fields of learning," including the busy intersection of natural history and lexicography.

50. The previous literature, however, focuses on the limits of the state action as "the art of the possible," not how the *Siku quanshu* project also selectively interpreted scholarship produced by the Qing court in earlier reigns. See Guy, *Emperor's Four Treasuries*, 201–2; Elman, *Philosophy to Philology*, 15–18.

Bibliography

A-dun 阿頓. *Tongwen guanghui quanshu* 同文廣彙全書 | *Tung wen guwang lei ciowan šu* (Broadly collected complete text in the standard script). 1693. Reprint, Nanjing: Tingsong lou, 1702.

Ai Rulüe 艾儒略 (Giulio Aleni). *Xifang dawen* 西方答問 (Questions and answers on the Western regions). Facsimile. Vol. 2. *Ai Rulüe Hanwen zhushu quanji* 艾儒畧漢文著述全集. 1637. Reprint, Nanning: Guangshi shifan daxue chubanshe, 2011.

——. *Zhifang waiji* 職方外紀 (Annals of foreign lands). Facsimile. Vol. 1. *Ai Rulüe Hanwen zhushu quanji* 艾儒畧漢文著述全集. 1623. Reprint, Nanning: Guangshi shifan daxue chubanshe, 2011.

Allen, Barry. *Vanishing into Things: Knowledge in Chinese Tradition*. Cambridge, MA: Harvard University Press, 2015.

Allsen, Thomas. *The Royal Hunt in Eurasian History*. Philadelphia: University of Pennsylvania Press, 2006.

Almonte, Victoria. "The Perception of Exotic Features in Some Animals, Mentioned by Zhou Qufei in the *Lingwai Daida* (1178)." *Sulla via Del Catai* 15, no. 26 (2022): 33–63.

Alt'aiŏ yŏn'guso 阿爾泰語研究所. *"Ŏje Ch'ŏngmun'gam" saegin* 御製清文鑑索引 (An index to *Imperially Commissioned Mirror of the Manchu Language*). Vol. 2. Taegu: Hyosŏng yŏja taehakkyo ch'ulp'anbu, 1982.

Amiot, Joseph-Marie. *Dictionnaire Tartare-Mantchou François, composé d'après un Dictionnaire Mantchou-Chinois: Rédigé et publié avec des additions et l'alphabet de cette langue*. Edited by Louis-Mathieu Langlès. Paris: Didot l'aîné, 1789.

——. *Éloge de la Ville de Moukden et de ses environs; poème composé par Kien-long, Empereur de la Chine & de la Tartarie, actuellement régnant*. Paris: N. M. Tilliard, 1770.

———. *Hymne Tartare-Mantchou chanté à l'occasion de la conquête du Kin-tchouen*. Edited by Louis-Mathieu Langlès. Paris: Didot l'aîné, 1792.

An Dawei 安大伟. "Nalan Chang-an yu *Shenshui sanchun ji*" 纳兰常安与《沈水三春集》 (Nalan Chang-an and *Spending the Beginning of Spring, the Greeting of Spring, and the End of Spring by the Shen* [Wanquan 萬泉] *River*). *Lantai shijie* 兰台世界, no. 19 (2016): 98–102.

Bābur, Ẓahīr al-Dīn Muḥammad. *Bābur-nāma (vaqāyi')*. Edited by Eiji Mano. Critical edition. Kyoto: Syokado, 1995.

Batsaki, Yota, Sarah Burke Cahalan, and Anatole Tchikine, eds. *The Botany of Empire in the Long Eighteenth Century*. Cambridge, MA: Harvard University Press, 2017.

Behr, Wolfgang. "In the Interstices of Representation: Ludic Writing and the Locus of Polysemy in the Chinese Sign." In *The Idea of Writing: Play and Complexity*, edited by Alex Voogt and Irving Finkel, 281–314. Leiden: Brill, 2010.

Bello, David A. *Across Forest, Steppe and Mountain: Environment, Identity, and Empire in Qing China's Borderlands*. Cambridge: Cambridge University Press, 2016.

Bi Yuan 畢沅, ed. *Shanhai jing xin jiaozheng* 山海經新校正 (Classic of mountains and seas, newly collated and corrected). Privately owned copy. Chongde tang, 1783.

Bian, He. *Know Your Remedies: Pharmacy and Culture in Early Modern China*. Princeton, NJ: Princeton University Press, 2020.

———. "Re-Collecting the Glorious Age: Yang Fuji and the Disciplining of Zhaodai Congshu, 1772–1844." *Late Imperial China* 40, no. 1 (2019): 1–41.

Biggerstaff, Knight. "Some Notes on the *Tung-Hua Lu* 東華錄 and the *Shih-Lu* 實錄." *Harvard Journal of Asiatic Studies* 4 (1939): 101–15.

Biot, Édouard, trans. *Le Tcheou-li ou Rites des Tcheou*. 2 vols. Paris: Imprimerie Nationale, 1851.

Black, Lydia. "The Nivkh (Gilyak) of Sakhalin and the Lower Amur." *Arctic Anthropology* 10, no. 1 (1973): 1–110.

Bocci, Chiara. "The Animal Section in Boym's (1612–1659) *Flora Sinensis*: Portentous Creatures, Healing Stones, Venoms, and Other Curiosities." *Monumenta Serica* 59, no. 1 (2011): 353–81.

Boivin, Jean. *Mémoires pour l'histoire de la Bibliothèque du Roy*. Digitized manuscript, call no. NAF 1328. Bibliothèque Nationale de France, Paris.

Bossiere, Yves de Thomas. *Jean-François Gerbillon, S.J. (1654–1707): Un des cinq mathématiciens envoyés en Chine par Louis XIV*. Leuven: Ferdinand Verbiest Foundation, 1994.

Bottéro, Françoise. "Ancient China." In *The Cambridge World History of Lexicography*, edited by John Considine, 51–66. Cambridge: Cambridge University Press, 2019.

Bramao-Ramos, Sarah. "Manchu-Language Books in Qing China." PhD diss., Harvard University, 2023.

Bray, Francesca. "Essence and Utility: The Classification of Crop Plants in China." *Chinese Science* 9 (1989): 1–13.

Breen, Benjamin. *The Age of Intoxication: Origins of the Global Drug Trade*. Philadelphia: University of Pennsylvania Press, 2019.

Bretschneider, Emil. *Botanicon Sinicum*. London: Trübner & Co., 1882.

Brook, Timothy. "Native Identity under Alien Rule: Local Gazetteers of the Yuan Dynasty." In *Pragmatic Literacy, East and West, 1200–1330*, edited by Richard Britnell, 235–46. Woodbridge: Boydell Press, 1997.

Brunnert, H. S., and V. V. Hagelstrom. *Present Day Political Organization of China*. Edited by N. Th. Kolessoff. Shanghai: Kelly and Walsh, Limited, 1912.

B·Sod 波·索德. "Cong *Wuti Qingwen jian* Manyu gouci fujia chengfen -*tu* kan Mengguyu dui Manyu de yingxiang" 从《五体清文鉴》满语构词附加成分 -tu 看蒙古语对满语的影响 (A consideration of Mongolian influence on Manchu from the point of view of the use of the Manchu derivational suffix -*tu* in the *Pentaglot Mirror of the Manchu Language*). *Neimenggu minzu daxue xuebao (shehui kexue ban)* 内蒙古民族大学学报 (社会科学版) 34, no. 6 (2008): 39–41.

Cai Xiang 蔡襄. *Lizhi pu* 荔枝譜 (Catalog of lychees). Vol. 845. *Wenyuan Ge Siku quanshu* 文淵閣四庫全書. 1782. Reprint, Taipei: Taiwan Shangwu yinshu guan, 1983.

Cams, Mario. *Companions in Geography: East-West Collaboration in the Mapping of Qing China (c. 1685–1735)*. Leiden: Brill, 2017.

Cao Liya 曹莉亚. "*Honglou meng* 'xingxing' kaobian" 《红楼梦》"猩猩" 考辨 (An investigation of *xingxing* [a kind of simian] in *Dream of the Red Chamber* [i.e., *Story of the Stone*]). *Cao Xueqin yanjiu* 曹雪芹研究, no. 4 (2017): 73–77.

Carr, Michael Edward. "A Linguistic Study of the Flora and Fauna Sections of the *Erh-Ya*." PhD diss., University of Arizona, 1979.

Catalogue des livres imprimez de la Bibliothèque du Roy. Vol. 1. Paris: Imprimerie royale, 1739.

Chang, Michael G. *A Court on Horseback: Imperial Touring and the Construction of Qing Rule, 1680–1785*. Cambridge, MA: Harvard University Asia Center, 2007.

Chang Te-Ch'ang. "The Economic Role of the Imperial Household in the Ch'ing Dynasty." *Journal of Asian Studies* 31, no. 2 (1972): 243–73.

Chang-an 常安. *Shenshui sanchun ji* 潘水三春集 (Spending the beginning of spring, the greeting of spring, and the end of spring by the Shen [i.e., Wanquan 萬泉] river). Facsimile. Vol. 233. *Qingdai shiwen ji zhenben congkan* 清代詩文集珍本叢刊. 1740. Reprint, Beijing: Guojia Tushuguan chubanshe, 2017.

Chen Jingzhang 陳敬璋. *Zha Shenxing nianpu* 查慎行年譜 (Yearly chronicle of Zha Shenxing's life). Originally titled *Zha Tashan xiansheng nianpu* 查他山先生年譜. 1992. Reprint, Beijing: Zhonghua shuju, 2006.

Chen, Kai Jun. *Porcelain for the Emperor: Manufacture and Technocracy in Qing China*. Seattle: University of Washington Press, 2023.

Chen Menglei 陳夢雷. *Songhe shanfang wenji* 松鶴山房文集 (Collected prose from Pine Crane Mountain Lodge). Facsimile. Vol. 1416. *Xuxiu Siku quanshu* 續修四庫全書. Reprint of Kangxi edition, Shanghai: Shanghai guji chubanshe, 1995–2002.

Chen Menglei 陳夢雷 and Jiang Tingxi 蔣廷錫. *Qinding Gujin tushu jicheng* 欽定古今圖書集成 (Imperially authorized consummate collection of texts and illustrations from past and present). 1726. Reprint, Shanghai: Zhonghua shuju, 1934.

Chen Renxi 陳仁錫, ed. *Qianque ju leishu* 潛確居類書 (Classified writings from the Hidden and Firm Dwelling). Wumen: Xu Guanwo Daguan tang, 1630–1632.

Chen Shou-yi 陳受頤. "*Kangxi Ji xia gewu bian* de Fawen jieyi ben" 康熙幾暇格物編的法文節譯本 (A selection of French translations from *Collection of the Investigation of Things during Times of Leisure from the Myriad Affairs [of State] in the Kangxi Period*). *Lishi yuyan yanjiu suo jikan* 歷史語言研究所集刊 28, no. 2 (1957): 841–51.

Chen Yuanjing 陳元靚. "Suishi guangji" 歲時廣記 (Expanded account of the seasonal cycle). In *Xuehai leibian* 學海類編 (Categorized compilation from the sea of learning), edited by Cao Rong 曹溶. Facsimile. *Zhongguo fengtu zhi congkan* 中國風土志叢刊 7. Yangzhou: Guangling shushe, 2003.

Cheng Fangyi 程方毅. "Mingmo-Qingchu Hanwen Xishu zhong 'haizu' wenben zhishi suyuan: Yi *Zhifang waiji*, *Kunyu tushuo* wei zhongxin" 明末清初汉文西书中"海族"文本知识溯源—以《职方外纪》《坤輿图说》为中心 (Tracing the origins of the "sea creatures" in Chinese-language Western books from the late Ming and early Qing: With a focus on *Annals of Foreign Lands* and *Illustrated Explanation of the World*). *Anhui daxue xuebao (zhexue, shehui kexue ban)* 安徽大学学报（哲学社会科学版）, no. 6 (2019): 88–96.

Chiu, Elena Suet-Ying. *Bannermen Tales (Zidishu): Manchu Storytelling and Cultural Hybridity in the Qing Dynasty*. Harvard-Yenching Monograph Series 105. Cambridge, MA: Harvard University Asia Center, 2018.

Ch'oe Hak-kŭn 崔鶴根. "Manmun ŭro ssyŏjin Kŏllongje ŭi Ŏje Sŏnggyŏng bu e taehaesŏ" 滿文으로 쓰여진 乾隆帝의 「御製盛京賦」에 對해서 (On Qianlong's *Imperially Commissioned Ode to Mukden*, written in Manchu). In *Chŭngbo Alt'aiŏhak non'go: Munhŏn kwa munbŏp* 增補알타이語學論攷—文獻과 文法—, edited by Yi Hŭi-sŭng 李熙昇, 95–102. 1968. Reprint, Seoul: Pogyŏng munhwasa, 1989.

Chuang Chi-fa 莊吉發. "Guoli Gugong Bowu Yuan diancang de Man-Han hebi zouzhe" 國立故宮博物院典藏的滿漢合璧奏摺 (Bilingual Manchu-Chinese palace memorials in the collection of the National Palace Museum). *Manzu wenhua* 滿族文化, no. 3 (1981): 27–31.

———. Niaopu *Manwen tushuo jiaozhu*《鳥譜》滿文圖說校注 (A critical edition of the Manchu legends to the pictures in the *Catalog of Birds*). 6 vols. Taipei: Wen shi zhe chubanshe, 2017.

———. Shoupu *Manwen tushuo jiaozhu*《獸譜》滿文圖說校注 (Annotated edition of the Manchu legends to *Catalog of Beasts*). 2 vols. Taipei: Wen shi zhe chubanshe, 2018.

———. "Xiangxing huiyi: Manwen yu Qingdai yishu shi yanjiu" 象形會意—滿文與清代藝術史研究 ("Symbolizing physical shape" and "associating ideas": Studies on the Manchu language and Qing art history). In *Qingshi lunji* 清史論集, vol. 22: 239–304. Taipei: Wen shi zhe chubanshe, 2012.

———. *Xie Sui "Zhigong tu" Manwen tushuo jiaozhu* 謝遂《職貢圖》滿文圖說校註 (A critical edition of the Manchu legends to the images in Xie Sui's *Album of Tributary Peoples*). Taipei: Guoli Gugong bowuyuan, 1989.

Chunhua 春花. "Gugong cang zhenben *Yuzhi jian Han Qingwen jian*: Jian tan Qing Neifu kanke, shoucang de Man-, Mengwen cidian" 故宮藏珍本《御制兼汉清文鉴》—兼谈清内府刊刻、收藏的满蒙文词典 (*Imperially Commissioned Mirror of the Manchu Language with Chinese Included*, a rare book in the collection of the Palace [Museum Library]: With a discussion on the Manchu and Mongol dictionaries printed and held by the Qing Imperial Household Department). *Zijincheng* 紫禁城, no. 8 (2010): 72–79.

———. *Qingdai Man-, Mengwen cidian yanjiu* 清代满蒙文词典研究 (Research on Manchu- and Mongol-language dictionaries of the Qing period). Shenyang: Liaoning minzu chubanshe, 2008.

———. "*Yuzhi Qingwen jian* leimu tixi laiyuan kao"《御制清文鉴》类目体系来源考 (Examination of the origin of the system of categories in the *Imperially Commissioned Mirror of the Manchu Language*). *Shenyang Gugong Bowuyuan yuankan* 沈阳故宫博物院院刊 3 (2007): 107–15.

Cibot, Pierre-Martial. "Observations de Physique et d'Histoire naturelle de l'Empereur Kang-hi." In *Mémoires concernant les Chinois*, vol. 4: 452–83. Paris: Nyon l'aîné, 1779.

Coblin, Weldon South. "An Introductory Study of Textual and Linguistic Problems in Erh-Ya." PhD diss., University of Washington, 1972.

Cook, Harold J. *Matters of Exchange: Commerce, Medicine, and Science in the Dutch Golden Age*. New Haven, CT: Yale University Press, 2007.

Cooley, Mackenzie. "Animal Empires: The Perfection of Nature between Europe and the Americas, 1492–1630." PhD diss., Stanford University, 2018.

Corff, Oliver, Kyoko Maezono, Wolfgang Lipp, Dorjpalam Dorj, Görööchin Gerelmaa, Aysima Mirsultan, Réka Stüber, Byambajav Töwshintögs, and Xieyan Li, eds. *Auf kaiserlichen Befehl erstelltes Wörterbuch des Manjurischen in fünf Sprachen "Fünfsprachenspiegel": Systematisch angeordneter Wortschatz auf Manjurisch, Tibetisch, Mongolisch, Turki und Chinesisch*. Wiesbaden: Harrassowitz, 2013.

Couvreur, Séraphin, trans. *Li ki: Ou, mémoires sur les bienséances et les cérémonies*. 2 vols. Ho Kien Fou: Imprimerie de la Mission catholique, 1913.

Crossley, Pamela Kyle. "An Introduction to the Qing Foundation Myth." *Late Imperial China* 6, no. 2 (1985): 13–24.

———. *A Translucent Mirror: History and Identity in Qing Imperial Ideology*. Berkeley: University of California Press, 1999.

Crossley, Pamela Kyle, and Evelyn S. Rawski. "A Profile of the Manchu Language in Ch'ing History." *Harvard Journal of Asiatic Studies* 53, no. 1 (1993): 63–102.

Csillag, Eszter. "Natural History Illustrations in Michael Boym's Chinese Atlas (*Borg. Cin. 531*) and *Flora Sinensis*." *Miscellanea Bibliothecae Apostolicae Vaticanae* 26 (2020): 115–41.

Da Dai liji 大戴禮記 (Record of rites of the elder Dai). Vol. 128. *Wenyuan Ge Siku quanshu* 文淵閣四庫全書. 1782. Reprint, Taipei: Taiwan Shangwu yinshu guan, 1983.

Dai Jianguo 戴建国. Yuanjian leihan *yanjiu*《渊鉴类函》研究 (A study of the *Categorized Boxes of the Profound Mirror Studio*). Shanghai: Dongfang chuban zhongxin, 2014.

Daicing gurun-i abkai wehiyehe-i gūsin duici aniya, sohon ihan, erin forgon-i ton-i bithe (Book of calculations of propitious times for the 34th year of

Qianlong of the Great Qing state, the year of the yellow cow). Digitized blockprint, call no. mandchou 192, 19769. Bibliothèque nationale de France, Paris.

Daicing gurun-i elhe taifin-i juwan uyuci aniya-i forgon-i yargiyan ton (True calculations of propitious times for the 19th year of Kangxi of the Great Qing state). Digitized blockprint, call. no. Phillipps 1986, 1680. Staatsbibliothek zu Berlin.

Daicing gurun-i g'aodzung yongkiyangga hūwangdi-i yargiyan kooli (Veritable records for the Lofty Progenitor, Emperor Chun of the Great Qing state). Manuscript, call no. 故宮012976. National Palace Museum Library, Taipei.

Daicing gurun-i yooningga dasan-i jakūci aniya sohon meihe erin forgon-i ton-i bithe (Book of calculations of propitious times for the 8th year of Tongzhi, the year of the yellow snake). Digitized blockprint, call no. Moellendorff 88, 1869. Staatsbibliothek zu Berlin.

Daigu 戴穀. *Manju gisun-i yongkiyame toktobuha bithe | Qingwen beikao* 清文備考 (Complete and definite book of the Manchu language). First edition. Microfilm. Tenri University Library, 1722.

Daston, Lorraine, and Glenn W. Most. "History of Science and History of Philologies." *Isis* 106, no. 2 (2015): 378–90.

De Troia, Paolo. "Real and Unreal Animals in Jesuit Geographical Works in China: A Research Note." *Sulla via del Catai* 15, no. 26 (2022): 123–55.

Delbourgo, James. *Collecting the World: Hans Sloane and the Origins of the British Museum*. Cambridge, MA: Harvard University Press, 2019.

Dennis, Joseph. "Early Printing in China Viewed from the Perspective of Local Gazetteers." In *Knowledge and Text Production in an Age of Print: China, 900–1400*, edited by Lucille Chia and Hilde De Weerdt, 105–34. Leiden: Brill, 2011.

Dennis, Joseph R. *Writing, Publishing, and Reading Local Gazetteers in Imperial China, 1100–1700*. Cambridge, MA: Harvard University Asia Center, 2015.

Di Cosmo, Nicola. "The Qing and Inner Asia: 1636–1800." In *The Cambridge History of Inner Asia: The Chinggisid Age*, edited by Nicola Di Cosmo, Allen J. Frank, and Peter B. Golden, 333–62. Cambridge: Cambridge University Press, 2009.

Doerfer, Gerhard. *Mongolo-Tungusica*. Wiesbaden: Harrassowitz, 1985.

———. "Terms for Aquatic Animals in the Wu T'i Ch'ing Wên Chien." In *Proceedings of the International Symposium on B. Pilsudski's Phonographic Records and the Ainu Culture*, edited by the Executive Committee of the International Symposium, Hokkaido University, 190–202.

Sapporo: Executive Committee of the International Symposium, Hokkaido University, 1985.

Duan Yucai 段玉裁. Shuowen jiezi zhu 說文解字注 (Commentary to *Explain the Graphs to Unravel the Written Words*). Edited by Xu Weixian 許惟賢. Critical edition. 2 vols. *Duan Yucai quanji* 段玉裁全集 1. 1815. Reprint, Nanjing: Fenghuang chubanshe, 2007.

Dunnell, Ruth W., and James A. Millward. Introduction to *New Qing Imperial History: The Making of Inner Asian Empire at Qing Chengde*, edited by James A. Millward, Ruth W. Dunnell, Mark C. Elliott, and Philippe Forêt, 1–12. London: RoutledgeCurzon, 2004.

Durand, Pierre-Henri. *Lettrés et pouvoirs: Un procès littéraire dans la Chine impériale*. Paris: Éditions de l'École des hautes études en sciences sociales, 1992.

Durrant, Stephen. "Sino-Manchu Translations at the Mukden Court." *Journal of the American Oriental Society* 99, no. 4 (1979): 653–61.

Edmonds, Richard L. "The Willow Palisade." *Annals of the Association of American Geographers* 69, no. 4 (1979): 599–621.

Elliott, Mark C. "Abel-Rémusat, la langue mandchoue et la sinologie." *Comptes rendus des séances de l'Académie des Inscriptions et Belles-Lettres*, April 2014, 973–93.

———. *Emperor Qianlong: Son of Heaven, Man of the World*. New York: Longman, 2009.

———. "The Limits of Tartary: Manchuria in Imperial and National Geographies." *Journal of Asian Studies* 59, no. 3 (2000): 603–46.

———. *The Manchu Way: The Eight Banners and Ethnic Identity in Late Imperial China*. Stanford, CA: Stanford University Press, 2001.

———. "Manwen dang'an yu Xin Qingshi" 滿文檔案與新清史 (The Manchu archives and New Qing History). *Gugong xueshu jikan* 故宮學術季刊 24, no. 2 (2006): 1–18.

———. "Whose Empire Shall It Be? Manchu Figurations of Historical Process in the Early Seventeenth Century." In *Critical Readings on the Manchus in Modern China (1616–2012)*, edited by Lars Laamann, vol. 1: 251–86, 2005. Reprint, Leiden: Brill, 2013.

Elliott, Mark C., and Ning Chia. "The Qing Hunt at Mulan." In *New Qing Imperial History: The Making of Inner Asian Empire at Qing Chengde*, edited by James A. Millward, Ruth W. Dunnell, Mark C. Elliott, and Philippe Forêt, 66–83. London: RoutledgeCurzon, 2004.

Elman, Benjamin A. *Classicism, Politics, and Kinship: The Ch'ang-Chou School of New Text Confucianism in Late Imperial China*. Berkeley: University of California Press, 1990.

———. "Collecting and Classifying: Ming Dynasty Compendia and Encyclopedias (*Leishu*)." In *Qu'était-ce qu'écrire une encyclopédie en Chine?*, edited by Florence Bretelle-Establet and Karine Chemla, 131–57. *Extrême-Orient, Extrême-Occident,* extra number. Paris: Presses Universitaires de Vincennes, 2007.

———. *A Cultural History of Civil Examinations in Late Imperial China.* Berkeley: University of California Press, 2000.

———. *From Philosophy to Philology: Intellectual and Social Aspects of Change in Late Imperial China,* rev. ed. UCLA Asian Pacific Monograph Series. 1984. Reprint, Los Angeles: Asia-Pacific Institute, University of California, 2001.

———. "Rethinking 'Confucianism' and 'Neo-Confucianism' in Modern Chinese History." In *Rethinking Confucianism: Past and Present in China, Japan, Korea, and Vietnam,* edited by Benjamin A. Elman, John B. Duncan, and Herman Ooms, 518–54. UCLA Asian Pacific Monograph Series. Los Angeles: University of California, 2002.

———. *On Their Own Terms: Science in China, 1550–1900.* Cambridge, MA: Harvard University Press, 2005.

Elvin, Mark. *The Retreat of the Elephants: An Environmental History of China.* New Haven, CT: Yale University Press, 2004.

Etō Toshio 衛藤利夫. *Kenryū gyosei "Seikyō fu" ni tsuite* 乾隆御製「盛京賦」に就いて (On the *Ode to Mukden,* imperially commissioned by Qianlong). Shenyang: Hōten toshokan [Fengtian tushuguan], 1931.

Fan, Fa-ti. *British Naturalists in Qing China: Science, Empire, and Cultural Encounter.* Cambridge, MA: Harvard University Press, 2004.

Fang Chao-ying. "Hung-Li." In *Eminent Chinese of the Ch'ing Period (1644–1912),* edited by Arthur W. Hummel, vol. 1: 369–73. Washington, DC: U.S. Government Printing Office, 1944.

———. "Kao Shi-Ch'i." In *Eminent Chinese of the Ch'ing Period (1644–1912),* edited by Arthur W. Hummel, vol. 1: 413–15. Washington, DC: U.S. Government Printing Office, 1944.

———. "Yü Min-Chung." In *Eminent Chinese of the Ch'ing Period (1644–1912),* edited by Arthur W. Hummel, vol. 2: 942–44. Washington, DC: U.S. Government Printing Office, 1944.

Fang Chongding 方崇鼎 and He Yingsong 何應松. *Xiuning xian zhi* 休寧縣志 (Gazetteer of Xiuning County). Facsimile. *Zhongguo fangzhi congshu* 中國方志叢書 627, 1815. Reprint, Taipei: Chengwen chubanshe, 1985.

Fang Gongqian 方拱乾. *Ningguta zhi* 寧古塔志 (Treatise on Ningguta). Facsimile of the 1844 Shikai Tang edition. *Zhaodai congshu* 昭代叢書, *bing* 丙 series. Shanghai: Shanghai guji chubanshe, 1990.

Fang Jianchang 房建昌. "Yi-liu-ba-ba nian Faguo Yesuhui shi Zhang Cheng Menggu xingcheng kaolüe: Yi chongjian tongxing Zhang Penghe, Qian Liangze lüxing luxian tu wei zhongxin" 1688年法国耶稣会士张诚蒙古行程考略—以重建同行张鹏翮、钱良择旅行路线图为中心 (Cursory investigation into the route taken by the French Jesuit Jean-François Gerbillon in Mongolia in 1688, with a focus on reestablishing a route map for his fellow travelers Zhang Penghe and Qian Liangze). *Neimenggu shifan daxue xuebao (zhexue, shehui kexue ban)* 内蒙古师范大学学报（哲学社会科学版）49, no. 5 (2020): 34–49.

Fang Shiji 方式濟. *Longsha jilüe* 龍沙紀略 (Cursory record of the area beyond the pass). Facsimile. Vol. 2. *Zhaodai congshu* 昭代叢書, *ji* 己 series. Shanghai: Shanghai guji chubanshe, 1990.

Fèvre, Francine, and Georges Métailié. *Dictionnaire Ricci des plantes de Chine*. Paris: Association Ricci pour le grand dictionnaire de la langue chinoise & Cerf, 2005.

FHA (First Historical Archives, Diyi lishi dang'an guan 第一历史档案馆), ed. *Kangxi qiju zhu* 康熙起居注 (Record of rise and repose of the Kangxi reign). Typeset edition. 3 vols. Beijing: Zhonghua shuju, 1984.

———, ed. *Qianlong chao shangyu dang* 乾隆朝上諭檔 (Imperial edicts of the Qianlong reign). Facsimile. 18 vols. Beijing: Dang'an chubanshe, 1991.

Fortescue, Michael. *Comparative Nivkh Dictionary*. Munich: LINCOM, 2016.

Foster, Christopher John. "Study of the *Cang Jie Pian*: Past and Present." PhD diss., Harvard University, 2017.

Foucault, Michel. *The Order of Things: An Archaeology of the Human Sciences*. Reprint, New York: Vintage Books, 1984. First published in translation by Pantheon Books, 1971.

Fracasso, Riccardo. "*Shan hai ching* 山海經." In *Early Chinese Texts: A Bibliographical Guide*, edited by Michael Loewe, 357–67. Berkeley: The Society for the Study of Early China & The Institute for East Asian Studies, University of California, 1993.

Fraser, Michie Forbes Anderson. *Tanggu Meyen and Other Reading Lessons*. London: Luzac, 1924.

Fuchs, Walter. "Eine unbeachtete Mandju-Übersetzung der Vier Bücher von 1741." In *Collectanea Mongolica. Festschrift für Professor Dr. Rintchen zum 60. Geburtstag*, edited by Walther Heissig, 59–64. Wiesbaden: Harrassowitz, 1966.

Führer, Bernhard. "Seers and Jesters: Predicting the Future and Punning by Graph Analysis." *East Asian Science, Technology, and Medicine* 25 (2006): 47–68.

Gabelentz, Hans Conon von der. *Sse-schu, Schu-king, Schi-king in mandschuischer Uebersetzung mit einem Mandschu-Deutschen Wörterbuch.* 2 vols. Abhandlungen für die Kunde des Morgenlandes 3. Leipzig: Brockhaus, 1864.

Gabelentz, Hans Georg Conon von der. "Mandschu-Bücher." *Zeitschrift der Deutschen Morgenländischen Gesellschaft* 16 (1862): 538–46.

Gandolfo, Stefano. "To Collect and to Order: The *Siku Quanshu* 四庫全書 and Its Organization." *Ming Qing Yanjiu* 24 (2020): 11–45.

Gang Song and Paola Demattè. "Mapping an Acentric World: Ferdinand Verbiest's *Kunyu Quantu*." In *China on Paper: European and Chinese Works from the Late Sixteenth to the Early Nineteenth Century*, edited by Paola Demattè and Marcia Reed. Los Angeles: Getty Research Institute, 2007.

Gao Lian 高濂. "Zunsheng bajian" 遵生八牋 (Eight discourses on the respect for life). In *Gao Lian ji* 高濂集, edited by Wang Dachun 王大淳, Typeset. Hangzhou: Zhejiang guji chubanshe, 2015.

Gao Shiqi 高士奇. *Hu cong dongxun rilu* 扈從東巡日錄 (Daily notes on the journey to the East following [the emperor]). New edition. Vol. 3. *Liaohai congshu* 遼海叢書 1. 1682. Reprint, Dalian: Liaohai shushe, 1934.

———. *Saibei xiaochao* 塞北小鈔 (Short notes from north of the border). *Xiao fanghu zhai yudi congchao zhengbian* 小方壺齋輿地叢鈔正編. Shanghai: Zhuyi tang, 1877.

Gao Wangling 高王凌. *Qianlong shisan nian* 乾隆十三年 (The first thirteen years of the Qianlong reign). Beijing: Jingji kexue chubanshe, 2012.

Gaozong 高宗. *Yuzhi shi er ji* 御製詩二集 (The imperially authored poems, second installment). Vols. 1303–4. *Wenyuan Ge Siku quanshu* 文淵閣四庫全書. 1782. Reprint, Taipei: Taiwan Shangwu yinshu guan, 1983.

———. *Yuzhi shi san ji* 御製詩三集 (The imperially authored poems, third installment). Vols. 1305–6. *Wenyuan Ge Siku quanshu* 文淵閣四庫全書. 1782. Reprint, Taipei: Taiwan Shangwu yinshu guan, 1983.

———. *Yuzhi shi si ji* 御製詩四集 (The imperially authored poems, fourth installment). Vols. 1307–8. *Wenyuan Ge Siku quanshu* 文淵閣四庫全書. 1782. Reprint, Taipei: Taiwan Shangwu yinshu guan, 1983.

Gaozong Chun huangdi shilu 高宗純皇帝實錄 (Veritable records of the Lofty Progenitor, Emperor Chun). *Qing shilu* 清實錄. Beijing: Zhonghua shuju, 1986.

Giles, Herbert Allen. *A Chinese-English Dictionary*, rev. ed. London: B. Quaritch, 1912.

Gimm, Martin. *Kaiser Qianlong (1711–1799) als Poet: Anmerkungen zu seinem schriftstellerischen Werk*. Stuttgart: Franz Steiner, 1993.

Golvers, Noël. *Letters of a Peking Jesuit: The Correspondence of Ferdinand Verbiest, SJ (1623–1688)*, rev. ed. Leuven: Ferdinand Verbiest Institute, 2017.

———. *Libraries of Western Learning for China: Circulation of Western Books between Europe and China in the Jesuit Mission (ca. 1650–1750): Of Books and Readers*. Vol. 3. Leuven: Ferdinand Verbiest Institute, 2015.

Goode, Walter. "On the *Sanbao Taijian Xia Xiyang-Ji* and Some of Its Sources." PhD diss., Australian National University, 1976.

Goodrich, L. Carrington. *The Literary Inquisition of Ch'ien-Lung*. Baltimore, MD: Waverly Press, 1935.

Gorelova, Lilya M. *Manchu Grammar*. Leiden: Brill, 2002.

Graham, Angus Charles. *Disputers of the Tao: Philosophical Argument in Ancient China*. La Salle, IL: Open Court, 1989.

Greenberg, Daniel M. "Taxonomy of Empire: The Compendium of Birds as an Epistemic and Ecological Representation of Qing China." *Journal18*, no. 7 (2019). https://www.journal18.org/3710.

———. "Weird Science: European Origins of the Fantastic Creatures in the Qing Court Painting, the Manual of Sea Oddities." In *The Zoomorphic Imagination in Chinese Art and Culture*, edited by Jerome Silbergeld and Eugene Y. Wang, 379–400. Honolulu: University of Hawai'i Press, 2016.

———. "Yuan cang *Haiguai tuji* chutan: Qing gong hua zhong de Xifang qihuan shengwu" 院藏《海怪圖記》初探—清宮畫中的西方奇幻生物 (A preliminary investigation of the Museum's *Illustrated Record of Sea Monsters*: Fantastic creatures of the West in Qing court paintings). Translated by Kang Shujuan 康淑娟. *Gugong wenwu* 故宮文物, no. 297 (2007): 38–51.

Griffin, Clare. "Russia and the Medical Drug Trade in the Seventeenth Century." *Social History of Medicine* 31, no. 1 (2018): 2–23.

Gu Songjie 顾松洁 and Gao Xi 高晞. "Guanyu Manwen chaoben *Geti quanlu* de ji ge wenti" 关于满文抄本《格体全录》的几个问题 (A few issues regarding the Manchu manuscript versions of the *Manchu Anatomy*). *Qingshi yanjiu* 清史研究, no. 3 (2021): 143–50.

Gulik, R. H. *The Gibbon in China: An Essay in Chinese Animal Lore*. Leiden: Brill, 1967.

Guo Weisen 郭維森 and Xu Jie 許結. *Zhongguo cifu fazhan shi* 中國辭賦發展史 (A history of the development of the Chinese rhapsody). Nanjing: Jiangsu jiaoyu chubanshe, 1996.

Guy, R. Kent. *The Emperor's Four Treasuries: Scholars and the State in the Late Ch'ien-lung Era*. Cambridge, MA: Council on East Asian Studies, Harvard University, 1987.

―――. *Qing Governors and Their Provinces: The Evolution of Territorial Administration in China, 1644–1796.* Seattle: University of Washington Press, 2010.

Haenisch, Erich. "Die Abteilung 'Jagd' im fünfsprachigen Wörterspiegel." *Asia Major*, first series, 10, no. 1 (1934): 59–93.

Halde, J. E., ed. *Description géographique, historique, chronologique, politique, et physique de l'empire de la Chine et de la Tartarie Chinoise.* 4 vols. The Hague: Henri Scheurleer, 1736.

Hammers, Roslyn L. "Assimilating the Classics: The Qing Imperium's Reworking of Cotton Textile Production." Presented at the Association for Asian Studies Annual Conference, online, March 24, 2021.

―――. *The Imperial Patronage of Labor Genre Paintings in Eighteenth-Century China.* New York: Routledge, 2021.

Han Qi 韩琦. *Kangxi huangdi—Yesu hui shi—kexue chuanbo* 康熙皇帝·耶稣会士·科学传播 (The Kangxi emperor, the Jesuits, and the transmission of science). Beijing: Zhongguo dabaike quanshu chubanshe, 2019.

Han-i araha manju gisun-i buleku bithe (Imperially commissioned mirror of the Manchu language). Beijing: Wuying dian, 1708.

Han-i araha manju gisun-i buleku bithe | Qayan-i bičigsen manju ügen-ü toli bičig (Imperially commissioned mirror of the Manchu [and Mongolian] language[s]). Beijing: Wuying dian, 1717.

Han-i araha manju monggo gisun-i buleku bithe | Qayan-i bičigsen manju mongyol-u gen-ü toli bičig (Imperially commissioned mirror of the Manchu and Mongolian languages). Beijing: Wuying dian, 1743. Call no. 19951.a.39. The British Library, London.

Han-i araha Mukden-i fu bithe | Yuzhi Shengjing fu 御製盛京賦 (Imperially commissioned ode to Mukden). Blockprint, 1743. Digitized copy, call no. FC.99.25. Cambridge University Library.

Han-i araha Mukden-i fujurun bithe (Imperially commissioned ode to Mukden). Blockprint, 1748. Digitized copy, call no. Moellendorff 53–1. Staatsbibliothek zu Berlin.

Han-i araha Ši ging bithe (Imperially commissioned edition of the *Poetry Classic*). Blockprint, 20 vols. Beijing, 1655. Call no. 故滿001016–001035. National Palace Museum Library, Taipei.

Hanson, Marta E. "Manchu Medical Manuscripts and Blockprints: An Essay and Bibliographic Survey." *Saksaha* 8 (2003): 1–32.

―――. "The Significance of Manchu Medical Sources in the Qing." In *Proceedings of the First North American Conference on Manchu Studies,* edited by Stephen Wadley and Carsten Naeher, vol. 1: 131–75. Wiesbaden: Harrassowitz, 2006.

Hao Yixing 郝懿行. *Erya yishu* 爾雅義疏 (Sub-commentary on the meaning of *Approaching Perfection*). Edited by Wang Qihe 王其和, Wu Qingfeng 吳慶峰, and Zhang Jinxia 張金霞. 2 vols. *Shisan jing Qingren zhu shu* 十三經清人注疏. Beijing: Zhonghua shuju, 2017.

Harbsmeier, Christoph. "On the Very Notions of Language and of the Chinese Language." *Histoire Épistémologie Langage* 31, no. 2 (2009): 143–61.

Hargett, James M. "Whales in Ancient China." In *Maritime Animals in Traditional China*, edited by Roderich Ptak, 93–119. Wiesbaden: Harrassowitz, 2010.

Hasbagan and Chen Shan. "The Cultural Importance of Animals in Traditional Mongolian Plant Nomenclature." In *Culture and Environment in Inner Asia*, edited by Caroline Humphrey and David Sneath, vol. 2: 25–29. Cambridge: The White Horse Press, 1996.

Hauer, Erich. "Ein Thesaurus der Mandschusprache." *Asia Major*, first series, 7 (1932): 629–41.

———. *Handwörterbuch der Mandschusprache*. 3 vols. Wiesbaden: Harrassowitz, 1952.

———. *Handwörterbuch der Mandschusprache*. 2nd ed. Edited by Oliver Corff. Wiesbaden: Harrassowitz, 2007.

Hayata Teruhiro 早田輝洋 and Teramura Masao 寺村政男, eds. *Daishin zensho: Zōho kaitei, tsuketari Manshūgo, Kango sakuin* 大清全書：增補改訂・附満洲語漢語索引 (Expanded and emended edition of *Complete Book of the Great Qing*, with appended Manchu and Chinese indexes). 3 vols. Fuchū: Tōkyō gaikokugo daigaku Ajia Afurika gengo bunka kenkyūjo, 2004.

He Qiutao 何秋濤. *Shuofang beisheng* 朔方備乘 (Historical records for the defense of the North). Facsimile. Vols. 740–42. *Xuxiu Siku quanshu* 續修四庫全書. 1881. Reprint, Shanghai: Shanghai guji chubanshe, 1995–2002.

He, Yingtian. "Well-Ordered Textures: The Book of Odes and the Study of Wu in Mid-Qing China." PhD diss., Princeton University, 2023.

He, Yuming. *Home and the World: Editing the "Glorious Ming" in Woodblock-Printed Books of the Sixteenth and Seventeenth Centuries*. Cambridge, MA: Harvard-Yenching Institute, 2013.

Hiowan yei 玄燁. Kangxi ji xia gewu bian *yizhu* 康熙几暇格物编译注 (Collection of the Investigation of Things during Times of Leisure from the Myriad Affairs [of State] in the Kangxi Period, translated and annotated). Edited by Li Di 李迪. Shanghai: Shanghai guji chubanshe, 1993.

———. *Yuzhi wenji* 御製文集 (Collection of imperial prose). Facsimile. *Qingdai shiwenji huibian* 清代詩文集彙編 194. Shanghai: Shanghai guji chubanshe, 2009.

Hirayama Teruo 平山輝男. "Ko Uehara Hisashi kyōju wo shinobu" 故上原久教授をしのぶ (Remembering the late Professor Uehara Hisashi). *Onsei kenkyū* 音声研究 2, no. 1 (1998): 96.

Hostetler, Laura. *Qing Colonial Enterprise: Ethnography and Cartography in Early Modern China*. Chicago, IL: University of Chicago Press, 2001.

Hu Zengyi 胡增益. *Xin Man-Han da cidian* 新满汉大词典 (New and comprehensive Manchu-Chinese dictionary). Urumqi: Xinjiang renmin chubanshe, 1994.

Huainan honglie 淮南鴻烈 (The radiant light of Huainan). Vol. 848. *Wenyuan Ge Siku quanshu* 文淵閣四庫全書. 1782. Reprint, Taipei: Taiwan Shangwu yinshu guan, 1983.

Huang Aiping 黃爱平. *Siku quanshu zuanxiu yanjiu* 四库全书纂修研究 (A study on the compilation of the *Complete Books of the Four Repositories*). Beijing: Zhongguo renmin daxue chubanshe, 1989.

Huang Juan 黃娟. "'Qingshu shujishi' kaoxi" "清书庶吉士"考析 (Examination and analysis of the "Manchu-language Hanlin Bachelors"). In *Guoji qingnian xuezhe Manxue yanjiu lunji (2015)* 国际青年学者满学研究论集 (2015), edited by Liu Xiaomeng 刘小萌, 114–33. Beijing: Zhongguo shehui kexue chubanshe, 2017.

Huang Sin-yu 黃莘瑜. "Yuan, pu, wenxue yu Huaxu guo: Renwen shiyu xia de *Er ru ting qunfang pu*" 園、譜、文學與華胥國—人文視域下的《二如亭群芳譜》(Garden, catalogs, literature and the ancient kingdom: The *Catalog of Myriad Flowers from the Pavilion of Double Resemblance* examined from a humanistic perspective). *Tai Da Zhongwen xuebao* 臺大中文學報 55 (2016): 53–96.

Huang Wei-Jen 黃韋仁. "*Yuzhi zengding Qingwen jian* Man-Han duiyin yanjiu" 《御製增訂清文鑑》滿漢對音研究 (A study of the Manchu-Chinese transcriptions in the *Imperially Commissioned Expanded and Emended Mirror of the Manchu Language*). Master's thesis, National Taiwan Normal University, 2020.

Hucker, Charles O. *A Dictionary of Official Titles in Imperial China*. Stanford, CA: Stanford University Press, 1985.

Hūng Jeo 弘晝, Maci 馬奇, and Ortai 鄂爾泰, eds. *Baqi tongzhi [chuji]* 八旗通志 [初集] (Comprehensive treatises of the Eight Banners [first installment]). Blockprint. Beijing: Wuying dian, 1739. Digitized copy, call no. T 4718 2008. Harvard-Yenching Library, Cambridge, MA.

Hūng Jeo, Maci, and Ortai, eds. *Han-i araha jakūn gūsai tung j'i bithe* (Comprehensive treatises of the Eight Banners). Blockprint, Beijing:

Wuying dian, 1739. Digitized copy, call no. Libri. sin. N.S. 1916. Staatsbibliothek zu Berlin.

Ianaccone, Isaia. "Lo zoo dei Gesuiti: La transmissione scientifica del bestiario rinascimentale europeo alla Cina dei Qing in *Kunyu tushuo* di Ferdinand Verbiest (1674)." In *Studi in onore di Lionello Lanciotti*, edited by S. M. Carletti, M. Sacchetti, and P. Santangelo, 739–64. Naples: Istituto Universitario Orientale, 1996.

Idema, Wilt L., trans. *Two Centuries of Manchu Women Poets: An Anthology*. Seattle: University of Washington Press, 2017.

Ihing 宜興. *Qingwen buhui* 清文補彙 | *Manju gisun be niyeceme isabuha bithe* (Manchu collected, supplemented). Blockprint of 2nd edition, 1802. Digitized copy, 8 vols. Waseda University Library, Japan.

Ikegami Jirō 池上二良, ed. *Uiruta go jiten* ウイルタ語辞典 (A dictionary of the Uilta language). Sapporo: Hokkaido daigaku toshokan kankōkai, 1997.

Ilire tere be ejehe dangse (Archival records of rise and repose). Manuscript. Call nos. 故官 008526, 故官009456. National Palace Museum Library, Taipei.

Imanishi Shunjū 今西春秋. "*Shinbunkan*: Tantai kara gotai made" 清文鑑—単体から五体まで (The *Mirrors of the Manchu Language*: From monolingual to pentaglot). *Chōsen gakuhō* 朝鮮学報, no. 39/40 (1966): 121–63.

———. "*Zōtei Shinbunkan* no ihan ni tsuite" 増訂清文鑑の異版に就いて (On the different editions of the *Expanded and Emended Mirror of the Manchu Language*). *Shilin* 史林 23, no. 4 (1938): 219–26.

Imanishi Shunjū 今西春秋, Tamura Jitsuzō 田村実造, and Satō Hisashi 佐藤長, eds. *Gotai Shinbunkan yakukai* 五體清文鑑譯解 (*Mirror of the Manchu Language in Five Scripts*, translated and explained). 2 vols. Kyōto: Kyōto daigaku bungakubu nairiku Ajia kenkyūjo, 1966.

'Jam-dpal-rdo-rje. *An Illustrated Tibeto-Mongolian Materia Medica of Āyurveda*. Edited by Lokesh Chandra. Facsimile. Delhi: Jayyed Press, 1971.

———. *Mengyao zhengdian* 蒙药正典 (True canon of Mongolian medicines). Edited by Liubaiyila 柳白乙拉. Chinese translation. Beijing: Minzu chubanshe, 2006.

Jami, Catherine. *The Emperor's New Mathematics: Western Learning and Imperial Authority during the Kangxi Reign (1662–1722)*. Oxford: Oxford University Press, 2012.

———. "Imperial Science Written in Manchu in Early Qing China: Does It Matter?" In *Looking at It from Asia: The Process that Shaped the Sources of History of Science*, edited by Florence Bretelle-Establet, 371–91. Dordrecht: Springer, 2010.

———. "Investigating Things under Heaven: Imperial Mobility and the Kangxi Emperor's Construction of Knowledge." In *Individual Itineraries and the Spatial Dynamics of Knowledge: Science, Technology, and Medicine in China, 17th–20th Centuries*, edited by Catherine Jami, 173–205. Paris: Collège de France, Institut des hautes études chinoises, 2017.

———. "Western Learning and Imperial Scholarship: The Kangxi Emperor's Study." *East Asian Science, Technology, and Medicine* 27 (2007): 146–72.

Janhunen, Juha. "Unicorn, Mammoth, Whale: Mythological and Etymological Connections of Zoonyms in North and East Asia." In *Linguistics, Archaeology and the Human Past*, edited by Toshiki Osada and Hitoshi Endo, 189–222. Kyoto: Research Institute for Humanity and Nature, 2011.

Jarring, Gunnar. *An Eastern Turki-English Dialect Dictionary*. Lund: Håkan Ohlssons Boktryckeri, 1964.

Ji Yun 紀昀, ed. *Qinding Siku quanshu zongmu* 欽定四庫全書總目 (General catalog of the *Imperially Authorized Complete Books of the Four Repositories*). Facsimile. Vols. 1–5. *Wenyuan Ge Siku quanshu* 文淵閣四庫全書. 1782. Reprint, Taipei: Taiwan Shangwu yinshu guan, 1983.

Jiang Qiao 江桥. *Kangxi Yuzhi Qingwen jian yanjiu* 康熙《御制清文鉴》研究 (Research on Kangxi's *Imperially Commissioned Mirror of the Manchu Language*). Beijing: Beijing Yanshan chubanshe, 2009.

Jin Taofang 金泰方. "Manwen shi *Shengjing fu, songci* de yishu tese" 满文诗《盛京赋·颂词》的艺术特色 (The artistic qualities of the Manchu-language poem *Ode to Mukden, Hymn of Praise*). *Manzu yanjiu* 满族研究, no. 1 (1985): 41, 71–75.

Juntu 屯圖. *Yi xue san guan Qingwen jian* 一學三貫清文鑑 (Mirror of the Manchu language, which will direct you to three things when you consult only one). Facsimile. *Gugong zhenben congkan* 故宮珍本叢刊 723. 1746. Reprint, Haikou: Hainan chubanshe, 2001.

Kam, Tak-Sing. "The Romanization of the Early Manchu Regnal Names." *Studia Orientalia* 87 (1999): 133–48.

Kane, Daniel. *The Sino-Jurchen Vocabulary of the Bureau of Interpreters*. Bloomington: Indiana University Research Institute for Inner Asian Studies, 1989.

Karlgren, Bernhard. *The Book of Odes*. Stockholm: Museum of Far Eastern Antiquities, 1950.

———. "Loan Characters in Pre-Han Texts." *Bulletin of the Museum of Far Eastern Antiquities*, no. 35 (1963): 1–128.

Keliher, Macabe. "Administrative Law and the Making of the First *Da Qing Huidian*." *Late Imperial China* 37, no. 1 (2016): 55–107.

Kicengge 承志 (Chengzhi). *Daichin Gurun to sono jidai: Teikoku no keisei to hakki shakai* ダイチン・グルンとその時代―帝国の形成と八旗社会― (*Daicing gurun* [the Great Qing state] and its time: The formation of the empire and Eight Banner society). Nagoya: Nagoya daigaku shuppankai, 2009.

———. "Manwen *Wula deng chu difang tu kao*" 滿文《烏喇等處地方圖》考 (The Manchu-language *Ula Region Map*). *Central Asiatic Journal* 59, no. 1/2 (2016): 179–238.

Kicentai 庄声 (Zhuangsheng). *Teikoku o tsukutta gengo seisaku: Daichin gurun shoki no gengo seikatsu to bunka* 帝国を創った言語政策―ダイチン・グルン初期の言語生活と文化 (The language policies that created an empire: Linguistic life and culture in the early period of the Qing state). Kyōto: Kyōto daigaku gakujutsu shuppankai, 2016.

Kim, Loretta. *Ethnic Chrysalis: China's Orochen People and the Legacy of Qing Borderland Administration*. Cambridge, MA: Harvard University Asia Center, 2019.

Klaproth, Julius. *Chrestomatie mandchou, ou recueil de textes mandchou*. Paris: Imprimerie Royale, 1828.

Kleutghen, Kristina. *Imperial Illusions: Crossing Pictorial Boundaries in the Qing Palaces*. Seattle: University of Washington Press, 2015.

Knechtges, David R. *The Han Rhapsody: A Study of the Fu of Yang Hsiung (53 B.C.–A.D. 18)*. Cambridge: Cambridge University Press, 1976.

———. "Problems of Translating Descriptive Binomes in the *Fu*." *Tamkang Review* 15 (1984): 329–47.

Knechtges, David R., and Taiping Chang, eds. *Ancient and Early Medieval Chinese Literature: A Reference Guide, Part Two*. Leiden: Brill, 2014.

Ko, Dorothy. *The Social Life of Inkstones: Artisans and Scholars in Early Qing China*. Seattle: University of Washington Press, 2017.

Köhle, Natalie. "Why Did the Kangxi Emperor Go to Wutai Shan?: Patronage, Pilgrimage, and the Place of Tibetan Buddhism at the Early Qing Court." *Late Imperial China* 29, no. 1 (2008): 73–119.

Kornicki, Peter Francis. *Languages, Scripts, and Chinese Texts in East Asia*. Oxford: Oxford University Press, 2018.

Koroloff, Rachel. "In the Business of Gardens: The Management of the Tsar's Vineyards along the Volga River at the Turn of the Eighteenth Century." Presented at the History of Science Graduate Workshop, "Trading Objecthood: Global Business and the Language of Natural History in the Long Nineteenth Century," Princeton University, February 8–9, 2019.

Kowalewski, Joseph Étienne. *Dictionnaire mongol-russe-français*. 3 vols. 1844. Reprint, Tientsin, 1941.

Krempien, Rainer. "Wolfgang Seuberlich zum 75. Geburtstag." *NOAG* 129 (1981): 6–10.
Kroll, Paul W. "The Significance of the *fu* in the History of T'ang Poetry." *Tang Studies* 18–19 (2000): 87–105.
Küçük, Harun. *Science without Leisure: Practical Naturalism in Istanbul, 1660–1732.* Pittsburgh, PA: University of Pittsburgh Press, 2020.

Lai, Yu-chih. "Domesticating the Global and Materializing the Unknown: A Study of the *Album of Beasts* at the Qianlong Court." In *EurAsian Matters: China, Europe, and the Transcultural Object, 1600–1800*, edited by Anna Grasskamp and Monica Juneja, 125–71. Cham, CH: Springer, 2018.
———. "Images, Knowledge and Empire: Depicting Cassowaries in the Qing Court." Translated by Philip Hand. *Transcultural Studies*, no. 1 (2013): 7–100.
Lai Yu-chih 賴毓芝. "Qianlong chao *Niaopu* de qianshen: Jiang Tingxi hui *Niaopu*" 乾隆朝《鳥譜》的前身：蔣廷錫繪《鳥譜》(The forerunner of the Qianlong-court *Album of Birds* [i.e., *Catalog of Birds*]: Jiang Tingxi's *Album of Birds* [*Catalog of Birds*]). Chapter draft presented at the Institute of Modern History colloquium, Academia Sinica, Taipei, October 7, 2021.
———. "Qing gong dui Ouzhou ziran shi tuxiang de zaizhi: Yi Qianlong chao *Shoupu* wei li" 清宮對歐洲自然史圖像的再製：以乾隆朝《獸譜》為例 (Reproducing Renaissance naturalist images and knowledge at the Qianlong court: A study of the *Album of Beasts* [i.e., *Catalog of Beasts*]). *Zhongyang yanjiu yuan jindaishi yanjiusuo jikan* 中央研究院近代史研究所集刊, no. 80 (2013): 1–75.
———. "Qinggong, Ouzhou yu Riben: Qinggong *Niaopu* de chengli yu liubo" 清宮、歐洲與日本：清宮《鳥譜》的成立與流播 (The Qing Court, Europe, and Japan: Webbing the production of the *Album of Birds* [i.e., *Catalog of Birds*] and its proliferation in Asia). Taipei, 2015.
———. "Zhishi, xiangxiang yu jiaoliu: Nan Huairen *Kunyu quantu* zhi shengwu chahui yanjiu" 知識、想像與交流：南懷仁《坤輿全圖》之生物插繪研究 (Knowledge, imagination, and exchange: A study of the animal illustrations in Ferdinand Verbiest's *Complete Map of the World*). In *Gantong shenshou: Zhong-Xi wenhua jiaoliu beijing xia de ganguan yu ganjue* 感同身受—中西文化交流背景下的感官與感覺, edited by Dong Shaoxin 董少新, 141–82. Shanghai: Fudan daxue chubanshe, 2018.
Lasari, ed. *Inenggi giyangnaha sy šu-i jurgan be suhe bithe* (Annotations to the daily lectures on the meaning of the Four Books). Digitized blockprint, call no. Libri sin. N.S. 1546. Beijing, 1677. Staatsbibliothek zu Berlin.
Lau, D. C., trans. *The Analects.* 1979. Reprint, Hong Kong: The Chinese University of Hong Kong Press, 1992.

Le Blanc, Charles. "*Huai nan tzu* 淮南子." In *Early Chinese Texts: A Bibliographical Guide*, edited by Michael Loewe, 189–95. Berkeley: Society for the Study of Early China & Institute for East Asian Studies, University of California, 1993.

Legge, James, trans. *The Chinese Classics*. 1893. Facsimile. Reprint, Taipei: SMC Publishing, 1991.

———. *The Sacred Books of China: The Texts of Confucianism*. Oxford: Clarendon Press, 1885.

Lessing, Ferdinand D., Mattai Haltod, John Gombojab Hanggin, and Serge Kassatkin, eds. *Mongolian-English Dictionary*. Berkeley: University of California Press, 1960.

Li Shizhen 李時珍. *Bencao gangmu* 本草綱目 (Systematic materia medica). Vols. 772–74. *Wenyuan Ge Siku quanshu* 文淵閣四庫全書. 1782. Reprint, Taipei: Taiwan Shangwu yinshu guan, 1983.

Li Xian 李賢. *Ming yitong zhi* 明一統志 (Unified gazetteer of the Great Ming). Facsimile. Vols. 472–73. *Wenyuan Ge Siku quanshu* 文淵閣四庫全書. 1782. Reprint, Taipei: Taiwan Shangwu yinshu guan, 1983.

Li Xiongfei 李雄飞. "*Yuzhi zengding Qingwen jian* keben chutan"《御制增订清文鉴》刻本初探 (Tentative exploration of the blockprinted editions of the *Imperially Commissioned Expanded and Emended Mirror of the Manchu Language*). *Manyu yanjiu* 满语研究, no. 1 (2013): 37–43.

Li Yanji 李延基. *Qingwen huishu* 清文彙書 | *Manju isabuha bithe*. Facsimile. Vol 719. *Gugong zhenben congkan* 故宫珍本叢刊. 1750. Reprint, Haikou: Hainan chubanshe, 2001.

Lin Shih-hsuan 林士鉉. "Huang yi pei du, shi wei di xiang: Qianlong huangdi yu Man-, Hanwen *Yuzhi Shengjing fu*" 皇矣陪都，實惟帝鄉：乾隆皇帝與滿、漢文《御製盛京賦》("August, the secondary capital; it truly is the imperial homeland": The Qianlong emperor and the Manchu-Chinese bilingual *Imperially Commissioned Ode to Mukden*). *Gugong wenwu yuekan* 故宮文物月刊, no. 367 (2013): 52–67.

———. "Meiguo Huashengdun daxue tushuguan diancang Man-, Hanwen zhuanzi *Yuzhi Shengjing fu* banben tanjiu" 美国华盛顿大学图书馆典藏满汉文篆字《御制盛京赋》版本探究 (An investigation into the Manchu-Chinese seal script edition of the *Imperially Commissioned Ode to Mukden* held by the University of Washington Library). In *Qingqian lishi yu Shengjing wenhua: Qingqian shi yanjiu zhongxin chengli ji jinian Shengjing dingming sanbai bashi zhounian xueshu yantao hui* 清前历史与盛京文化：清前史研究中心成立暨纪念盛京定名380周年学术研讨会, edited by Bai Wenyu 白文煜, 1:373–88. Shenyang: Liaoning minzu chubanshe, 2015.

———. "Qianlong nianjian duiyu Dongbei xiangcao de kaocha yu Yu Xing Jiachan jianxiang ce" 乾隆年間對於東北香草的考察與余省〈嘉產薦馨〉冊 (The investigations of northeastern fragrant herbs in the Qianlong period and Yu Xing's album *Auspicious Products of Commendable Fragrance*). Presented at the Exploring Ecological Art History 探索生態藝術史 workshop, National Taiwan University, September 10, 2021.

———. "Qianlong shidai de gongma yu Manzhou zhengzhi wenhua" 乾隆時代的貢馬與滿洲政治文化 (Tribute horses in the Qianlong period and Manchu political culture). *Gugong xueshu jikan* 故宮學術季刊 24, no. 2 (2006): 51–108.

———. *Qingdai Menggu yu Manzhou zhengzhi wenhua* 清代蒙古與滿洲政治文化 (The Mongols in the Qing period and Manchu political culture). Taipei: Guoli Zhengzhi daxue lishi xue xi, 2009.

Linke, Bernd-Michael. *Zur Entwicklung des mandjurischen Khanats zum Beamtenstaat: Sinisierung und Bürokratisierung der Mandjuren während der Eroberungszeit*. Wiesbaden: Harrassowitz, 1982.

Liu, Cary Y. "Archive of Power: The Qing Dynasty Imperial Garden-Palace at Rehe." *Guoli Taiwan daxue meishu shi yanjiu jikan* 國立臺灣大學美術史研究集刊 28 (2010): 43–82.

Liu Chao-ming 劉昭民. "Qingchu *Ji xia gewu bian* zhong de kexue shiliao" 清初幾暇格物編中的科學史料 (Scientific sources in the early Qing *Collection of the Investigation of Things during Times of Leisure from the Myriad Affairs [of State]*). *Si yu yan* 思與言 24, no. 2 (1986): 167–78.

Liu, Shi-yee. "Containing the West in the Manchu Realm? Emperor Qianlong's Deer Antler Scrolls." *Orientations* 46, no. 6 (2015): 58–69.

Liu, Yan. *Healing with Poisons: Potent Medicine in Medieval China*. Seattle: University of Washington Press, 2021.

Long Yun 龙云. "Qian Deming fayi *Shengjing fu*: Banben, jiazhi, zhengwu" 钱德明法译《盛京赋》：版本、价值、证误 (Joseph-Marie Amiot's French translation of the *Ode to Mukden*: Editions, value, and errata). *Guoji Hanxue* 国际汉学, no. 4 (2021): 169–78.

Lu Dian 陸佃. *Piya* 埤雅 (Supplement to *Approaching Perfection*). Edited by Wang Minhong 王敏紅. Critical edition. Hangzhou: Zhejiang daxue chubanshe, 2008.

Lu Dingpu 陸定圃. *Lenglu yihua* 冷廬醫話 (Medical talk from the cold hut). Facsimile. *Zhongguo yixue dacheng* 中國醫學大成 39. 1859. Reprint, Shanghai: Shanghai kexue jishu chubanshe, 1990.

Lu Ji 陸機. *Mao Shi cao mu niao shou chong yu shu* 毛詩草木鳥獸蟲魚疏 (Flora and fauna in the Mao tradition of the *Poetry Classic*). Facsimile.

Vol. 70. *Wenyuan Ge Siku quanshu* 文淵閣四庫全書. 1782. Reprint, Taipei: Taiwan Shangwu yinshu guan, 1983.

Lu, Tina. *Accidental Incest, Filial Cannibalism, and Other Peculiar Encounters in Late Imperial Chinese Literature*. Cambridge, MA: Harvard University Asia Center, 2008.

Ma Zimu 马子木 and Borjigidai Oyunbilig 乌云毕力格. "'Tongwen zhi zhi': Qingchao duo yuwen zhengzhi wenhua de gouni yu shijian" "同文之治"：清朝多语文政治文化的构拟与实践 ("Rule in the standardized script": Structure and practice of the Qing dynasty's multilingual political culture). *Minzu yanjiu* 民族研究, no. 4 (2017): 82–94.

Magnani, Arianna. "Searching for *Sirenes* in the 17th and 18th Centuries: Fantastic Taxonomies of Anthropomorphic Fish in Chinese and Jesuit Texts." *Sulla via del Catai* 15, no. 26 (2022): 87–105.

Magnus, Olaus. *A Description of the Northern Peoples, 1555*. Edited by Peter G. Foote and John Granlund. Translated by Peter Fisher and Humphrey Higgens. 3 vols. London: The Hakluyt Society, 1998.

———. *Historia de gentibus septentrionalibus*. Rome, 1555.

Marcon, Federico. "The Critical Promises of the History of Knowledge: Perspectives from East Asian Studies." *History and Theory* 59 (2020): 19–47.

———. *The Knowledge of Nature and the Nature of Knowledge in Early Modern Japan*. Chicago, IL: University of Chicago Press, 2015.

Mathieu, Rémi. *Étude sur la mythologie et l'ethnologie de la Chine ancienne*. Vol. 2. Paris: Collège de France, Institut des Hautes Études Chinoises, 1983.

Matsuura Shigeru 松浦茂. *Shinchō no Amūru seisaku to shōsū minzoku* 清朝のアムール政策と少数民族 (The Qing court's Amur policy and the ethnic minorities). Kyōto: Kyōto daigaku gakujutsu shuppankai, 2006.

McMahon, Keith. *Causality and Containment in Seventeenth-Century Chinese Fiction*. Leiden: Brill, 1988.

Meiguo Pulinsidun daxue Dongya tushuguan 美國普林斯頓大學東亞圖書館. *Pulinsidun daxue tushuguan cang Zhongwen shanben shumu* 普林斯頓大學圖書館藏中文善本書目 (Catalog of Chinese rare books in the Princeton University Library collection). Beijing: Guojia tushuguan chubanshe, 2017.

Menegon, Eugenio. "New Knowledge of Strange Things: Exotic Animals from the West." *Gujin lunheng* 古今論衡 15 (2006): 40–48.

Menon, Minakshi. "Making Useful Knowledge: British Naturalists in Colonial India, 1784–1820." PhD diss., University of California, San Diego, 2013.

Menzies, Nicholas K. *Ordering the Myriad Things: From Traditional Knowledge to Scientific Botany in China*. Seattle: University of Washington Press, 2021.

Meserve, Ruth I. "The Expanded Role of Mongolian Domestic Livestock Classification." *Acta Orientalia Academiae Scientiarum Hungaricae* 53, no. 1–2 (2000): 23–45.

Métailié, Georges. "Aux sources d'une vocation: L'herbier chinois d'Abel-Rémusat." In *Jean-Pierre Abel-Rémusat et ses successeurs: Deux cents ans de sinologie française en France et en Chine*, edited by Pierre-Étienne Will and Michel Zink, 116–26. Paris: Académie des Inscriptions et Belles-Lettres, 2020.

———. *Traditional Botany: An Ethnobotanical Approach*. Translated by Janet Lloyd. *Science and Civilisation in China*, vol. 6, *Biology and Biological Technology*, part 4. Cambridge: Cambridge University Press, 2015.

Meyer, Iben Raphael. "Das schamanistische Begriffsinventar des manjurischen Wörterspiegels von 1708." *Oriens Extremus* 29, no. 1/2 (1982): 173–208.

Miles, Steven B. *The Sea of Learning: Mobility and Identity in Nineteenth-Century Guangzhou*. Cambridge, MA: Harvard University Asia Center, 2006.

Miller, Roy Andrew. "*Shih ming* 釋名." In *Early Chinese Texts: A Bibliographical Guide*, edited by Michael Loewe, 424–28. Berkeley: Society for the Study of Early China & Institute for East Asian Studies, University of California, 1993.

Millward, James A., Ruth W. Dunnell, Mark C. Elliott, and Philippe Forêt, eds. *New Qing Imperial History: The Making of Inner Asian Empire at Qing Chengde*. London: RoutledgeCurzon, 2004.

Mingdo 明鐸. *Yin Han Qingwen jian* 音漢清文鑑 | *Nikan hergen-i ubaliyambuha manju gisun-i buleku bithe* (The Mirror of the Manchu language translated into Chinese). Blockprint. Beijing: Hongwen ge, 1735. Microfilm. Tenri Library, Japan.

Miu Zhijin 繆之晉, ed. *Da Qing shixian shu jianshi* 大清時憲書箋釋 (Annotations to the Great Qing calendar). Facsimile. Vol. 1040. *Xuxiu Siku quanshu* 續修四庫全書. 1723. Reprint, Shanghai: Shanghai guji chubanshe, 1995–2002.

Monnet, Nathalie. "Abel-Rémusat (1788–1832): Un autodidacte et ses livres." In *Jean-Pierre Abel-Rémusat et ses successeurs: Deux cents ans de sinologie française en France et en Chine*, edited by Pierre-Étienne Will and Michel Zink, 71–116. Paris: Académie des Inscriptions et Belles-Lettres, 2020.

———. "Le livre impérial de la dynastie des Qing : Quelques éléments d'appréciation." In *Imprimer sans profit ? Le livre non commercial dans la Chine impériale*, edited by Michela Bussotti and Jean-Pierre Drège, 425–530. Geneva: Librairie Droz, 2015.

Morrison, Robert. *A Dictionary of the Chinese Language*. 2 vols. 1815. Reprint, Shanghai: London Mission Press, 1865.

Mosca, Matthew. *From Frontier Policy to Foreign Policy: The Question of India and the Transformation of Geopolitics in Qing China*. Stanford, CA: Stanford University Press, 2013.

Mosca, Matthew W. "The Literati Rewriting of China in the Qianlong-Jiaqing Transition." *Late Imperial China* 32, no. 2 (2011): 89–132.

Murakami Nobuaki 村上信明. *Shinchō no Mōko kijin: Sono jitsuzō to teikoku tōchi ni okeru yakuwari* 清朝の蒙古旗人: その実像と帝国統治における役割 (The Mongol bannermen of the Qing dynasty: Their actual situation and their role in the administration of the empire). Tōkyō: Fūkyōsha, 2007.

Nappi, Carla S. "Making 'Mongolian' Nature: Medicinal Plants and Qing Empire in the Long Eighteenth Century." In *The Botany of Empire in the Long Eighteenth Century*, edited by Yota Batsaki, Sarah Burke Cahalan, and Anatole Tchikine, 337–47. Cambridge, MA: Harvard University Press, 2017.

———. *The Monkey and the Inkpot: Natural History and Its Transformations in Early Modern China*. Cambridge, MA: Harvard University Press, 2009.

———. *Translating Early Modern China: Illegible Cities*. Oxford: Oxford University Press, 2021.

Needham, Joseph, and Lu Gwei-Djen. *Biology and Biological Technology*, part 1, *Botany*. Science and Civilisation in China, vol. 6. Cambridge: Cambridge University Press, 1986.

Needham, Joseph, and Wang Ling. *History of Scientific Thought*. Science and Civilisation in China, vol. 2. Cambridge: Cambridge University Press, 1956.

Nie Huang 聂璜. *Qinggong haicuo tu* 清宫海错图 (The illustrations of the miscellaneous things from the sea held at the Qing palace). Edited by Gugong Bowu Yuan 故宫博物院 and Wen Jinxiang 文金祥. Beijing: Gugong chubanshe, 2014.

Norman, Jerry. *A Comprehensive Manchu–English Dictionary*. Cambridge, MA: Harvard University Asia Center, 2013.

———. "A Sketch of Sibe Morphology." *Central Asiatic Journal* 18, no. 3 (1974): 159–74.

Obringer, Frédéric. "Jean-Pierre Abel-Rémusat, médicin et sinologue." In *Jean-Pierre Abel-Rémusat et ses successeurs: Deux cents ans de sinologie française en France et en Chine*, edited by Pierre-Étienne Will and Michel Zink, 127–50. Paris: Académie des Inscriptions et Belles-Lettres, 2020.

Omont, Henri. *Missions archéologiques françaises en Orient.* Vol. 2. Paris: Imprimerie nationale, 1902.
Ōno Nobutane 大野延胤. "Matsumoto Tokizō to sono chojutsu, josetsu" 松本斗機蔵とその著述・序説 (Matsumoto Tokizō and his writings and introductions). *Gakushūin joshi daigaku kiyō* 学習院女子大学紀要 2 (2000): 51–67.
Owen, Stephen, ed. *Du Fu Shi (The Poetry of Du Fu).* Translated by Stephen Owen. Boston, MA: De Gruyter, 2016.

Pang, Tatiana A. "Three Versions of a Poem Composed by Emperor Qianlong." In *Florilegia Altaistica: Studies in Honour of Denis Sinor on the Occasion of His 90th Birthday*, edited by Elena V. Boikova and Giovanni Stary, 85–91. Wiesbaden: Harrassowitz, 2006.
Pang, Tatiana A., and M. P. Volkova. *Descriptive Catalogue of Manchu Manuscripts and Blockprints in the St. Petersburg Branch of the Institute of Oriental Studies, Russian Academy of Sciences: Issue 2.* Wiesbaden: Harrassowitz, 2001.
Parrenin, Dominique. "Lettre du Père Parennin ... à M. de Fontenelle." In *Lettres édifiantes et curieuses écrites des missions étrangères*, 19:208–42. 1738. Reprint, Toulouse: Sens & Gaudé, 1811.
Perdue, Peter C. *China Marches West: The Qing Conquest of Central Eurasia.* Cambridge, MA: Harvard University Press, 2005.
Peterson, Willard J. "Advancement of Learning in Early Ch'ing: Three Cases." In *The Ch'ing Empire to 1800, Part Two*, edited by Willard J. Peterson, vol. 9: 513–70. *The Cambridge History of China.* Cambridge: Cambridge University Press, 2016.
———. "Dominating Learning from Above during the K'ang-Hsi Period." In *The Ch'ing Empire to 1800, Part Two*, edited by Willard J. Peterson, vol. 9: 571–605. *The Cambridge History of China.* Cambridge: Cambridge University Press, 2016.
———. "Introduction: The Ch'ing Dynasty, the Ch'ing Empire, and the Great Ch'ing Integrated Domain." In *The Ch'ing Empire to 1800, Part 2*, edited by Willard J. Peterson, vol. 9: 1–15. *The Cambridge History of China.* Cambridge: Cambridge University Press, 2016.
Pino, Angel. "RÉMUSAT (Abel-Rémusat) Jean-Pierre Abel de." In *Dictionnaire des orientalistes de langue française*, edited by François Pouillon, expanded ed., 858–59. Paris: Karthala, 2012.
Plaks, Andrew H., trans. *"Ta Hsüeh" and "Chung-Yung": The Highest Order of Cultivation, and, On the Practice of the Mean.* London: Penguin, 2003.
Poppe, Nicholas. "On Some Mongolian Names of Wild Beasts." *Central Asiatic Journal* 9, no. 3 (1964): 161–74.

Ptak, Roderich. *Birds and Beasts in Chinese Texts and Trade: Lectures Related to South China and the Overseas World*. Wiesbaden: Harrassowitz, 2011.

Pu Xian 浦銑. *Lidai fuhua jiaozheng* 歷代賦話校證 (Rhapsody criticism throughout the ages, critical edition). Edited by He Xinwen 何新文 and Lu Chengwen 路成文. 1788. Reprint, Shanghai: Shanghai guji chubanshe, 2007.

Pulleyblank, Edwin G. *Lexicon of Reconstructed Pronunciation in Early Middle Chinese, Late Middle Chinese, and Early Mandarin*. Vancouver: University of British Columbia Press, 1991.

Qian Liangze 錢良擇. *Chusai jilüe* 出塞紀略 (Cursory record of a trip outside our border). Facsimile. Vol. 8. Ming-Qing shiliao huibian chuji 明清史料彙編初集. 1840. Reprint, Taipei: Wenhai chubanshe, 1967.

Qianlong qi qiju zhuce (dingben) 乾隆期起居注冊（定本）(Registers of rise and repose from the Qianlong period [definite copy]). Photocopied edition. Taipei: Guoli Gugong bowuyuan, n.d.

Qianlong yuzhi sanshier ti zhuanshu Shengjing fu 乾隆御製三十二體篆書盛京賦 (Qianlong's imperially commissioned *Ode to Mukden* in thirty-two kinds of seal script). Facsimile. 1748. Reprint, Changchun: Jilin renmin chubanshe, 2000.

Qinding Huangchao tongzhi 欽定皇朝通志 (Imperially authorized Comprehensive treatises of our august dynasty). Facsimile. Vols. 644–45. Wenyuan Ge Siku quanshu 文淵閣四庫全書. 1782. Reprint, Taipei: Taiwan Shangwu yinshu guan, 1983.

Qinding Huangyu Xiyu tuzhi 欽定皇輿西域圖志 (Imperially approved treatises on the Western territories under imperial rule). Vol. 251. Wenyuan Ge Siku quanshu 文淵閣四庫全書, 1782. Reprint, Taipei: Taiwan Shangwu yinshu guan, 1983.

Qinding Qing-Han duiyin zishi 欽定清漢對音字式 (Imperially authorized Manchu and Chinese characters presented in corresponding sounds). Blockprint, 1773. Reprint, Jinggu Tang, 1836. Call no. A940 338.1. National Taiwan Normal University Library, Taipei.

Qinding Shengjing tongzhi 欽定盛京通志 (Imperially authorized comprehensive gazetteer of Mukden). Facsimile. 3 vols. 1736. Reprint, Taipei: Wenhai chubanshe, 1965.

Qinggong Neiwu fu zaoban chu dang'an zonghui 清宮內務府造辦處檔案總匯 (Assembled archives of the Workshops of the Imperial Household in the Qing palace). 55 vols. Beijing: Renmin chubanshe, 2005.

Räsänen, Martti. *Versuch eines etymologischen Wörterbuchs der Türksprachen*. Helsinki: Suomalais-ugrilainen seura, 1969.

Read, Bernard E. *Chinese Medicinal Plants from the Pen Ts'ao Kang Mu* 本草綱目 *A.D. 1596 of Botanical, Chemical and Pharmacological Reference List*. Facsimile. 1936. Reprint, Taipei: Southern Materials Center, 1982.

Rémusat, Jean-Pierre Abel. "Notice sur le dictionnaire intitulé: Miroir des langues mandchoue et mongole." *Notices et extraits des manuscrits de la Bibliothèque du Roi* 13 (1838): 1–125.

———. *Nouveaux mélanges asiatiques*. Paris: Schubart et Heideloff, 1829.

———. "Observations sur l'état des sciences naturelles chez les peuples de l'Asie orientale." *Mémoires de l'Institut de France* 10 (1833): 116–67.

———. *Table alphabétique de l'Encyclopédie japonaise*. Digitized manuscript, call no. Japonais 391. Bibliothèque nationale de France, Paris.

———. *Table alphabétique du Thsing-wen-kien*. Digitized manuscript, call no. Japonais 392. Bibliothèque nationale de France, Paris.

Riegel, Jeffrey K. "*Ta Tai Li chi* 大戴禮記." In *Early Chinese Texts: A Bibliographical Guide*, edited by Michael Loewe, 456–59. Berkeley: Society for the Study of Early China & The Institute for East Asian Studies, University of California, 1993.

Rikugunshō 陸軍省. *Mōkogo daijiten* 蒙古語大辭典 (Unabridged dictionary of Mongolian). 3 vols. Tōkyō: Kaikōsha hensanbu, 1933.

Rogaski, Ruth. *Knowing Manchuria: Environments, the Senses, and Natural Knowledge on an Asian Borderland*. Chicago, IL: University of Chicago Press, 2022.

Rondelet, Guillaume. *Histoire entière des poissons*. Translated by Laurent Joubert. Lyon: Mace Bonhome, 1558.

Ross, E. Denison. *A Polyglot List of Birds in Turki, Manchu and Chinese*. Calcutta: Baptist Mission Press, 1909.

Rowe, William T. *China's Last Empire: The Great Qing*. Cambridge, MA: Harvard University Press, 2009.

Rozycki, William. *Mongol Elements in Manchu*. Bloomington: Indiana University Research Institute for Inner Asian Studies, 1994.

Ruan Yuan 阮元, ed. *Erya zhu shu* 爾雅注疏 (*Approaching Perfection*, with commentary and subcommentary). Facsimile. *Yingyin Ruan ke* Shisan jing zhu shu *fu jiaokan ji* 影印阮刻十三經注疏附校勘記. Reprint, Taipei: Qiming shuju, 1959.

———, ed. *Liji zhengyi* 禮記正義 (Correct meaning of the *Record of Rites*). Facsimile. *Yingyin Ruan ke* Shisan jing zhu shu *fu jiaokan ji* 影印阮刻十三經注疏附校勘記. Reprint, Taipei: Qiming shuju, 1959.

———, ed. *Mao Shi zhengyi* 毛詩正義 (Correct meaning of the Mao tradition of the *Poetry Classic*). Facsimile. *Yingyin Ruan ke* Shisan jing zhu shu *fu jiaokan ji* 影印阮刻十三經注疏附校勘記. Reprint, Taipei: Qiming shuju, 1959.

———, ed. *Zhou Yi zhengyi* 周易正義 (Correct meaning of the *Zhou Changes*). Facsimile. *Yingyin Ruan ke Shisan jing zhu shu fu jiaokan ji* 影印阮刻十三經注疏附校勘記. Reprint, Taipei: Qiming shuju, 1959.

Rudolph, Richard C., and Hartmut Walravens. "Comprehensive Bibliography of Manchu Studies (1909–2003)." *Monumenta Serica* 57 (2009): 231–494.

Rusk, Bruce. *Critics and Commentators: The "Book of Poems" as Classic and Literature*. Cambridge, MA: Harvard University Asia Center, 2012.

Sangge 桑額, ed. *Man-Han leishu* 滿漢類書 | *Man han lei šu bithe* (Manchu and Chinese, divided into sections). Blockprint. Zixing zhai, 1700.

Saunders, John B. de C. M., and Francis R. Lee, eds. *The Manchu Anatomy and Its Historical Origin*. Taipei: Li Ming Cultural Enterprise, 1981.

Schäfer, Dagmar. *The Crafting of the 10,000 Things: Knowledge and Technology in Seventeenth-Century China*. Chicago, IL: University of Chicago Press, 2011.

Schiebinger, Londa. *Plants and Empire: Colonial Bioprospecting in the Atlantic World*. Cambridge, MA: Harvard University Press, 2007.

Schlesinger, Jonathan. "The Qing Invention of Nature: Environment and Identity in Northeast China and Mongolia, 1750–1850." PhD diss., Harvard University, 2012.

———. *A World Trimmed with Fur: Wild Things, Pristine Places, and the Natural Fringes of Qing Rule*. Stanford, CA: Stanford University Press, 2017.

——— [Xie Jian 谢健]. *Diguo zhi qiu: Qingchao de shanzhen, jindi yiji ziran bianjiang* 帝国之裘：清朝的山珍、禁地以及自然边疆 (The empire's furs: Delicacies of the land, restricted areas, and the natural frontiers of the Qing dynasty). Translated by Guan Kang 关康. Beijing: Beijing daxue chubanshe, 2019.

Schmidt, Peter. "Chinesische Elemente im Mandschu. Mit Wörterverzeichnis." *Asia Major*, first series, 7 (1932): 573–628.

———. "Chinesische Elemente im Mandschu. Wörterverzeichnis (Fortsetzung und Schluss)." *Asia Major*, first series, 8 (1933): 353–436.

Schuessler, Axel. *ABC Etymological Dictionary of Old Chinese*. Honolulu: University of Hawai'i Press, 2007.

Sela, Ori. *China's Philological Turn: Scholars, Textualism, and the Dao in the Eighteenth Century*. New York: Columbia University Press, 2018.

Shapin, Steven. "The House of Experiment in Seventeenth-Century England." *Isis* 79, no. 3 (1988): 373–404.

Shen Qiliang 沈啟亮. *Da Qing quanshu* 大清全書 (Complete book of the Great Qing). Facsimile. 1683. Reprint, Shenyang: Liaoning minzu chubanshe, 2008.

———. *Sishu yaolan* 四書要覽 | *Sye šu oyonggo tuwara bithe* (Essential readings from the *Four Books*). Chongli Tang, 1686. Digitized, call no. mandchou 18. Bibliothèque Nationale de France, Paris.

Shengjing tongzhi 盛京通志 (Comprehensive gazetteer of Mukden). Blockprint. 12 vols. 1684. Digitized. Call no. T 3116/0.81. Harvard-Yenching Library, Cambridge, MA.

Shengzu Ren huangdi shilu 聖祖仁皇帝實錄 (Veritable records of the Sage Progenitor, Emperor Ren). *Qing shilu* 清實錄. Beijing: Zhonghua shuju, 1986.

Shengzu Ren huangdi yuzhi wenji 聖祖仁皇帝御製文集 (Collection of imperially authored prose by the Sagacious Progenitor, Emperor Ren). Facsimile. Vols. 1298–99. *Wenyuan Ge Siku quanshu* 文淵閣四庫全書. 1782. Reprint, Taipei: Taiwan Shangwu yinshu guan, 1983.

Shi You 史游. *Jijiu pian* 急就篇 (Slips for swift employ). Facsimile of a Ming-era manuscript. *Sibu congkan xubian* 四部叢刊續編. Reprint, Shanghai: Shangwu yinshu guan, 1934.

Shih Ching-fei 施靜菲. "Cong *Huoji dang* zhong de qiwu zhizuo kan Qianlong huangdi yaoqiu de Manzhou tese" 從《活計檔》中的器物製作看乾隆皇帝要求的滿洲特色 (A look at the Manchu characteristics demanded by Qianlong on the basis of the object production seen in the *Work-related Archives of the Imperial Workshops*). Presented at the Academia Sinica International Conference on Ming-Qing studies 2021中央研究院明清研究國際學術研討會, Taipei, December 15, 2021.

Shinneman, Vicki M. "Horse Colors of the [*Wuti Qingwen jian*] 五體清文鑑." Master's thesis, University of Washington, 1995.

Shirokogoroff, S.M. "Reading and Transliteration of Manchu Lit." *Rocznik Orjentalistyczny* 10 (1934): 122–30.

Shizhuan daquan 詩傳大全 (Poetry [classic], complete tradition). Vol. 78. *Wenyuan Ge Siku quanshu* 文淵閣四庫全書. Facsimile. 1782. Reprint, Taipei: Taiwan Shangwu yinshu guan, 1983.

Siebert, Martina. "Animals as Text: Producing and Consuming Text Animals." In *Animals through Chinese History: Earliest Times to 1911*, edited by Roel Sterckx, Martina Siebert, and Dagmar Schäfer, 139–59. Cambridge: Cambridge University Press, 2018.

———. "Klassen und Hierarchien, Kontrastpaare und Toposgruppen Formen struktureller Eroberung und literarischer Vereinnahmung der Tierwelt im alten China." *Zeitschrift der Deutschen Morgenländischen Gesellschaft* 162 (2012): 171–95.

———. *Pulu: "Abhandlungen und Auflistungen" zu materieller Kultur und Naturkunde im traditionellen China*. Wiesbaden: Harrassowitz, 2006.

Siebert, Martina, Kai Jun Chen, and Dorothy Ko. Introduction to *Making the Qing Palace Machine Work: Mobilizing People, Objects, and Nature in the Qing Empire*, edited by Martina Siebert, Kai Jun Chen, and Dorothy Ko, 23–35. Amsterdam: Amsterdam University Press, 2021.

———, eds. *Making the Palace Machine Work: Mobilizing People, Objects, and Nature in the Qing Empire*. Amsterdam: Amsterdam University Press, 2021.

Sinor, Denis. "Some Altaic Names for Bovines." *Acta Orientalia Academiae Scientiarum Hungaricae* 15, no. 1–3 (1962): 315–24.

Söderblom Saarela, Mårten. "Alphabets *avant la lettre*: Phonographic Experiments in Late Imperial China." *Twentieth-Century China* 41, no. 3 (2016): 234–57.

———. *The Early Modern Travels of Manchu: A Script and Its Study in East Asia and Europe*. Philadelphia: University of Pennsylvania Press, 2020.

———. "Lexicography." In *Plurilingualism in Traditional Eurasian Scholarship: Thinking in Many Tongues*, edited by Glenn W. Most, Dagmar Schäfer, and Mårten Söderblom Saarela, 229–39. Leiden: Brill, 2023.

———. "Linguistic Compartmentalization and the Palace Memorial System in the Eighteenth Century." *Late Imperial China* 41, no. 2 (2020): 131–79.

———. "Manchu and the Study of Language in China (1607–1911)." PhD diss., Princeton University, 2015.

———. *The Manchu Language at Court and in the Bureaucracy under the Qianlong Emperor*. Leiden: Brill, 2024.

———. "On the Manchu Names for Grasshoppers, Locusts, and a Few Other Bugs in the Seventeenth and Eighteenth Centuries." In *Insect Histories of East Asia*, edited by David A. Bello and Daniel Burton-Rose, 41–63. Seattle: University of Washington Press, 2023.

———. "Public Inscriptions and Manchu Language Reform in the Early Qianlong Reign (1740s–60s)." *Saksaha* 16 (2019): 31–53.

Sŏng Paek-in 成百仁. "Ŏje chŭngjŏng Ch'ŏngmun'gam ŭi ip'anbon sikpyŏl ŭl wihan t'ŭkching chosa" 『御製增訂清文鑑』의 異版本 識別을 위한 特徵 調査 (A survey for the purpose of distinguishing the different editions of the *Imperially Commissioned Expanded and Emended Mirror of the Manchu Language*). *Han'gugŏ yŏn'gu* 韓國語研究 1 (2003): 149–75.

Spence, Jonathan. *The Question of Hu*. New York: Vintage Books, 1988.

Standaert, Nicolas. "Comprehensive Histories in Late Ming and Early Qing: The Genealogy of the *Gangjian* 綱鑑 Texts." *Bulletin of the Museum of Far Eastern Antiquities* 79/80 (2018): 245–334.

Stary, Giovanni. *Die chinesischen und mandschurischen Zierschriften: Mit 64 Tafeln und 12 Abbildungen im Text.* Hamburg: Helmut Buske, 1980.

———. "Fundamental Principles of Manchu Poetry." In *Selected Manchu Studies: Contributions to History, Literature, and Shamanism of the Manchus,* edited by Hartmut Walravens, 76–109. Berlin: Klaus Schwarz, 2013.

———. "Linguistic and Cultural Limits of Manchu Poetry in Comparison with Chinese." In *Selected Manchu Studies: Contributions to History, Literature, and Shamanism of the Manchus,* edited by Hartmut Walravens, 292–99. Berlin: Klaus Schwarz, 2013.

———. "'L'Ode di Mukden' dell'imperatore Ch'ien-lung: Nuovi spunti per un'analisi della tecnica versificatoria mancese." *Cina,* no. 17 (1981): 235–51.

———. "'L'Ode di Mukden' di Qianlong: La versione mancese versificata et la sua traduzione in prosa." In *Caro Maestro . . . Scritti in onore di Lionello Lanciotti per l'ottantesimo compleanno,* edited by Maurizio Scarpari and Tiziana Lippiello, 1095–104. Venice: Libreria Editrice Cafoscarina, 2005.

———. "A Manchu Word List from 1682." In *Opuscula Altaica: Essays Presented in Honor of Henry Schwarz,* edited by Edward H. Kaplan and Donald W. Whisenhunt, 577–86. Bellingham: Center for East Asian Studies, Western Washington University, 1994.

———. "Mandschurische Miszellen." In *Florilegia Manjurica: in Memoriam Walter Fuchs,* edited by Michael Weiers and Giovanni Stary, 76–86. Wiesbaden: Harrassowitz, 1982.

Statman, Alexander. "A Forgotten Friendship: How a French Missionary and a Manchu Prince Studied Electricity and Ballooning in Late Eighteenth Century Beijing." *East Asian Science, Technology, and Medicine* 46 (2017): 89–118.

Sterckx, Roel. "Animal Classification in Ancient China." *East Asian Science, Technology, and Medicine* 23 (2005): 26–53.

Sturman, Peter C. "Cranes above Kaifeng: The Auspicious Image at the Court of Huizong." *Ars Orientalis* 20 (1990): 33–68.

Su, Jui-lung. "The Origins of the Term 'Fu' as a Literary Genre of Recitation." In *The Fu Genre of Imperial China: Studies in the Rhapsodic Imagination,* edited by Nicholas Morrow Williams, 19–38. Amsterdam: Amsterdam University Press, 2019.

Su Ya-fen 蘇雅芬. "Jiang Tingxi *Hua qunfang xiu xiu ce* yanjiu" 蔣廷錫《畫群芳擷秀冊》研究 (A Study on Jiang Tingxi's *Picked Beauties of Beautiful Flowers*). Master's thesis, National Taiwan Normal University, 2018.

Sukita Tomohiko 鋤田智彦. "Manshūji hyōki no Kango ni motozuku kinsei Chūgokugoon no kenkyū: *Manbun sangokuji* shiryō to shite" 満洲字表記の漢語に基づく近世中国語音の研究—『満文三国志』資料として— (A study of the early modern Chinese pronunciations reflected in Manchu transcription, using the *Manchu-language Record of the Three Kingdoms* as a source). PhD diss., Waseda University, 2013.

Sun Ruozhi 孫柔之. *Ruiying tu* 瑞應圖 (Illustrated compendium of auspicious responses). Facsimile. Vol. 4. *Yuhan shanfang ji yishu* 玉函山房輯佚書. 1871. Kyōto: Chūbun shuppansha, n.d.

Taiping yulan 太平御覽 (Imperial digest [from the era of rousing the state through] great tranquility). Facsimile. Vols. 893–901. *Wenyuan Ge Siku quanshu* 文淵閣四庫全書, 1782. Reprint, Taipei: Taiwan Shangwu yinshu guan, 1983.

Takahashi Kageyasu 高橋景保. *Yakugo shō* 譯語抄 (Excerpted translations). Manuscript. 6 vols. Call no. 和 18555. Naikaku Bunko, Tokyo.

———. *Zōtei Manbun shūin* 增訂滿文輯韻 (Compiled Manchu rhymes, expanded and emended). Manuscript, 1828. Imperial Household Agency Library, Tokyo. Photocopies. Call no. 401 75. Princeton University Library, NJ.

Talpe, Lode. "The Manchu Text of the *Hsi-Tu-Shih* or *Lapis Serpentinus*." *Orientalia Lovaniensia Periodica* 22 (1991): 215–34.

Terajima Ryōan 寺島良安. *Wakan sansai zue* 和漢三才圖會 (Illustrated dictionary of the three realms in Japanese and Chinese). Facsimile. 1713. Reprint, Tōkyō: Tōkyō bijutsu, 2004.

Tojin 托津. *Da Qing huidian shili* 大清會典事例 (Cases and precedents of the collected statutes of the Great Qing). Blockprint, Jiaqing edition. 1801–1818.

Tong Yonggong 佟永功 and Guan Jialu 关嘉禄. "Qianlong chao 'Qinding xin Qingyu' tanxi" 乾隆朝《钦定新清语》探析 (Exploratory analysis of the *Imperially Authorized New Manchu Words* of the Qianlong period). *Zhongguo minzu gu wenzi yanjiu* 中国民族古文字研究, no. 4 (1994): 35–42.

Torbert, Preston M. *The Ch'ing Imperial Household Department: A Study of Its Organization and Principal Functions, 1662–1796.* Cambridge, MA: Council on East Asian Studies, Harvard University, 1977.

Tsai Ming-che 蔡名哲. "Huacheng tianxia: *Yuzhi Qingwen jian* xuyan fanying de wenzhi guan" 化成天下—《御制清文鑑》序言反映的文治观 (To successfully transform the world through instruction: The view of cultural attainment reflected in the preface to the *Imperially Commissioned Mirror of the Manchu Language*). In *Manxue luncong* 满学论丛,

edited by Chang Yuenan 常越男, 67–79. Shenyang: Liaoning minzu chubanshe, 2021.

———. "Qing qianqi Manzhou tezhi zhi zai jiangou yanjiu" 清前期滿洲特质之再建构研究 (A study of the reconstruction of Manchu specificity in the early Qing period). PhD diss., Minzu University of China, 2021.

Tsintsius, V. I., ed. *Sravnitel'nyĭ slovar' tunguso-manchzhurskikh iazykov: Materialy k ėtimologicheskomu slovariu* (Comparative dictionary of the Manchu-Tungus languages: Materials toward an etymological dictionary). Leningrad: Nauka, 1975.

Tu, Chenyu. "On the Source Text of Ši Ging Ni Bithe (1654)." *Manchu Studies Group* (blog), April 22, 2021. https://www.manchustudiesgroup.org/2021/04/22/on-the-source-text-of-si-ging-ni-bithe–1654.

Tu Kai 屠楷. *Jingyang xian zhi* 涇陽縣志 (Gazetteer of Jingyang County). Vol. 5. *Shaanxi sheng tushuguan cang xijian fangzhi congkan* 陝西省圖書館藏稀見方志叢刊. 1732. Reprint, Beijing: Beijing tushuguan chubanshe, 2006.

Tu Lien-chê. "Kung Ting-Tzŭ." In *Eminent Chinese of the Ch'ing Period (1644–1912)*, edited by Arthur W. Hummel, vol. 1: 431. Washington, DC: U.S. Government Printing Office, 1944.

Tulišen 圖麗琛 and Chuang Chi-fa 莊吉發. *Man-Han Yiyu lu jiaozhu* 滿漢異域錄校注 (Critical edition of the Manchu and Chinese versions of the *Book Recording the Mission to Distant Territories*). Taipei: Wen shi zhe chubanshe, 1983.

Tulišen 圖麗琛 and Imanishi Shunjū 今西春秋. *Kōchū Iikiroku* 校注異域錄 (*Book Recording the Mission to Distant Territories*, critical edition). Tenri: Tenri daigaku oyasato kenkyūjo, 1964.

Uehara Hisashi 上原久. *Takahashi Kageyasu no kenkyū* 高橋景保の研究 (Studies on Takahashi Kageyasu). Tōkyō: Kōdansha, 1977.

Uray-Kőhalmi, Käthe. "Die Farbbezeichnungen der Pferde in den Mandschu-Tungusischen Sprachen." *Acta Orientalia Academiae Scientiarum Hungaricae* 19, no. 1 (1966): 45–55.

———. "Von woher kamen die Iče Manju?" In *Tumen jalafun jecen akū: Manchu Studies in Honor of Giovanni Stary*, edited by Alessandra Pozzi, Juha Antero Janhunen, and Michael Weiers, 235–43. Wiesbaden: Harrassowitz, 2006.

Valenti, Federico. "Biological Classification in Early Chinese Dictionaries and Glossaries: From Fish to Invertebrates and Vice Versa." PhD diss., Università degli Studi di Sassari, 2015.

Van Duzer, Chet. *Sea Monsters on Medieval and Renaissance Maps*. London: The British Library, 2013.

Vedal, Nathan. *The Culture of Language in Ming China: Sound, Script, and the Redefinition of Boundaries of Knowledge*. New York: Columbia University Press, 2022.

———. "The Manchu Reading of *Jinpingmei*: Commentary, Encyclopedism, and Translingual Practices in Early Eighteenth-Century China." *Late Imperial China* 42, no. 2 (2021): 1–48.

———. "Preferring Omission over Falsity: The Politics of Compilation in the Kangxi 'Classic of Characters' 康熙字典." *Historiographia Linguistica* 40, no. 1–2 (2013): 3–37.

Verbiest, Ferdinand. *Kunyu tushuo* 坤輿圖說 (Illustrated explanation of the world). Facsimile. *Congshu jicheng chubian* 叢書集成初編 3266. Shanghai: Shangwu yinshuguan, 1937.

Vovin, Alexander. "Some Thoughts on the Origins of the Old Turkic 12-Year Animal Cycle." *Central Asiatic Journal* 48, no. 1 (2004): 118–32.

Wadley, Stephen A. "A Preliminary Investigation of Manchu Tree Names in the *Wuti Qingwenjian*." *Central Asiatic Journal* 36, no. 1/2 (1992): 107–22.

Wakeman, Frederic. "High Ch'ing: 1683–1839." In *Modern East Asia: Essays in Interpretation*, edited by James B. Crowley, 1–28. New York: Harcourt, Brace & World, 1970.

Waley-Cohen, Joanna. "The New Qing History." *Radical History Review* 88 (2004): 193–206.

Walravens, Hartmut. "Das Huang Ch'ing chih-kung t'u als Werk der mandjurischen Literatur." In *Tumen jalafun jecen akū: Manchu Studies in Honor of Giovanni Stary*, edited by Alessandra Pozzi, Juha Antero Janhunen, and Michael Weiers, 267–308. Wiesbaden: Harrassowitz, 2006.

———. "Die Deutschland-Kenntnisse der Chinesen (bis 1870): Nebst einem Exkurs über die Darstellung fremder Tiere im K'un-yü t'u-shuo des P. Verbiest." PhD diss., Universität zu Köln, 1972.

———. *Ivan Il'ič Zacharov (1817–1885): Russischer Diplomat und Sinologe, eine biobibliographische Skizze*. Hamburg: Ch. Bell, 1982.

———. *Jean Pierre Abel Rémusat (1788–1832): Zu Leben und Werk eines Wegbereiters der Ostasienwissenschaften*. Norderstedt: Books on Demand, 2020.

———. "Konrad Gessner in chinesischem Gewand." *Gesnerus* 30, no. 3–4 (1973): 87–98.

———. "Medical Knowledge of the Manchus and the *Manchu Anatomy*." *Études Mongoles et Sibériennes* 27 (1996): 359–74.

———. "'Tribute-Bearers in Manchu and Chinese': A Unique 18th-Century Source for East and Central Asian History." *Acta Orientalia Academiae Scientiarum Hungaricae* 49, no. 3 (1996): 395–406.

Walravens, Hartmut, Larry V. Clark, John R. Krueger, Manfred Taube, and Michael L. Walter. *Bibliographies of Mongolian, Manchu-Tungus, and Tibetan Dictionaries*. Wiesbaden: Harrassowitz, 2006.

Walravens, Hartmut, and Martin Gimm, eds. *Wei jiao zi ai* 為教自愛 «*Schone dich für die Wissenschaft»: Leben und Werk des Kölner Sinologen Walter Fuchs (1902–1979) in Dokumenten und Briefen*. Wiesbaden: Harrassowitz, 2010.

Wang Cheng-hua 王正華. "*Ting qin tu* de zhengzhi yihan: Huizong chao de yuanhua fengge yu yiyi wangluo" 《聽琴圖》的政治意涵：徽宗朝院畫風格與意義網絡 (The political import of *Painting of Listening to the Zither*: The academy style of painting at the Huizong court and its web of meanings). *Guoli Taiwan daxue meishushi yanjiu jikan* 國立台灣大學美術史研究集刊 5 (1998): 77–122.

Wang Guoliang 王國良. *Shenyi jing yanjiu* 神異經研究 (A study on the *Classic of Divine Marvels*). Taipei: Wen shi zhe chubanshe, 1985.

Wang Hao 汪灝. *Sui luan ji en* 隨鑾紀恩 (An account of grace during my trip accompanying His Majesty's carriages). Edited by Wang Xiqi 王錫祺. *Xiao fanghu zhai yudi congchao zheng bian* 小方壺齋輿地叢鈔正編. Shanghai: Zhuyi tang, 1877.

———, ed. *Yuding Peiwen zhai Guang Qunfang pu* 御定佩文齋廣群芳譜 (*Imperially Commissioned Expanded Catalog of Myriad Flowers* from the Studio of Literary Admiration). Facsimile. Vols. 845–47. *Wenyuan Ge Siku quanshu* 文淵閣四庫全書. 1782. Reprint, Taipei: Taiwan Shangwu yinshu guan, 1983.

Wang Shizhen 王士禛. *Chibei outan* 池北偶談 (Occasional conversations at the pond's north shore). 2 vols. Beijing: Zhonghua shuju, 1982.

———. *Fen gan yu hua* 分甘餘話 (Superfluous words while sharing sweets). Typeset. Beijing: Zhonghua shuju, 1989.

Wang Ting 王頲. *Xiyu, Nanhai shidi yanjiu* 西域南海史地研究 (Research on the history and geography of the Western regions and Southern seas). Shanghai: Shanghai guji chubanshe, 2005.

Wang Xiangjin 王象晋. *Er ru ting Qunfang pu* 二如亭群芳譜 (*Catalog of Myriad Flowers* from the Pavillon of Double Resemblance). Facsimile. Vol. 80. *Siku quanshu cunmu congshu bubian* 四庫全書存目叢書補編. Jinan: Qi-Lu shushe, 2001.

Wang Xianqian 王先謙, ed. *Donghua lu* 東華錄 (Records from [within the] Eastern Resplendent Gate). Facsimile. *Xuxiu Siku quanshu* 續修四庫全書 369–75. Reprint, Shanghai: Shanghai guji chubanshe, 1995.

Wang Zhao 王釗. "Dixiang qingfen: *Jiachan jiansheng* zhong xiangliao zhiwu kao" 帝鄉清芬：《嘉產薦聲》中香料植物考 (The noble fragrance of the imperial homeland: An examination of the herbs in *Auspicious Products of Commendable Fragrance*). *Gugong wenwu yuekan* 故宮文物月刊, no. 408 (2017): 64–75.

———. "Guan hu dong zhi: Kang-Qian shidai de Qing gong bowu huihua yanjiu" 观乎动植：康乾时代的清宫博物绘画研究 (To observe animals and plants: A study of natural-historical paintings at the Qing court during the Kangxi and Qianlong reigns). PhD diss., Peking University, 2018.

———. "Zhong shen zhi xiang: Yi *Jiachan jianxin* wei zhongxin tanjiu Qingdai Manzu saman jisi yong xiang" 众神之飨—以《嘉产荐馨》为中心探究清代满族萨满祭祀用香 (A banquet for the many spirits: Exploratory study of the use of fragrant herbs in Manchu shamanist offerings, centered on *Auspicious Products of Commendable Fragrance*). *Manzu yanjiu* 满族研究, no. 4 (2017): 61–68.

Weiers, Michael. "Der Mandschu-Khortsin Bund von 1626." In *Documenta Barbarorum: Festschrift für Walther Heissig zum 70. Geburtstag*, edited by Klaus Sagaster and Michael Weiers, 412–35. Wiesbaden: Harrassowitz, 1983.

Weng Lianxi 翁連溪, ed. *Qing Neifu keshu dang'an shiliao huibian* 清內府刻書檔案史料彙編 (Compilation of historical sources from archives relating to the printing of books at the Qing Imperial Household Department). 2 vols. Yangzhou: Guangling shushe, 2007.

———. *Qingdai Neifu keshu tulu* 清代內府刻書圖錄 (Illustrated catalog of the imprints of the Imperial Household Department in the Qing period). Beijing: Beijing chubanshe, 2004.

———. *Qingdai Neifu keshu yanjiu* 清代内府刻书研究 (Studies on the imprints of the Imperial Household Department in the Qing period). 2 vols. Beijing: Beijing chubanshe, 2013.

Wilhelm, Richard. *The I Ching or Book of Changes*. Translated by Cary F. Baynes. 1950. Reprint, New York: Pantheon Books, 1961.

Williams, Nicholas Morrow. "Introduction: The Rhapsodic Imagination." In *The Fu Genre of Imperial China: Studies in the Rhapsodic Imagination*, edited by Nicholas Morrow Williams, 1–16. Amsterdam: Amsterdam University Press, 2019.

Winter, Marc. "Conrad Gessner und die Entwicklung der Zoologie in China." In *Conrad Gessner (1516–1565): Die Renaissance der Wissenschaften/The Renaissance of Learning*, edited by Urs Leu and Peter Opitz, 639–52. Berlin: De Gruyter Oldenbourg, 2019.

Wu, Guo. "New Qing History: Dispute, Dialog, and Influence." *Chinese Historical Review* 23, no. 1 (2016): 47–69.

Wu Huey-Fang 吳蕙芳. *Wanbao quanshu: Ming-Qing shiqi de minjian shenghuo shilu* 萬寶全書：明清時期的民間生活實錄 (*Complete Book of Myriad Treasures*: A faithful record of popular life in the Ming-Qing period), rev. ed. Taipei: Huamulan, 2005.

Wu Huiyi and Zheng Cheng. "Transmission of Renaissance Herbal Images to China: The Beitang Copy of Mattioli's Commentaries on Dioscorides and Its Annotations." *Archives of Natural History* 47, no. 2 (2020): 236–53.

Wu Siyu 吳思雨 and Wang Le 王乐. "Qianlong shiqi (1736–1795 nian) zaoban chu *Huoji dang* suo zai xingxing zhan tanxi" 乾隆时期（1736—1795年）造办处《活计档》所载猩猩毡探析 (An exploratory examination of the mentions of *xingxing* [a kind of simian] felt in the *Archive on the Products of the Imperial Workshop* in the Qianlong period [1736–1795]). *Sichou sichou* 丝绸 57, no. 4 (2020): 94–100.

Wu Weiping 吳偉蘋. "Qianlong huangdi yu yu banzhi" 乾隆皇帝與玉扳指 (The Qianlong emperor and archery thumb rings made from jade). *Gugong wenwu yuekan* 故宮文物月刊, no. 367 (2013): 94–109.

Wu, Yulian. *Luxurious Networks: Salt Merchants, Status, and Statecraft in Eighteenth-Century China*. Stanford, CA: Stanford University Press, 2017.

Wu Zhenchen 吳桭臣. *Ningguta jilüe* 寧古塔紀略 (Cursory record of Ningguta). Facsimile. *Zhaodai congshu* 昭代叢書, *geng* 庚 series. Shanghai: Shanghai guji chubanshe, 1990.

Xiang Xuan 項旋. *Qingdai* Gujin tushu jicheng *guan yanjiu* 清代古今图书集成馆研究 (A study of the compilation office for *Consummate Collection of Texts and Illustrations from Past and Present* in the Qing period). Zhengzhou: Henan renmin chubanshe, 2019.

Xiao Tong and David R. Knechtges. *Wen xuan, or Selections of Refined Literature*. Vol. 1. Princeton, NJ: Princeton University Press, 1982.

Xi-qing 西清. *Heilongjiang waiji* 黑龍江外記 (An external account of Heilongjiang). Blockprint. Guangzhou: Guangya shuju, 1877.

Xu Baoguang 徐葆光. *Zhongshan chuanxin lu* 中山傳信錄 (Record based on letters transmitted from the Ryūkyū kingdom). *Taiwan wenxian congkan* 台灣文獻叢刊 306. Taipei: Taiwan yinhang jingji yanjiushi, 1972.

Xu Jie 许结. "Qianlong *Shengjing fu* de xiezuo yu chuanbo" 乾隆《盛京賦》的写作与传播 (The writing and circulation of Qianlong's *Ode to Mukden*). *Gudian wenxue zhishi* 古典文学知识, no. 3 (2017): 125–30.

Yan, Zinan 颜子楠. "Routine Production: Publishing Qianlong's Poetry Collections." *T'oung Pao*, 2nd series, 103, no. 1–3 (2017): 206–45.

Yanagisawa Akira 柳澤明. "Kokyū Hakubutsuin zō Man-Kan gōheki *Chōfu* ni tsuite" 故宮博物院蔵満漢合璧『鳥譜』について (On the Manchu-Chinese bilingual *Catalog of Birds* held at the Palace Museum). *Manzokushi kenkyū* 滿族史研究 3 (2004): 18–39.

Yang Bin 楊賓. *Liubian jilüe* 柳邊紀略 (Cursory record from the land by the willow palisade). 1690. New edition. Vol. 3. *Liaohai congshu* 遼海叢書 1. Dalian: Liaohai shushe, 1934.

Yang Shen 楊慎. *Yiyu tu zan* 異魚圖贊 (Guide to the illustrations of exotic fish). Facsimile. Vol. 847. *Wenyuan Ge Siku quanshu* 文淵閣四庫全書. 1782. Reprint, Taipei: Taiwan Shangwu yinshu guan, 1983.

Yeh Kao-shu 葉高樹. "Man-Han hebi *Qinding fanyi wujing sishu* de wenhua yihan: Cong 'yin guoshu yi tong jingyi' dao 'yin jingyi yi tong guoshu'" 滿漢合璧《欽定繙譯五經四書》的文化意涵：從「因國書以通經義」到「因經義以通國書」 (The cultural import of the Manchu-Chinese bilingual *Imperially Authorized Translations of the Five Classics and Four Books*: From "Grasping the meaning of the classics through the dynastic language" to "Grasping the dynastic language through the meaning of the classics"). In *Jingxue yanjiu luncong* 經學研究論叢, edited by Lin Chingchang 林慶彰, vol. 13, 1–42. Taipei: Taiwan xuesheng shuju, 2006.

———. *Manwen "Qinding Manzhou jishen jitian dianli" yizhu* 滿文《欽定滿洲祭神祭天典禮》譯註 (An annotated translation of the Manchu version of *Imperially Commissioned Manchu Rituals for the Worship of Gods and Heaven*). Taipei: Xiuwei zixun keji gufen youxian gongsi, 2018.

Yu Wentao 郁文韜. "Huangdi de bowutu: Yu Xing, Zhang Weibang hui *Mo Jiang Tingxi niaopu, Shoupu* yanjiu" 皇帝的博物图—余省、张为邦绘《摹蒋廷锡鸟谱》《兽谱》研究 (The emperor's natural history drawings: A study of Yu Xing's and Zhang Weibang's *Catalog of Birds Traced on the Basis of Jiang Tingxi* and *Catalog of Beasts*). *Zhongguo meishu* 中国美术, no. 3 (2016).

Yu Xing 余省 and Zhang Weibang 张为邦. *Qinggong shoupu* 清宫兽谱 (*Catalog of Beasts* held at the Qing palace). Edited by Gugong Bowu Yuan 故宫博物院. Beijing: Gugong chubanshe, 2014.

Yu Yue 于越. "*Yuzhi Qingwen jian* yongli yanjiu" 《御制清文鉴》用例研究 (A study on the example sentences in the *Imperially Commissioned Mirror of the Manchu Language*). Master's thesis, Heilongjiang University, Harbin, 2018.

Yuan Ke 袁珂, ed. *Shanhai jing jiaozhu* 山海經校注 (Critical edition of the *Classic of Mountains and Seas*). Taipei: Li ren shuju, 1981.

Yuding Kangxi zidian 御定康熙字典 (Imperially established Character standard of the reign of secure peace). Facsimile. Vol. 229–231. *Wenyuan Ge Siku quanshu* 文淵閣四庫全書, 1782. Reprint, Taipei: Taiwan Shangwu yinshu guan, 1983.

Yunggui 永貴. Qingwen jian *wai xinyu* 清文鑑外新語 (New words that were not included in the *Mirror of the Manchu Language*). Blockprint. Minzu University of China Library.

Yuzhi fanyi Mengzi 御製繙譯孟子 (Imperially commissioned translation of *Mencius*). Facsimile. Vol. 189. *Wenyuan Ge Siku quanshu* 文淵閣四庫全書. 1782. Reprint, Taipei: Taiwan Shangwu yinshu guan, 1983.

Yuzhi fanyi Shijing 御製繙譯詩經 (Imperially commissioned translation of the *Poetry Classic*). Facsimile. Vol. 185. *Wenyuan Ge Siku quanshu* 文淵閣四庫全書. 1782. Reprint, Taipei: Taiwan Shangwu yinshu guan, 1983.

Yuzhi Manzhu-, Menggu-, Hanzi sanhe qieyin Qingwen jian 御製滿珠蒙古漢字三合切音清文鑑 | *Han-i araha manju monggo nikan hergen ilan hacin-i mudan acaha buleku bithe* | *Qayan-u bičigsen manju mongyol kitad üsüg yurban jüil-ün ayalyu neilegsen toli bičig* (Imperially commissioned Manchu, Mongolian, and Chinese script mirror with tripartite spellings). Facsimile. Vol. 234. *Wenyuan Ge Siku quanshu* 文淵閣四庫全書. 1782. Reprint, Taipei: Taiwan Shangwu yinshu guan, 1983.

Yuzhi zengding Qingwen jian 御製增訂清文鑑 | *Han-i araha nonggime toktobuha manju gisun-i buleku bithe* (Imperially commissioned mirror of the Manchu language, expanded and emended). Blockprint. Beijing: Wuying dian, 1772. Call no. 19951.a.31. The British Library, London.

Yuzhi zengding Qingwen jian 御製增訂清文鑑 | *Han-i araha nonggime toktobuha manju gisun-i buleku bithe* (Imperially commissioned mirror of the Manchu language, expanded and emended). Facsimile. Vols. 232–33. *Wenyuan Ge Siku quanshu* 文淵閣四庫全書. 1782. Reprint, Taipei: Taiwan Shangwu yinshu guan, 1983.

Yuzhi zengding Qingwen jian 御製增訂清文鑑 | *Han-i araha nonggime toktobuha manju gisun-i buleku bithe* (Imperially commissioned mirror of the Manchu language, expanded and emended). Facsimile. *Qinding Siku quanshu huiyao* 欽定四庫全書薈要. 1778. Reprint, Changchun: Jilin chuban jituan, 2005.

Zach, Erwin Ritter von. *Lexicographische Beiträge*. Vol. 3. Beijing, 1905.
———. "Über Wortzusammensetzungen im Mandschu." *Wiener Zeitschrift für die Kunde des Morgenlandes* 11 (1897): 242–48.

Zakharov, Ivan Il'ich. *Grammatika man'chzhurskogo ïazyka* (Grammar of the Manchu language). 1879. Reprint, Folkestone, Kent: Global Oriental, 2010.
———. *Polnyĭ Man'chzhursko-Russkiĭ slovar'* (Unabridged Manchu-Russian dictionary). St. Petersburg: Tip. I. Akademīi nauk, 1875.

Zha Shenxing 查慎行. "Peilie biji" 陪獵筆記 (Jottings during my trip accompanying the imperial hunt). In *Zha Shenxing quanji* 查慎行全集, edited by Fan Daoji 范道濟, vol. 3. Beijing: Zhonghua shuju, 2017.

———. *Renhai ji* 人海記 (My account in the sea of humankind). Beijing: Beijing guji chubanshe, 1989.

Zhang Hong 张虹, Cheng Dakun 程大鲲, and Tong Yonggong 佟永功, eds. "Qianlong chao 'Qinding Xin Qingyu' (er)" 乾隆朝"钦定新清语"(二) (The "Imperially authorized new Manchu words," part 2). *Manyu yanjiu* 满语研究, no. 2 (1994): 68–77.

———, eds. "Qianlong chao 'Qinding Xin Qingyu' (san)" 乾隆朝"钦定新清语"(三) (The "Imperially authorized new Manchu words," part 3). *Manyu yanjiu* 满语研究, no. 2 (1995): 51–58.

———, eds. "Qianlong chao 'Qinding Xin Qingyu' (qi)" 乾隆朝"钦定新清语"(七) (The "Imperially authorized new Manchu words," part 7). *Manyu yanjiu* 满语研究, no. 2 (2000): 26–33.

———, eds. "Qianlong chao 'Qinding Xin Qingyu' (ba)" 乾隆朝"钦定新清语"(八) (The "Imperially authorized new Manchu words," part 8). *Manyu yanjiu* 满语研究, no. 2 (2001): 83–88.

Zhang Jinyan 张缙彦. *Ningguta shan shui ji; Yu wai ji* 宁古塔山水记；域外集 (An account of mountains and rivers in Ningguta; Collection of the outer domains). Edited by Li Xingsheng 李兴盛. Typeset. Harbin: Heilongjiang renmin chubanshe, 1984.

Zhang Penghe 張鵬翮. *Fengshi Eluosi riji* 奉使俄羅斯日記 (Daily record of an imperial embassy to Russia). Edited by Cheng Yansheng 程演生. Typeset. *Zhongguo neiluan waihuo lishi congshu* 中國內亂外禍歷史叢書. Shanghai: Shenzhou guoguang she, 1946.

———. *Fengshi Eluosi riji* 奉使俄羅斯日記 (Daily record of an imperial embassy to Russia). Facsimile. Vol. 9. *Biji xiaoshuo daguan* 筆記小說大觀 3. Taipei: Xinxing shuju, 1974.

———. *Fengshi Eluosi xingcheng lu* 奉使俄羅斯行程錄 (Record of the route during an imperial embassy to Russia). Facsimile. Vol. 97. *Congshu jicheng xinbian* 叢書集成新編. Reprint, Taipei: Xin wenfeng chuban gongsi, 1985.

Zhang Qingwen 张清文. "Zhuwei de qiyuan ji gongyong deng zhu shi zakao" 麈尾的起源及功用等诸事杂考 (Assorted examinations of the origin, function, etc. of the "*zhu* tail"-whisk). *Meishu yu sheji* 美术与设计, no. 1 (2019): 43–46.

Zhang, Qiong. "The Infrastructure of Science Making in Early Modern China." *Asian Review of World Histories* 11, no. 1 (2023): 90–129.

Zhang Tianqi 張天祈. *Lianzhu ji* 聯珠集 (Collection of stringed pearls). Edited by Liu Shun 劉順. 1700. Digitized microfilm, call no. manchou 95. Bibliothèque nationale de France, Paris.

Zhang, Xue. "Imperial Maps of Xinjiang and Their Readers in Qing China, 1660–1860." *Journal of Chinese History* 4, no. 1 (2020): 111–33.

Zhang Yichi 张一弛. "Yuzhi *Shengjing fu* yu Qingdai zhengzhi wenhua: Yi *Shengjing fu* de fabu, chuanbo yu yuedu wei zhongxin" 御制《盛京赋》与清代政治文化—以《盛京赋》的发布、传播与阅读为中心 (The imperially commissioned *Ode to Mukden* and Qing political culture, with a focus on the distribution, circulation, and reading of the *Ode*). *Qingshi yanjiu* 清史研究, no. 1 (2018): 37–51.

Zhang Yichi 张一弛 and Liu Fengyun 刘凤云. "Qingdai 'da yi tong' zhengzhi wenhua de goujian: Yi *Shengjing tongzhi* de zuanxiu yu chuanbo wei li" 清代"大一统"政治文化的构建—以《盛京通志》的纂修与传播为例 (The construction of the Qing-period political culture of the "integrated domain": The example of the *Comprehensive Gazetteer of Mukden*). *Zhongguo renmin daxue xuebao* 中国人民大学学报, no. 6 (2018): 159–69.

Zhang, Ying. *Confucian Image Politics: Masculine Morality in Seventeenth-Century China*. Seattle: University of Washington Press, 2016.

Zhang Ying 張英 and Wang Shizhen 王士禛, eds. *Yuding yuanjian leihan* 御定淵鑑類函 (Imperially commissioned categorized boxes of the Profound Mirror Studio). Facsimile. 1701? Reprint, Taipei: Xinxing shuju, 1960.

Zhang Zilie 張自烈. *Zhengzi tong* 正字通 (Mastery of correct characters). Blockprint, 1671. Reprint, Xiushui: Wu Yuanqi Qingwei Tang, 1685. Digitized, call no. T 5172 1321. Harvard-Yenching Library, Cambridge, MA.

———. *Zhengzi tong* 正字通 (Mastery of correct characters). Edited by Liao Wenying 廖文英. Vols. 234–35. *Xuxiu Siku quanshu* 續修四庫全書. 1671. Reprint, Shanghai: Shanghai guji chubanshe, 1995–2002.

Zhao Erxun 趙爾巽, ed. *Qingshi gao* 清史稿 (Draft history of the Qing). 1927. Reprint, Beijing: Zhonghua shuju, 1977.

Zhao Hong'en 趙弘恩, ed. *Jiangnan tongzhi* 江南通志 (Comprehensive gazetteer of Jiangnan). Vols. 507–12. *Wenyuan Ge Siku quanshu* 文淵閣四庫全書, 1782. Reprint, Taipei: Taiwan Shangwu yinshu guan, 1983.

Zhao Huanlin 赵焕林 and Liaoning sheng dang'an guan 辽宁省档案馆, eds. *Heitu dang: Kangxi chao* 黑图档：康熙朝 (*Hetu dangse* [archive]: Kangxi reign). 56 vols. Beijing: Xianzhuang shuju, 2017.

———, eds. *Heitu dang: Qianlong chao* 黑图档：乾隆朝 (*Hetu dangse* [archive]: Qianlong reign). 24 vols. Beijing: Xianzhuang shuju, 2016.

Zhao Xuemin 趙學敏. *Bencao gangmu shiyi* 本草綱目拾遺 (Supplement to *Systematic Materia Medica*). Facsimile. 1871. Vols. 994–95. *Xuxiu Siku quanshu* 續修四庫全書. 1881. Reprint, Shanghai: Shanghai guji chubanshe, 1995–2002.

Zhao Zhiqiang 赵志强. "Lun Qingdai de Nei Fanshu Fang" 论清代的内翻书房 (On the Manchu-Chinese Translation Office in the Qing period). *Qingshi yanjiu* 清史研究, no. 2 (1992): 22–28, 38.

———. *Qingdai zhongyang juece jizhi yanjiu* 清代中央决策机制研究 (A study on the policymaking mechanisms of the Qing central government). Beijing: Kexue chubanshe, 2007.

Zheng, Bingyu. "Experimenting with the National Language: Use of Manchu in Bannermen Poetry and Songs in the Nineteenth Century." *CHINOPERL: Journal of Chinese Oral and Performing Literature* 39, no. 1 (2020).

Zhi-kuan 志寬 and Pei-kuan 培寬. *Duiyin jizi* 對音輯字 / *Dui yen ji dzi bithe* (Collection of characters for transcription). Blockprint. 2 vols. Jingzhou: Jingzhou Fanyi Zongxue, 1890. Capital Library, Beijing.

Zhu Pengshou 朱彭壽. *Anle kangping shi suibi* 安樂康平室隨筆 (Random jottings at the House of Peaceful Joy and Healthful Tranquility). Beijing: Zhonghua shuju, 1982.

Zhu Su 朱橚, ed. *Jiuhuang bencao* 救荒本草 (Materia medica for famine relief). Facsimile. Vol. 730. *Wenyuan Ge Siku quanshu* 文淵閣四庫全書. 1782. Reprint, Taipei: Taiwan Shangwu yinshu guan, 1983.

Zou Zhenhuan 邹振环. "*Haicuo tu* yu Zhong-Xi zhishi zhi jiaoliu" 《海错图》与中西知识之交流 (*Illustrations of the Miscellaneous Things from the Sea* and Sino-Western knowledge exchange). *Zijincheng* 紫禁城, no. 3 (2017): 124–31.

———. "Jiaoliu yu hujian: *Qinggong haicuo tu* yu Zhong-Xi haiyang dongwu de zhishi ji huayi" 交流与互鉴：《清宫海错图》与中西海洋动物的知识及画艺 (Exchange and Cross-Consultation: *The Qing Palace's Illustrations of the Miscellaneous Things from the Sea* and knowledge and painting of marine animals from China and the West). *Huadong shifan daxue xuebao (zhexue shehui kexue ban)* 华东师范大学学报 (哲学社会科学版), no. 3 (2020): 96–106.

———. "*Shoupu* zhong de wailai 'yiguo shou'" 《兽谱》中的外来"异国兽" (The foreign "beasts from strange countries" in the *Catalog of Beasts*). *Zijincheng* 紫禁城, no. 10 (2015): 142–49.

———. *Zai jian yishou: Ming-Qing dongwu wenhua yu Zhong-wai jiaoliu* 再见异兽：明清动物文化与中外交流 (Exotic beasts seen anew: Animal culture in the Ming-Qing period and China's interaction with the outside world). Shanghai: Shanghai guji chubanshe, 2021.

Zucker, Arnaud. "Zoologie et philologie dans les grands traités ichtyologiques renaissants." *Kentron* 29 (2013): 135–74.

Index

Locators in *italics* refer to figures.

Aleni, Giulio (Ai Rulüe 艾儒略): on "a fish called the swordfish" (*jianyu* 劍魚, *xiphias*), 53, 53n93; Ming period scholarship on natural history published by, 41

Amiot, Joseph-Marie
—*Dictionnaire Tartare-Mantchou François*: *shenyu* rendered as oyster (*huître*), 47; *ucika* associated with dorsal fins, 44n67
—translation of the *Ode to Mukden*: on Ch. *zun* 鱒 and Ma. *jelu* (trout [?]), 225; the *Mirror of the Manchu Language* consulted for, 196, 201, 225; tree identified as Ch. *duanmu* 椴木 and Ma. *nunggele* (Tilia), 171, 225

Analects. See under *Four Books* (*Sishu*)

aquatic creatures: carp (Ma. *mujuhu*; Ch. *li* 鯉), 202, 225; creatures identified as strange and unknown to locals in *Shengjing tongzhi*, 130; *deyengge nimaha* (flying fish; Ch. *feiyu* 飛魚) introduced in the new *Mirror*, 239; eels (Ma. *horo*; Ch. *man* 鰻), 202, 215, 226; eels (Ma. *hūwara*; Ch. *li* 鱧), 225–26; *erin nimaha* (time fish) coined for (Ch. *shiyu* 鰣魚 reeves shad), 300; *Haicuo tu* 海錯圖 (Illustrations of the miscellaneous things from the sea) (Nie Huang 聶璜), 42, 42n62, 51n88; *Haiguai tuji* 海怪圖記 (Illustrated record of sea monsters), 42; *haizha* 海蜇 (sea jellyfish) in the *Ode*, 218; "human fish" (*niyalma nimaha* [mermaid]), 43, 55–56; immature locusts identified as born from fish and shrimp eggs, by Qianlong, 255–56; *miantiao yu* 麵條魚 (noodle fish), 296; on the need for the study of words for aquatic creatures in the *Pentaglot Mirror* suggested by Doerfer, 40, 233n7, 236, 236n19; *niomošon* (Siberian salmon) translated as *eyu* 鮧魚, 281; *niomošon* used to translate *xilin* 細鱗 (narrow-scaled fish), 222–23, 281; a note in the Kangxi *Mirror* regarding "sea fish" possibly related to an encounter with European natural histories or Chinese pictorial albums, 241; *sangguji* used for *haizha* 海蜇 (sea jellyfish) in the second edition of the *Ode*, 218, 236; transcribed

aquatic creatures (*continued*)
Chinese loan words for aquatic animals instead of Manchu words in the *Lianzhu ji*, 202; turtles (Ma. *aihūma*; Ch. *bie* 鱉), 32, 202, 258. See also *Mirror of the Manchu Language* (Kangxi *Mirror*) (1708)—fauna chapters (*šošohon*)—"Creatures with scales and shells" (*esihengge hurungge*); *Mirror of the Manchu Language* (Kangxi *Mirror*) (1708)—fauna chapters (*šošohon*)—"Creatures with scales and shells" (*esihengge hurungge*)—section on "River fish" (*birai nimaha*); *Mirror of the Manchu Language* (Kangxi *Mirror*) (1708)—fauna chapters (*šošohon*)—"Creatures with scales and shells" (*esihengge hurungge*)—section on "Sea fish" (*mederi nimaha*)

—*edeng* (water tiger [sawfish? swordfish? orca?]): definition in *Qingwen huishu* (Li Yanji), 48, 48n78, 51, 52n90, 53; definition in the Kangxi *Mirror*, 43, 44–46, 57; "old people" or "old men" (*sakdasa*) consulted about it, 43, 49n81, 50–51, 51n88, 54, 57; translated as *shuihu* 水虎 (water tiger), 46–47n70; *xiphias* described in his *Historia de gentibus septentrionalibus* compared to the *edeng* in the Kangxi *Mirror*, 52–53, 52n90. See also *yasa jerkišembi* (dazzling eyes)

Auf kaiserlichen Befehl erstelltes Wörterbuch des Manjurischen in fünf Sprachen "Fünfsprachenspiegel": Systematisch angeordneter Wortschatz auf Manjurisch, Tibetisch, Mongolisch, Turki und Chinesisch (Corff): on *forimbi* (to beat, to strike), 294n47; *gaha oton* (bitter melon) identified as *Momordica charantia*, 73n47; on the list of flower names in *Guangqun fangpu*, 305, 306–7; on *wanggua* as a variant of *huanggua* 黃瓜 (cucumber; Ger. *Salzgurke*), 70n39

Baqi tongzhi [chuji] 八旗通志[初集] (Comprehensive treatises of the Eight Banners [first installment]) (Hūng Jeo, Maci, and Ortai): Fudari's 傅達禮 (fl. 1671–1675) biography in, 26n19

Bencao gangmu 本草綱目 (Systematic materia medica) (Li Shizhen): on bitter melon (*gaha oton*), 72, 73–74; *cangji* 鶬雞 identified as a word used by southerners, 223; on *gaha hengke* (small red snake gourd, lit. crow tub), 72–74; information of *qianma* 蕁麻 (nettle) quoted in *Guang Qunfang pu*, 161–62; Li Shizhen's close attention to names in, 14; on the macaque (*mihou* 獼猴), 74

beyond-the-border plants (Ma. *jasei tulergi*): directly studied by Kangxi in his "investigation of things" (*gewu* 格物), 185; as a focus of scholar-officials in Kangxi's hunting entourages of the 1700s, 9–10; *isi* (larch) identified in the Kangxi *Mirror* as, 37. See also flora—golden lotus

birds: *elan* 阿藍 described as a generic name for small sparrows in *Shengjing tongzhi*, 134–35; *elan* (likely a crested lark) encountered by Qian Liangze in Mongolia, 134–35; *gasha* (bird) defined in the Kangxi *Mirror*, 67; *heturhen* (Ch. *lan hushou* 攔虎獸 [*lan* tiger beast ?] or *tugu* 兔鶻 [rabbit falcon]), 294; jay (Ma. *isha*; Ch. *songya* 松鴉), 248–50, 249n52, 282n18; *sama cecike* (small shaman bird) described as *wolan que* 窩藍雀, 135; specimens of birds in Qianlong's Grand Council's collection, 246–47, 247n46; *tojin* (peacock) as a loanword in Manchu, 234. See also *Catalog of Birds*; *Mirror of the Manchu Language* (Kangxi *Mirror*)

(1708)—fauna chapters (*šošohon*)—
"Large birds" (*gasha cecike*)
—ducks (Ch. *ya* 鴨): *alhacan niyehe* (mottled duck), 282n17; "domesticated geese, ducks, and other animals" (*boode ujiha niongniyaha, niyehe-i jergi ergengge jaka*) identified as "domestic animals" (*ujima*) in the Kangxi *Mirror*, 67; mandarin duck (Ch. *yuanyang* 鴛鴦) identified with lama duck (*lama ya*), 303; mandarin duck (Ch. *yuanyang* 鴛鴦) listed as *ijifun niyehe* in the original and expanded *Mirrors*, 67, 302–3
—in the Kangxi *Mirror* (*cecike-i hacin*): bull quail (*ihan mušu*), 283n19; *fengtou elan* 鳳頭阿蘭; Ma. *saman cecike* (Phoenix head elan) and *elan* 阿蘭 (Ma. *wenderhen*) in, 135, 135n28
—ostriches: naming in Manchu of, 2, 5; as *temege coko* in the revised *Mirror*, 255; *temen coko* (camel chicken) proposed in a memorial by Qianlong (Feb. 28, 1754), 254–55; as *to gi* (*tuoji* 駝雞 [camel chickens]) in the Kangxi *Mirror*, 39, 254–55; as *tuoji* (camel chickens) in Ming-period sources, 39
Bouvet, Joachim (1656–1730), 26
Bray, Francesca, 13n23
Bretschneider, Emil, 11n19, 106n44, 116n61
Broadly Collected Complete Text in the Standard Script (Ch. *Tongwen guanghui quanshu*; Ma. *Tung wen guwang lei ciowan* šu): categories of the Kangxi *Mirror* compared with those in it, 63–64, 66; *Lianzhu ji* 聯珠集 (Collection of stringed pearls) published as a supplement to later editions of, 201
Brook, Timothy, on local gazetteers predating the Qing, 127n8
B·Sod 波索德, 236
Buddhism: *ergengge jaka* (thing endowed with breath) identified as having Buddhist overtones, 67; 'Jam-dpal-rdo-rje (1792–1855) identified as a Buddhist monk, 322; Mount Wutai as a holy place in Mongolian Buddhism, 37, 194; names for birds that originate in the "*kanjur* canon" (*g'anjur nomun*) or "Buddhist canon" (*fucihi nomun*) in the 1772 *Mirror*, 240n31

Catalog of Beasts: bilingual legends in, as a continuation of late Kangxi-era encyclopedism, 231; *honin* (sheep, Ch. *yang* 羊) depicted in, 84; *kandatu* used to translate the name of a legendary Chinese bovine, 264; *mi* (Ma. *suwa*) depicted in, 259, 260, 261n80, 262, 296n58; *miyahūtu* not used in the place of *gi buhū* in, 219; *sabintu* (unicorn) included in, 295–96, 297; simians from, included in the supplement of the 1772 bilingual *Mirror*, 310; strange beast from, included in the supplement of the 1772 bilingual *Mirror*, 309; *sumaltu* coined as a name for the opossum, 247

Catalog of Birds: bilingual legends in, as a continuation of late Kangxi-era encyclopedism, 231; black-naped oriole (Ch. *huangli* 黃鸝; Ma. *galin cecike*), 248; *isha* as *songya* (pine crow) in, 249–50, 182n18; legendary jay (Ma. *isha*; Ch. *songya* 松鴉) in, related to the paradise flycatcher in the Kangxi *Mirror*, 248–49; words lifted from it for the new *Mirror*, 292–94

catalogs (*pu*): Chen Jiru's valorization of *pu* by associating it with family genealogies (*jiapu* 家譜), 92–93; cross-pollination with the scholarly study of words and characters, 331–32; transformation of the categorical learning in *leishu* into past and present marvels, 112–13. See also *Catalog of Birds*; *Guang Qunfang pu* (Expanded catalog of myriad flowers); *Qunfang pu* (Catalog of myriad flowers) (Wang Xiangjin)

Chang-an (Nalan Chang-an 納蘭常安, 1683–1748): background and bannerman status of, 204, 205; scorpion grass (*gabtama*) identified with *qianma*, 161–62; "Shengjing wuchan fu" 盛京物產賦 (Rhapsody on the natural products of Mukden), 201, 204

Change Classic: Kong Yingda's commentary on "*guo* is the fruit of trees, *luo* is the fruit of vines," 70–71; *Shuo gua* 說卦 commentary on the trigram *gen* 艮 cited in the definition of *jancuhūn hengke* (sweet melon) in the Kangxi *Mirror*, 70

Chen Jiru 陳繼儒 (1558–1639, or Chen Migong 陳麋公): inclusion of his biography in *Qunfang pu*, 97, 98n30; preface written for *Qunfang pu*, 89–90, 92; valorization of *pu* by associating it with family genealogies (*jiapu* 家譜), 92–93; on the wide scope of *Qunfang pu*, 95

Chen, Kai Jun, 7

Chen Menglei 陳夢雷 (1650–1741): acceptance of Yinchi's task to compile a new encyclopedia, 114–15, 173; exile in Manchuria, 115, 172; scholars recruited to work on *Guanjin tushu jicheng*, 173; on Yinchi's views of the limitations of current comprehensive references, 114. See also *Gujin tushu jicheng* 古今圖書集成 (Consummate collection of texts and illustrations from past and present) (Chen Menglei and Jiang Tingxi)

Chiu, Elena Suet-Ying, 7n12

Ch'oe Hak-kŭn 崔鶴根, 210n50

Chongzhen emperor (r. 1628–1644), Wang Xiangjin reinstated in office at the beginning of the reign of, 89

Chuang Chi-fa 莊吉發: on the coining of Manchu words based on corresponding Chinese expressions, 235;

ya 鴉 (crow) asserted as mistakenly used for *ya* 鴨 (duck) in the *Catalog of Birds*, 250

Chunhua 春花, 66, 280–81

Cibot, Pierre-Martial (Han Guoying 韓國英, 1727–1780), 175–76

"classified writings" (*leishu*) as a genre: Chen Menglei's interest in improving it by intense and exhaustive textual searching, 115; cross-pollination with the scholarly study of words and characters, 331–32; erudition derived from the *leishu* tradition evidenced by Chinese scholarship quoted in the Kangxi *Mirror*, 75; Kangxi's impatience with its focus on "literary composition," 117–18. See also *Gujin tushu jicheng* 古今圖書集成 (Consummate collection of texts and illustrations from past and present) (Chen Menglei and Jiang Tingxi); *Taiping yulan* 太平御覽 (Imperial digest [from the era of rousing the state through] great tranquility); *Yuanjian leihan* 淵鑑類函 (Categorized boxes of the Profound Mirror Studio)

Confucian classics: appending of quotes from to definitions in the *Mirror*, to anchor a Manchu word in the tradition of classical learning, 58, 164; Manchu versions published by the Qing court, 278. See also *Four Books* (*Sishu* 四書)

Cook, Harold J., 10

Corff, Oliver. See *Auf kaiserlichen Befehl erstelltes Wörterbuch des Manjurischen in fünf Sprachen "Fünfsprachenspiegel": Systematisch angeordneter Wortschatz auf Manjurisch, Tibetisch, Mongolisch, Turki und Chinesisch* (Corff)

Crossley, Pamela Kyle, on the *Ode to Mukden*, 213

INDEX

Da Dai liji 大戴禮記 (Record of rites of the elder Dai), 65

Da Ming yitong zhi 大明一統志 (Unified gazetteer of the Great Ming) (Li Xian 李賢): on Arhat's weed, 101–2; as a source for *Guang Qunfang pu*, 101; as a source for *Shengjing tongzhi*, 130

Da Qing quanshu 大清全書 (Complete book of the Great Qing) (Shen Qiliang 沈啟亮): arrangement determined by its Manchu lemmata rather than by topic, 64; on *ayan buhū* translated as *milu* 麋鹿 (elk) in, 260; *boo nimaha* translated as *fuyu* 鮄魚, 47n74; *fethe* (tail fin) in the sense of *yuchi* 魚翅 (fins), listed with *yusai* 魚鰓 (gills), 44n67; *hualu* (blossom deer) as a translation for *suwa buho* (i.e., buhū), 261n80, 296n57; *irgebun* (poem, poetry) associated with *fu* 賦, 209–10; on *malu* (elk, lit. horse deer), 260; Manchu lemmata left-untranslated in, 21–22; on *nunggila* (read: *nunggile*) *moo* and *nunggele moo* with the translation *jiamu* 椵木 (*jia* tree), 171; *qi* 杞 (goji berries) identified with *una*, 186; on Sika deer (*suwa buhū*), 296n57; words left untranslated in, 222

deer and moose: *arfu buhū* proposed as a new word for *mi* 麋 (deer), 258–59, 261–62; *ayan buhū* translated as *milu* 麋鹿 (elk) in *Da Qing quanshu*, 260; *ayan buhū* used for *malu* 馬鹿 (elk) in the new *Mirror*, 296n57; *buhekū* used for *buhū* in the 1769 official calendar, 264; *buhū* used for *lu* 鹿 (deer) in the new *Mirror*, 296n57; "camel deer" and *mi* deer equated in Lu Dian's *Piya*, 131–32, 179; fly whisks associated with the *mi* deer, 77; *gi buhū* (Red muntjac?), 219; *jolo buhū* (doe deer) in *Sishu yaolan* 四書要覽 (*Sye šu oyonggo tuwara bith*e [Essential readings from the *Four Books*]) (Shen Qiliang), 261n80; *kandahan* 堪打漢 (moose) in the Manchu *Mirror*, 31–32, 132, 132n19, 179, 264; *kjɛn* as the word for moose, 264–65; listings in the Kangxi *Mirror* chapter titled "Beasts," 31–32; *mafuta buhū* (buck deer) paired with *milu* in *Man-Han leishu*, 261n80; *malu* (elk, lit. horse deer) in *Da Qing quanshu*, 260; Père David's deer (*milu* 麋鹿), 258n72, 261n80, 263; Qianlong on *mi* (Ma. *suwa*) in his "Lujiao ji," 262–63; *si* 麖 ("a kind of muntjac deer [*jun*] with long fur and dog fee, hide suitable for shoes") (or *pao* 麃), 218–19, 220n91; Sika deer (*suwa buhū*), 258–59, 261n80, 296n57; *suwa* defined as a small-bodied and light pink deer, 260–61n80; translation of *mi* deer (*milu* 麋鹿) in the *Four Books*, 258, 261n80, 264n93; *uncehen golmin buhū* (long-tailed deer), 76–77, 258–59, 261

—*zhu* 麈 (long-tailed deer) (Ma. *uncehen golmin buhū*): *arfu* formed from *arfukū* to describe a fly whisk made from the long tail of a deer or mule, 262; *changwei lu* 長尾鹿 (long-tailed deer) associated with, by Qianlong, 263, 263n90; described alongside *lu* 鹿 (deer) and *mi* in *Gujin tushu jicheng*, 260, 261–62; *mi* changed to *zhu* in the phrase "the *zhu* deer sheds its antlers" (*zhujiao jie* 麈角解), 263; *uncehen golmin buhū* (long-tailed deer) used to translate *zhu*, 261; *zhu wei* 麈尾 (*zhu* tail) fly whisk, 260–61, 299n65

Doerfer, Gerhard (1920–2003): on the need for the study of words for aquatic creatures in the *Pentaglot Mirror*, 40, 233n7, 236, 236n19; *sahamha* identified

Doerfer, Gerhard (*continued*)
  as a neologism, 55, 55n98; on the word *kenggin* as a loan word from Tungusic, 55
domestic animals. *See* livestock and domestic animals
dragons and snakes: *boo nimaha* identified with *shenyu* (a legendary creature with the appearance of a snake and horns like a dragon), 47–48; the headword "snakes hibernate" (*meihe bulunambi*), 36, 38; phrasing of the Kangxi *Mirror* section on "Dragons and snakes" traced to *Da Dai liji*, 65; *teng* 螣 identified as a kind of dragon in *Erya*, 38n52
Dutch: *balsam appel manneken* (Momorica balsamina, male), in *Yakugo shō*, 328; orca called "swordfish" in seventeenth-century Dutch, 52
dynastic historiography (*guoshi* 國史): family genealogies (*jiapu* 家譜) viewed as analogous to, by Chen Jiru, 92–93; gazetteers as preparatory material for, 101–2; Manchu translation of the Liao, Jin, and Yuan dynastic histories, 24

Early Modern Travels of Manchu (Söderblom Saarela), 3, 26n19
Eight Banners: Fudari's 傅達禮 (fl. 1671–1675) biography in *Baqi tongzhi* [*chuji*] 八旗通志[初集] (Comprehensive treatises of the Eight Banners [first installment]) (Hūng Jeo, Maci, and Ortai), 26n19; as a key institution for entrenched Manchu identity, 7n12; sentencing of Wang Hao's household to servitude under, 119; spread of the Manchu language facilitated by, 33–34; study of the literary output by men and women of, 7
Elliott, Mark C., 7n12, 195, 209

Elman, Benjamin A., 11n19, 12–13n22, 108, 333 Elvin, Mark, 13n23
Er ru ting Qunfang pu (Catalog of myriad flowers from the Pavilion of Double Resemblance). See *Qunfang pu* (Catalog of myriad flowers) (Wang Xiangjin)
Erya 爾雅 (Approaching perfection): berry with the character *qian* 蒹 in, 185; entry *teng* 螣, 38n52; identification of "six domestic animals" (*liu chu* 六畜), 67; inclusion in Needham and Lu's survey of Chinese botany, 14; lexical material arranged according to subject matter, 65–67; *lu* 蘆 equated with *wei* 葦, 215; tacit use of its definitions in the Kangxi *Mirror*, 66–67; *xiao* 蕭 equated with *di* 荻, 215, 216
European natural history: a curious note in the Kangxi *Mirror* regarding "sea fish" possibly related to, 241; endeavor to find European equivalents to Manchu terms for natural products in Amiot's and Klaproth's translation of the *Ode to Mukden*, 194–95, 214; engagement of Catholic missionaries in China with it, 15, 309; influence on words for sea creatures in the Kangxi *Mirror*, 23, 57; lack of a term for it in late imperial Chinese sources, 11; observations of, in "Houses of Experiment" in seventeenth-century England, 188; orcas known to, 52; scant knowledge of Manchu dictionaries in the seventeenth century, 26n15
evidential learning (*kaozhengxue* 考證學): attention to plants and animals not yet examined in the scholarly narrative of, 12, 12–13n22; its rise related to interest in the investigation of historical facts and mundane aspects of life, 108–9, 117; the resulting professionalization of scholarship

marked by *Siku quanshu*, 321; role of *Guang Qunfang pu* in establishing the "best practice" of, 108–9

Expanded and Emended Manchu Mirror (*Yuzhi zengding Qingwen jian* 御製增訂清文鑑) (1772): additions, alterations, and distinctions that set it apart from the 1708 *Mirror*, 312; *arfukū* translated as Ch. *zhuwei* to describe a fly whisk made from the long tail of a deer or mule, 262; *ashangga singgeri* (winged mouse) given the Chinese translation *fei lei* 飛䴇 (flying squirrel), 302; basic categories in, 287–88; *buhū* used for *lu* 鹿 (deer) and *ayan buhū* used for *malu* 馬鹿 (elk) in, 296n57; *cao lizhi* 草荔支 (grass lychee) used to translate *ilhamuke* in, 265–67; changes based on Chinese encyclopedic works, pictorial albums, and discussions on neologisms at court, 305; Chinese classics as an undercurrent of knowledge in, 288–90; Chinese translations in *Shinbunkan meibutsu goshō* 清文鑑名物語抄 (Content words excerpted from the *Expanded and Emended Mirror of the Manchu Language*), 328; *deyengge nimaha* (flying fish; Ch. *feiyu* 飛魚) introduced in, 239; on *douban nan* 豆瓣楠 (bean-lobe nan), 291; effort to establish a one-to-one correspondence between nomenclature and taxonomy in Manchu, 244; *erin nimaha* (time fish) coined for Ch. *shiyu* 鰣魚 (reeves shad), 300; *hoošan* (paper), 325; *hoošari moo* (paper mulberry) introduced in, 325, 327; *hoto hengke* (lit. skull gourd) translated as *huzi* 瓠子 (squash) in, 285, 285; *huhucu* (Adenophera [bellflower], Ch. *mingzhe cai* 茗荠菜) given a new definition in, 304n84; *ica* (Ch. *miantiao yu* 麵條魚 [noodle fish]) given the synonym *honokta*, 301; on jasmine (Ma. *moli ilha*; Ch. *moli* 茉莉), 290; *koojiha* (fish called *latun* [cured piglet]) in, 304; Manchu neologisms invented to name plants and animals in, 4, 6, 237, 286, 295; on names for birds that originate in the "*kanjur* canon" (*g'anjur nomun*) or "Buddhist canon" (*fucihi nomun*), 240n31; *nenden ilha* (plum) (lit. first flower), 288; neologism *okjihad* ("root of sweet flag") (Ch. *cangzhu* 蒼朮 [rhizome of *Atractylodes lancea* or *Atractylodes chinensis*]), 291–92; new words in the revised translation of the *Poetry Classic* not included in, 239–40; *niomošon* (Siberian salmon), 281; on *tama* (sole, Ch. *xiedi yu* 鞋底魚) defined as "shaped similarly to the sole of a foot," 303; *tuopan guo* 托盤果 (plate-top fruit) used for goji berries, 186; "Ula grass" as a type of *foyo* added to, 143, 143–44n55; *wogua* 倭瓜 (Japanese gourd), 285, 285; words lifted from the *Catalog of Birds*, 292–94; Zach on neologisms in, 234–35

—"Beasts": *qilin* 麒麟 (or *kilin*, unicorn) replaced with *sabitun* (*qi*) and *sabintu* (*lin*), 295–96; simians from the *Catalog of Beasts* included in, 297–99, 298n63, 311; simians quoted in *Gujin tushu jicheng* included in, 299; *sirsing* placed first among the simians, 299n65; *xi shu* 磎鼠 (iced rodent, Ma. *juhen singgeri*) in, 296–97, 329, 329n42

—supplement (*bubian* 補編): creatures listed as fantastical in the *Catalog of Beasts* included in, 299; designing by Qianlong as a reservoir for plants and animals that did not appear in "everyday language," 305; *endurin soro* (divine jujube) and *soro* (jujubes) in, 307, 307n90; legendary creatures in Chinese texts listed in, 309; newly invented names for varieties of peacock

—supplement (*bubian* 補編) (*continued*)
included in, 253; simians from the *Catalog of Beasts* (that were left out of the main body of the dictionary) included in, 310–11; on the "terrestrial golden lotus" (Ch. *han jin lian* 旱金蓮; Ma. *aisin šu ilha*), 156n25; *wemburi* as a neologism for haw, to replace the Manchu word *umpu* (Ch. *wenpu*), 308

Fan, Fa-ti, 10
Fang Gongqian 方拱乾 (1596–1666, *jinshi* 1628): exile to Ningguta, 137. See also *Ningguta zhi* 寧古塔志 (Treatise on Ningguta) (Fang Gongqian 方拱乾)
Fang Shiji 方式濟 (*jinshi* 1709): detailed account of Manchuria included in *Siku quanshu*, 139, 139n42; on *uli* (*ou lizi* 歐李子) found in Ningguta and Aigun, 165–66
Fang Yuegong 方岳貢 (*jinshi* 1622), preface written for *Qunfang pu*, 90
fauna: *fen shu* 鼢鼠 (or zokor, dung rodent), 183, 329n42; *lefu* (bear) in *Da Qing quanshu*, 64; *xi shu* 磎鼠 (iced rodent, Ma. *juhen singgeri*), 183, 183n87, 296–97, 329, 329n42. See also aquatic creatures; birds; deer and moose; dragons and snakes; livestock and domestic animals; *Mirror of the Manchu Language* (Kangxi *Mirror*) (1708)—fauna chapters (*šošohon*); mythical animals and strange beasts (*yi shou* 異獸)
fish. See aquatic creatures; *Mirror of the Manchu Language* (Kangxi *Mirror*) (1708)—fauna chapters (*šošohon*)—"Creatures with scales and shells" (*esihengge hurungge*)—section on "River fish" (*birai nimaha*); *Mirror of the Manchu Language* (Kangxi

Mirror) (1708)—fauna chapters (*šošohon*)—"Creatures with scales and shells" (*esihengge hurungge*)—section on "Sea fish" (*mederi nimaha*)
flora: neologism *okjihad* ("root of sweet flag," Ch. *cangzhu* 蒼朮 [rhizome of *Atractylodes lancea* or *Atractylodes chinensis*]), 291–92; neologisms for flora and fauna evidenced in Grand Council documents, 2; revision of names for plants (sugarcane, apple, Chinese crabapple, pomegranate) during Qianlong's reform, 237; "yellow flower" (*huang hua* 黃花) in *Guang Qunfang pu*, praised by Kangxi, 149–52, 152n12, 154. See also fruits and nuts; grasses; melons and gourds; *Mirror of the Manchu Language* (Kangxi *Mirror*) (1708)—flora chapters (*šošohon*); mushrooms (Ma. *megu*; Ch. *mogu* 蘑菇), in Kangxi Mirror, "Vegetables and dishes" (*sogi booha*); vegetables
—golden lotus: Chinese rhapsody written about it, 194; "Golden lotus flower" (*jinlianhua* 金蓮花) in *Guang Quanfang pu*, 154–56, 156n25, 173, 307–8; iconic status in the supplement as *aisin su ilha*, 307–8; "terrestrial golden lotus" (Ch. *han jin lian* 旱金蓮; Ma. *aisin šu ilha*), 156, 156n25
Foucault, Michel, 332
Four Books (*Sishu* 四書): Chen Menglei on Yinchi's views of the *Extended Meaning* [of the Great Learning] ([*Daxue*] *yanyi* 大學衍義, 1229), 114; *jolo buhū* (doe deer) in *Sishu yaolan* 四書要覽 (*Sye šu oyonggo tuwara bithe* [Essential readings from the *Four Books*]) (Shen Qiliang), 261n80; Manchu editions of, 258–59n73; unsuccessful pre-conquest petitioning for a Manchu translation of, 23; Wang

INDEX

Xiaoling's paraphrasing of the *Doctrine of the Mean*, 91, 91nn11–12
—*Analects*, Confucius's deference to "Old farmers" and "old gardeners," 91
—*Mencius*, translation of *mi* deer (*milu* 麋鹿) in, 258, 261n80, 264n93
fruits and nuts: bush cherry (Ma. *uli*; Ch. *yuli* 郁李), 167; hazelnuts (Ch. *shanzha* 山楂; Ma. *umpu*), 138, 139, 187–88, 308; Kangxi's proposal of a tripartite classification of berries into vine-grown ones, upright-growth shrub berries, and herbaceous berries that simply crawl on the ground, 185; *kišimiši* (seedless green grape, Ch. *wuzi lü putao* 無子綠葡萄), 237–38; *mi-sun-wu-shi-ha* 米孫烏什哈 listed as a product from Ningguta but not translated as *misu hūsiha* in *Shengjing tongzhi*, 141, 147, 221; neologism for apple (*pingguri*), 237, 286; neologism for pomegranate (*useri*), 237, 286; pine nuts (*songzi* 松子), 138, 139, 204; *qi* 杞 (goji berries) identifed with *una* by Shen Qiliang, 186; *soro* (jujubes), 307, 307n90; *šulhe* (fragrant water pear) in Imperial Household Department documents, 128–28n14; *uli* (*ou lizi* 歐李子) found in Ningguta and Aigun, 165–66; *yi-er-ha-mu-ke* 衣而哈目克 fruit compared to a small *yangmei* (bayberry) by Wu Zhenchen, 58. See also melons and gourds
—crabapples: bifurcating interpretations between a crabapple (*nai* 柰) and a plum (*li* 李), 158, 162–67; entanglement between Chinese sources and those recorded in other languages revealed by, 158; in *Gujin tushu jicheng*, 173n64; neologism Chinese crabapple (*yonggari*), 237, 286; *ouli* 歐棃 ("Ula crabapple" [*Wula nai* 烏喇柰]), 162–67; pear-leafed crabapple (Ma. *mamugiya*), 244; "Ula crabapple" described and eaten by Wang Hao and Zha Shenxing, 162–63
—*ilhamuke* (flower water): *cao lizhi* 草荔支 (grass lychee) glossed as *tsuru reiishi* ツルレイシ (vine lychee), 328; *cao lizhi* 草荔支 (grass lychee) in Qianlong's poetry, 268–69; *cao lizhi* 草荔支 (grass lychee) paired with *ilhamuke* in a bilingual report submitted to the Grand Council (1765), 266–67, 269–70; *cao lizhi* 草荔支 (grass lychee) used to translate it in Qianlong's Manchu-Chinese *Mirror*, 265–66
—*yengge*: *yengge* associated with "Parrot pass" (*Yingge guan* 鸚哥關) in Gao Shiqi's Manchu word list, 179–80; *yengge* described as "astringently flavored" fruit that looked like small grapes (*boco sahaliyan, mucu de adalikan, amtan fekcuhun*) in the Kangxi *Mirror*, 140, 181; *yengge* (wild cherry, or bird cherry) in Tungusic, 180; *yengge* (wild cherry) in Manchu documents sent to the Mukden branch of the Imperial Household Department, 128; *ying-e* 櫻額 (Ma. *yengge*), described in *Kangxi ji xia gewu bian* as a smaller version of "wild black grapes," 179

fu 賦 (rhapsody): Ban Gu's 班固 (32–92 CE) "Liang du fu" 兩都賦 (Two capitals rhapsody; 58–75 CE), 198, 199; as the form of the *Ode to Mukden*, 194–95, 196, 197–99, 201, 211–12; in *Guang Qunfang pu*, 154; *irgebun* (poem, poetry) associated with, 209–10; as one of six Principles of Poetry (*liu yi* 六義) in the *Zhouli* 周禮 (Rites of Zhou), 197; Yuan Mei 袁枚 (1716–1798), on gazetteers and encyclopedias used for writing *fu*, 199; Zhang Heng's 張衡 (78–139 CE) "Er jing fu" 二京賦 (Two metropolises rhapsody), 198

INDEX

Fuchs, Walter (1902–1979): words for plants and animals in the Manchu *Mirrors* discussed with Wolfgang Seuberlich, 313–14, 316, 331

Gao Shiqi 高士奇 (1645–1703): sample of the "night-glow wood" shown to him, by Kangxi, 157; travels to Manchuria and to Mount Wutai with Kangxi in the 1680s, 124, 153, 153n16
—appendix (*fulu* 附錄): *ilhamuke* (flower water) rendered as *yi-er-ga-mu-ke* 一兒噶木克, 144; *mi* deer's tail mushroom (Ma. *gio ura*; Ch. *mizi wei* 麋子尾) listed in, 142; "names and facts" (*mingshi* 名實) as its focus, 125; names of local plants and animals with only Manchu names transcribed into Chinese characters, 221; Stary's study of the wordlist in, 125, 125n7; "Ula grass" (*Wula cao* 烏喇草) rendered as *hūwaitame gūlha foyo* (tied-up boot grass), 143; on *uli* (*wu-li* 烏立, in vernacular Ch. *laoya yan* 老鴉眼 [the old crow's eye]), 165; *yengge* associated with "Parrot pass" (*Yingge guan* 鸚哥關), 179–80

gazetteers: effort to grasp and describe flora and fauna across different languages, 9, 82; local gazetteers predating the Qing, 127; as preparatory material for dynastic historiography, 101–2; project to compile a Manchurian gazetteer, 83; provincial gazetteers ordered to be updated by Yongzheng, 128n11; as sources for *Guang Qunfang pu*, 101–3. See also *Da Ming yitong zhi* 大明一統志 (Unified gazetteer of the Great Ming) (Li Xian 李賢); *Shengjing tongzhi* 盛京通志 (Comprehensive gazetteer of Mukden)

Gessner, Conrad, engraved images of natural and wondrous creatures, 245, 246

Golvers, Noël, 42

Gong Dingzi 龔鼎孳 (1616–1673), on written characters (*wenzi*), 27, 278

Goodrich, L. Carrington, 8n15

gourds. *See* melons and gourds

Grand Council: *cao lizhi* 草荔支 (grass lychee) in a mixed-language report submitted to, by Hoiling and Horonggo, 266–70; Manchu word for locust discussed at, 255–58; request for information on "the word *guwalase* for *kegua* 客瓜" (*guwalase sere* 客瓜 *emu gisun*), 270–72; role in generating neologisms, 273n116

Grand Council documents: bilingual characteristics of, 1–2; the Kangxi *Mirror* referenced for *huaxin* 花心 (center of a flower) as *jilha*, 287; neologisms for flora and fauna evidenced in, 2; research efforts to find appropriate plant and animal names in Manchu revealed in, 15; terms for different kinds of deer in Manchu and Chinese discussed in, 257–64

grasses: *ayan hiyan* (Ch. *yunxiang*; Mo. *sülü* [incense]), 243, 244; *gabtama* (to shoot [an arrow]) (nettles), 35, 159–60, 161n35, 167; scorpion grass (*gabtama*) identified with *qianma*, 159, 160–62, 167, 182; "Ula grass" (*Wula cao* 烏喇草) rendered as *hūwaitame gūlha foyo* (tied-up boot grass), 143–44. See also *Mirror of the Manchu Language* (Kangxi *Mirror*) (1708)—flora chapters (*šošohon*)—"Grasses" (*orho-i hacin*)

Greenberg, Daniel M., 42

Gu Songjie 顾松洁 and Gao Xi 高晞, 322

Guang Qunfang pu (Expanded catalog of myriad flowers): *cao lizhi* mistakenly claimed to have been included in it, by Qianlong, 266–67; completion of, 85, 85n1; discussion of the list of flower names, 305307; editorial team of, 93, 94; emphasis on novelty and curiosity in *Yuanjian leihan* contrasted with, 113; format for presenting glosses of pronunciation and etymology, 106, *107*; *gabtama* (to shoot [an arrow]) (nettles), 159–60, 167; "Golden lotus flower" (*jinlianhua* 金蓮花) in, 154–56, 156n25, 173, 307–8; "Great efficacious bean" (*Daling dou* 大靈豆) in, 113; impact on Chinese-language encyclopedias and Manchu lexicography, 87, 269; as intentionally distinguished from *bencao* pharmacopeias, 104–5; introduction of Manchu and Mongolian flora traced to natural-historical observations during Kangxi's 1703 hunting tour, 150, 152, 153, 156, 160, 165, 167–68; Kangxi's imperial preface to, 95–96; lemmata from, carried over to the *Pentaglot Mirror*, 305–8; lexicographical erudition of, 87–88, 104, 106, 106n44, 108–11, 117, 119, 172, 174; local gazetteers as sources for, 101–3; natural-historical erudition of, 86, 87; new entries and sources not seen in the Ming original marked with the character "added" (*zeng* 增), 110–11, *111*; "night-glow wood" (*yeguang mu* 夜光木) described in, 157; office established for, 173–74; on *qianma* 蕁麻 (nettle) in *Bencao gangmu* quoted in, 161–62; role in establishing the "best practice" of evidential research, 108–9; scorpion grass (*gabtama*) identified with *qianma*, 159, 160–62, 167, 182; section from QFP titled "Fragrant traces of past luminaries" eliminated from, 98; structure of entries in, 104–6, *106*; voluntary donations supporting its publication, 94; "yellow flower" (*huang hua* 黃花) in, praised by Kangxi, 149–52, 152n12, 154

Gujin tushu jicheng 古今圖書集成 (Consummate collection of texts and illustrations from past and present) (Chen Menglei and Jiang Tingxi): the addition of pictures to the genre *pu* traced to Jiang's role as a chief editor, 241–42; contextualization of its emergence, 9, 88; crabapple in, 173n64; entry on *erin nimaha* (time fish) coined for Ch. *shiyu* 鰣魚 (reeves shad) possibly based on, 300; identified as the first draft of *Guang Qunfang pu*, 88; illustrations and descriptions lifted from *Kunyu tushuo*, 117; influence of *Guang Qunfang pu* on its natural erudition (*bowu* 博物), 116, 173, 173n64; as the pinnacle of the Chinese encyclopedic tradition (*leishu*), 174–75; plants and animals divided into minute entries rather than broadly defined "categories" (*lei*), 172; as a ploy in court politics, 172; search for *yi* 異 (unusual/strange), 113; strange beast from, included in the supplement of the 1772 bilingual *Mirror*, 309; *zhu* 麈 (long-tailed deer) (Ma. *uncehen golmin buhū*) described alongside *lu* 鹿 (deer) and *mi* in, 260

Guy, R. Kent, 7n14, 321, 333

Haicuo tu 海錯圖 (Illustrations of the miscellaneous things from the sea) (Nie Huang 聶璜), 42, 42n62, 51n88

Haiguai tuji 海怪圖記 (Illustrated record of sea monsters), 42

Hammers, Roslyn L., 7

Han-i araha manju gisun-i buleku bithe. See *Mirror of the Manchu Language* (Kangxi *Mirror*) (1708)

Han-i araha nonggime toktobuha manju gisun-i buleku bithe. See *Expanded and Emended Manchu Mirror* (*Yuzhi zengding Qingwen jian* 御製增訂清文鑑) (1772)

Hasegawa, Masato, 1

Hauer, Erich (1878–1936): on the coining of Manchu words based on corresponding Chinese expressions, 235, 314–15; *mahūntu* identified as simian neologism, 311n100; and *ukeci* (Ch. *penghou* 彭猴), 298; Walraven's criticism of his dictionary, 315n9; on yellow gibbon (Ma. *sobonio*; Ch. *rong* 狨) (*sohon* + *bonio*)

historical scholarship: bilingual natural-historical scholarship at the Qing court, 2; projects carried out by the Imperial Household Department, 24–25. *See also* European natural history; evidential learning (*kaozhengxue* 考證學); pharmacy and pharmacopeia; *Siku quanshu* 四庫全書 (Complete books of the four repositories)

horses. *See* livestock and domestic animals—horses

Hu Zengyi 胡增益, on *jerkišembi*, 45n67

Huainanzi 淮南子, on the ancient idea that the brains of fish grow with the moon, 68

Huang Aiping 黃爱平, 320n16

Huang Longmei 黄龍眷 (*jinshi* 1694), 94

Huang Wei-Jen 黄韋仁, 280, 283

Huang Qing zhigong tu 皇清職貢圖 (Album of tributary peoples of the august Qing), 43n66, 246

Ihing 宜興 (1747–1809). *See Qingwen buhui* 清文補彙 (*Manju gisun be niyeceme isabuha bithe* [Manchu collected, supplemented]) (Ihing 宜興)

Imanishi Shunjū 今西春秋 (1907–1979): bibliographical research on the Manchu-Chinese *Mirror*, 283

Imanishi Shunjū 今西春秋, Tamura Jitsuzō 田村実造, and Satō Hisashi 佐藤長, *edeng* translated by, 49n81

imperial diary: Kangxi on wanting the *Mirror* to imitate the Chinese *Zihui* 字彙, 25–26n14; no mention of Qianlong's commissioning of the *Ode* in the *Qiju zhuce* 起居注冊 (Register of rise and repose), 206–7

Imperial Household Department (Ch. Neiwu fu 內務府; Ma. Dorgi baita be uheri kadalara yamun): books printed by, during Shunzhi reign (1644–1661), 24–25; Chang-an on the estates managed by, 204; circulation of new words to the bureaucracy evidenced in documents sent between its counterpart in Mukden, 237; loan words in archival documents of, 218; natural products submitted to it from areas beyond the original Manchu homeland, 34; šulhe (fragrant water pear) in documents exchanged with captains in Mukden, 128–28n14; *yengge* (wild cherry) in Manchu documents sent to the Mukden branch of, 128

Imperially Commissioned Mirror of the Manchu Language. See *Mirror of the Manchu Language* (Kangxi *Mirror*) (1708)

insects: *dondon* defined as a "small butterfly" (*ajige gefehe*) in the Kangxi *Mirror*, 32; *usi umiyaha* (Ch. *huichong* 蛔蟲), 300

—locusts: immature locusts identified as born from fish and shrimp eggs, by Qianlong, 255–56; Manchu word *sabsaha* proposed for locust at the Grand Council, 255–57; *sebseheri* as the neologism chosen for locust, 257; *unika* proposed as a word for immature locust, 255–56

investigation of things (*gewu* 格物): articulation through an open-ended lexicographical encyclopedism, 119, 316; integrated taxonomic vision of, 177, 185; Kangxi's personal experience

INDEX

on his tours contrasted with the work of scholars isolated in the study, 4, 9–10, 123–24, 175, 177, 188; as a precept of the Learning-of-the-Way, 175. See also *Kangxi ji xia gewu bian* 康熙幾暇格物編 (Collection of the investigation of things during times of leisure from the myriad affairs [of state] in the Kangxi period)

Jami, Catherine, 175, 177

Japan and Japanese language glosses: citations of "classified writings" from the late Ming by Japanese scholars, 322; Haneda Tōru's 羽田亨 (1882–1955) *Man-Wa jiten* 滿和辭典 (Manchu-Japanese dictionary), 315; *honzōgaku* 本草学 adopted and reinvented by Japanese scholars, 12; insight into the dynamics of knowledge and state power afforded by the global scope of historical inquiry, 11; natural history featured in early Japanese studies of Manchu, 328; Rémusat's natural historical card catalog assembled from Manchu *Mirrors* and Japanese sources, 325, 326; Uehara Hisashi 上原久 (1908–1997), 327–28. *See also* Takahashi Kageyasu 高橋景保 (1785–1829); Yanagisawa Akira 柳澤明

Jarring, Gunnar, on čišmiškišmiš, 238n25

Jean-François Gerbillon (Zhang Cheng 張誠, 1654–1707), Mongolian travelogue, 123n1, 133

Jiachan jianxin 嘉產薦馨 (Auspicious products of commendable fragrance) (Yu Xing 余省): album leaf depicting *ayan hiyan* (Ch. *yunxiang*), 243; Yu Xing's production of (1747), 242

Jiang Fan 江藩 (1761–1831), 108

Jiang Qiao 江桥, *Kangxi* Yuzhi Qingwen jian *yanjiu* 康熙《御制清文鉴》研究 (Research on Kangxi's *Imperially Commissioned Mirror of the Manchu Language*), 25n13

Jiang Tingxi 蔣廷錫 (1669–1732): as editor of the *Gujin tushu jicheng*, 173; the golden lotus flower featured in his scroll paintings and albums, 156; wild poppy (*ye yingsu* 野罌粟) painting of a yellow flower, 152n12. *See also Gujin tushu jicheng* 古今圖書集成 (Consummate collection of texts and illustrations from past and present) (Chen Menglei and Jiang Tingxi)

Juntu 屯圖. See *Yi xue san guan Qingwen jian* 一學三貫清文鑑 (Juntu 屯圖)

Kangxi emperor (r. 1662–1722, Emperor Shengzu): identification of the grass lychee (*cao lizhi*) with the Fujian lychee, 268; imperial preface to *Guang Qunfang pu* (1708), 95–96; Manchu-Chinese Translation Office established (ca. 1671), 25; "night-glow wood" (*yeguang mu* 夜光木) praised in an eighteen-couplet rhapsody, 157; sea creature presented to him by Koreans, 42; southern tours, 93, 152; tours to Manchuria, 194; tours to Mount Wutai, 152, 153, 154, 194, 307; *xi* rodent identified as *fen shu* 鼢鼠, 183, 183n87, 329, 329n42. *See also Shengzu Ren huangdi yuzhi wenji* 聖祖仁皇帝御製文集 (Collection of imperially authored prose by the Sagacious Progenitor, Emperor Ren)

—hunting trips: beyond-the-border plants (Ma. *jasei tulergi*) as a focus of scholar-officials in Kangxi's hunting entourages of the 1700s, 9–10; introduction of Manchu and Mongolian flora traced to natural-historical observations during Kangxi's 1703 hunting tour (in the *Guang Qunfang pu*), 150, 152, 153, 156, 160, 165, 167–68. *See also* Wang Hao 汪灝

—hunting trips (*continued*)
(b. 1651)—*Suiluan jien* 隨鑾紀恩 (An account of grace during my trip accompanying His Majesty's carriages); Zha Shenxing 查慎行 (1650–1727, *jinshi* 1703)—*Ren hai ji* 人海記 (My account in the sea of humankind)

Kangxi ji xia gewu bian 康熙幾暇格物編 (Collection of the investigation of things during times of leisure from the myriad affairs [of state] in the Kangxi period): frontier flora and fauna's presence in court scholarship stabilized by, 148–49, 175–76; on the growth of auspicious millet in Manchurian wilderness, 178; the importance of travel and on-site observation articulated throughout it, 176–77; on Kangxi's desire to match names found in ancient Chinese sources with species in Manchuria, 182; publication of, 175–76; the *pupan* 普盤 named and described in accordance with imperial observations rather than in reference to the *Poetry Classic*, 185–86; the *qin-da-han* 秦達罕 (Ma. *cindahan*; Mo. *čindaya[n]*) related to hares (*tu* 兔), 178; on the *shanzha* 山楂 (haw), 187; sweeping taxonomic claims in, 187; *ying-e* 櫻額 (Ma. *yengge*), described as a smaller version of "wild black grapes," 179

Kangxi *Mirror*. See *Mirror of the Manchu Language* (Kangxi *Mirror*) (1708)

Kangxi zidian 康熙字典 (Character standard of the Kangxi reign): completion of (1716), 86; lexicographical erudition of, 119; Liu Hao as an "editor" (*zuanxiu* 纂修) for, 94n23; pronunciation of *zhi* 褆 in, 114n54; supplement not published for it until the nineteenth century, 28

Karlgren, Bernhard, *uli* (*qi* 杞) translated as a kind of willow, 164–65n42

Khingan Mountains: Kangxi's identification of the grass lychee (*cao lizhi*) growing in, with the Fujian lychee, 268; location in Manchuria, 37; snow rabbit (*cindahan*) living in, 37

Klaproth, Julius: endeavor to find European equivalents to Manchu terms for natural products in his translation of the *Ode to Mukden*, 194–95, 214; *haizha* 海蛇 (sea jellyfish) translated as sea snake (*serpent de mer*), 218

Know Your Remedies (He Bian), discussion of "pharmaceutical objecthood," 104n39, 105n42

Ko, Dorothy, 7

Köhle, Natalie, 152

Korea: "Koryŏ" (Gaoli 高麗) ditty in praise of ginseng, 170; sea creature presented to Kangxi by Koreans, 42–43

Kunyu tushuo 坤輿圖說 (Illustrated explanation of the world) (Verbiest): completion of, 41; illustrations and descriptions from, included in *Gujin tushu jicheng*, 117, 309; swordfish and mermaid (*xi-leng* 西楞), 53, 53n93, 54, 56

Lai Yu-chih 賴毓芝, 246, 247, 247n26, 283n19

Learning of the Way: and the "investigation of things" (*gewu* 格物), 175; Qianlong's ascription to, 277

leishu. See "classified writings" (*leishu*) as a genre

lexicographical encyclopedism: contribution of pictorial catalogs of birds and beasts to, 234, 250–52, 257, 308–10, 316; distinction between sources representing "facts and deeds" (*shishi* 事實) and "literary embellishments" (*jizao* 集藻), 108; shift away from the

INDEX

literati-centered tradition of natural erudition resulting from, 312, 332–33; Yuan Mei 袁枚 (1716–1798), on gazetteers and encyclopedias used for writing *fu*, 199. See also *Erya* 爾雅 (Approaching perfection); *Guang Qunfang pu* (Expanded catalog of myriad flowers); *Gujin tushu jicheng* 古今圖書集成 (Consummate collection of texts and illustrations from past and present) (Chen Menglei and Jiang Tingxi); *Shanhai jing* 山海經 (Classic of mountains and seas)
—at the Kangxi court: emergence of, 15–17, 87–88; the Manchu *Mirror* as a product of, 22–23; mediating role of language in discussions of Manchurian and Inner Asian flora and fauna, 125–27

Li Bai 李白 (701–762 CE), 210
Li Xiongfei 李雄飛, 283
Li Yanji. See *Qingwen huishu* 清文彙書 (*Manju isabuha bithe*) (Li Yanji)
Lianzhu ji 聯珠集 (Collection of stringed pearls) (Zhang Tianqi): eels (Ma. *hūwara*; Ch. *li* 鱧) in, 226; Liu Shan's preparation of the Manchu version of, 202; Liu Shun's preface to, 201–2n24; origins as a Chinese text by Zhang Tianqi, 201–2; publication as a supplement to later editions of *Broadly Collected Complete Text in the Standard Script*, 201; transcribed Chinese loan words for plants and animals instead of Manchu words in, 202
Liji 禮記 (Record of rites), "Monthly ordinances" (*yueling* 月令): on the collection of *micao* (delicate herbs), 289; cosmography of five evolutionary phases illustrated in, 69; as a source for plant and animal names, 69
Lin Shih-hsuan 林士鉉, 17n31, 206–7n40, 245

Liu, Cary Y., 177
Liu Hao 劉灝 (*jinshi* 1688), 94
Liu, Shi-yee, 262–63
livestock and domestic animals: *buka* (bull) in Mongolian, 251n60; *Erya*'s identification of "six domestic animals" (*liu chu* 六畜), 67; yaks in the "Bovines" (*ihan*) section of the chapter "Livestock and domestic animals" of the Kangxi *Mirror*, 32
—horses: "Horses" (Ma. *morin*; Ch. *ma* 馬) category of *Man-Han leishu*, 63; identified as one of "six domestic animals" in *Erya*, 67; names for horses in the "Horses and livestock" (*morin ulha*) section of Kangxi *Mirror* chapter "Livestock and domestic animals" (*ulha ujima*), 32; *yasa jerkišembi* (unclear vision when riding a horse at high speed), 45n67
—sheep: *haha honin* (male sheep, Ch. *gong yang* 公羊), 251, 304n85; *honin* (sheep, Ch. *yang* 羊), 67, 251
Lu Gwei-djen. See Needham, Joseph and Lu Gwei-djen

Magnus, Olaus: *xiphias* described in his *Historia de gentibus septentrionalibus* compared to the *edeng* in the Kangxi *Mirror*, 52–53, 52n90
Man-Han leishu (Manchu and Chinese classified writings) (Sangge 桑額): jay (Ma. *isha*) in, 282; *mafuta buhū* (buck deer) paired with *milu* in, 261n80; possible influence on the Kangxi *Mirror*, 63
Manchu-Chinese dictionary of Shen Qiliang. See *Da Qing quanshu* 大清全書 (Complete book of the Great Qing) (Shen Qiliang 沈啟亮)
Manchu-Chinese *Mirror*. See *Expanded and Emended Manchu Mirror* (*Yuzhi zengding Qingwen jian* 御製增訂清文鑑) (1772)

Manchu-Chinese Translation Office: establishment of (ca. 1671), 25; *Imperially Commissioned Mirror of the Manchu Language* compiled by, 25–26

Manchu Language at Court and in the Bureaucracy under the Qianlong Emperor (Söderblom Saarela), 3

Manchu *Mirrors*: evolving vision for a Qing order of things reflected in the study of, 4–5, 16, 307, 316, 331–32; iconic stature of, 2–3; move away from natural erudition, and toward languages, in scholarly studies, 331; narrow framing of, 15–16, 317, 319–20; Rémusat's card catalog as evidence of research on Asian plants in which Manchu *Mirrors* played a role, 324–25, 327; words for plants and animals in the Manchu *Mirrors* discussed by Wolfgang Seuberlich and Walter Fuchs, 313–14, 331. See also *Expanded and Emended Manchu Mirror* (*Yuzhi zengding Qingwen jian* 御製增訂清文鑑) (1772); *Mirror of the Manchu Language* (Kangxi *Mirror*) (1708); *Pentaglot Mirror of the Manchu Language*

Manchuria: Fang Shiji's account of Manchuria included in *Siku quanshu*, 139n42; the growth of auspicious millet in Manchurian wilderness noted in *Kangxi ji xia gewu bian*, 178; Kangxi's tours in the 1690s, 9, 42–43, 194, 198, 222; Lin Shih-hsuan 林士鉉 on Manchurian flora, 245; map of Manchuria in *Shengjing tongzhi*, 128, *129*; natural products submitted from it, to the Imperial Household Department, 34; project to compile a Manchurian gazetteer, 83, 126; Qianlong's tours of, 205, 206; Songhua stone from, 184. See also Khingan Mountains; Ningguta; *Ningguta zhi* 寧古塔志 (Treatise on Ningguta) (Fang Gongqian 方拱乾); *Shengjing tongzhi* 盛京通志 (Comprehensive gazetteer of Mukden)

Marcon, Federico, 12; on "metaphorical comparison" in the history of knowledge, 332

melons and gourds: distributions of gourds across three categories in the Kangxi *Mirror*, 61–63, 63n13; *gaha hengke* (small red snake gourd, lit. crow tub), 61, 62, 72–73; *hoto hengke* (lit. skull gourd) translated as *huzi* 瓠子 (squash) in the 1772 *Mirror*, 285, *285*; Kong Yingda's commentary on the *Change Classic*: "*guo* is the fruit of trees, *luo* is the fruit of vines," 70–71; *langgū* translated as *wogua* 倭瓜 (Japanese gourd) in the 1772 *Mirror*, 285, *285*; *lugiya hengke* (bitter melon *gaha oton*, lit. crow tub; Ch. *laigua* 癞瓜, *lai putao* 癞葡萄, *kugua* 苦瓜, and *jin lizhi* 錦荔枝), 60, 60n6, 61, 61n9, 62, 63, 72, 73, 73n47, 74; *lugiya hengke* (pickling melon, *shaogua* 稍瓜), 60, 60n6, 61n9, 73n47; muskmelon (Ma. *hengke*, Ch. *xianggua* 香瓜), 60, 204; *nasan hengke* (salting gourd, pickling gherkin), 60, 70–71. See also *Mirror of the Manchu Language* (Kangxi *Mirror*) (1708)—flora chapters (*šošohon*)—"Foodstuffs" (*jetere jaka*)—"Fruits" section

—*guwalase*: *guwalase* compared to a *hengke* (gourd) in the Kangxi *Mirror*, 60, 284–85; *guwalase* translated as *kegua* 客瓜 in *Yi xue san guan Qingwen jian*, 272n115; inquiry about *guwalase* from the Translation Office to the Grand Council, 270–71

Mencius. See under *Four Books* (*Sishu* 四書)

Métailié, Georges, 12–13, 13n23

Mingdo's 明鐸 translation of Kangxi *Mirror*. See *Yin Han Qingwen jian* 音漢清文鑑 (Mirror of the Manchu

INDEX

language translated into Chinese) (Mingdo)

Mirror of the Manchu Language (Kangxi *Mirror*) (1708): Amiot's consultation of, 225–26; chapters on medical conditions, 78, 78n56; a commercial Manchu-Chinese dictionary from the late seventeenth century compared with, 21–22; compilation by the Manchu-Chinese Translation Office, 25–26; *ko* (soot) and *bokida* (tassel) in, 78–79, 79n56; presentation of Inner Asian knowledge within individual entries, 34–37; *Taiping yulan* as a source for, 66; unevenness of chapters and sections in terms of names for plants and animals, 34. *See also* birds—in the Kangxi *Mirror* (*cecike-i hacin*); *Expanded and Emended Manchu Mirror* (*Yuzhi zengding Qingwen jian* 御製增訂清文鑑) (1772); *Yin Han Qingwen jian* 音漢清文鑑 (Mirror of the Manchu language translated into Chinese) (Mingdo)

—chapters (*šošohon*): arrangement of thirty-six chapters, 29; categories of Sangge's *Manchu and Chinese classified Writings* possibly influenced by, 63; principles behind the arrangement of, related to collections of "classical writings," 59; subdivided into sections (Ma. *meyen*; Ch. *duan* 段 [segment]), 29; subdivided into segments (*meyen*), 29, 30

—fauna chapters (*šošohon*)—"Beasts" (*gurgu*): on *arfukū* used to describe a fly whisk made from the long tail of a deer or mule, 262; on *ashangga singgeri* (winged mouse), 302; *gasha* (bird) defined in, 67; *gurgu* (beast) defined in, 67; *kandahan* (moose) listed in, 31–32, 132n19, 264; on *kilin* (unicorn) and *buhū* (deer), 72; on *mufuta*, 261n80; number of entries with translations in relation to the total number of entries in its section and to the number of translations overall, 77–79, 78, 79; a small monkey described as "a fickle troublemaker by nature" (*boco sohokon, uncehen foholon, hefeli dolo delihun akū, jeke jaka be yabuhai singgebumbi, banitai fiyen akū nungneku*), 74; snow rabbit (*cindahan*), 27; *suwa* defined as a small-bodied and light pink deer, 260–61n80; *uncehen golmin buhū* (long-tailed deer), 76–77; on *yuwan* (Ch. *yuan* 猿 [gibbon]), 74

—fauna chapters (*šošohon*)—"Creatures with scales and shells" (*esihengge hurungge*)—section on "Dragons and snakes" (*muduri meihe*), 32, 36, 38

—fauna chapters (*šošohon*)—"Creatures with scales and shells" (*esihengge hurungge*)—section on "River fish" (*birai nimaha*): number of entries with translations in relation to the total number of entries in its section and to the number of translations overall, 77–79, 78, 79; the sea fish *sohoco* said to be "somewhat similar to the *secu*" (*secu de adalikan*), 222n97

—fauna chapters (*šošohon*)—"Creatures with scales and shells" (*esihengge hurungge*)—section on "Sea fish" (*mederi nimaha*): *boo nimaha* (whale) described in, 43–44, 47; on *fethe* (tail fins), 44–45n67; on *giyaltu* (Ch. *bai daiyu* 白帶魚), 80; on *haga* (bones of fish), 44–45n67; on *ica* (Ch. *miantiao yu* 麵條魚 [noodle fish]), 81, 303; on *koojiha* (fish called *latun* [cured piglet]), 81, 82, 303, 304; on *kurce* (Ch. *baigao yu* 白膏魚 [white-fat fish]), 80; number of entries with translations in relation to the total number of entries in its section and to the number of translations overall,

393

—fauna chapters (*šošohon*) (*continued*) 77–81, *78*, *79*; ocean creatures listed with bare-bones definitions, 41; on *sotki* (Ch. *haiji yu* 海鯽魚 [seabass?]), 80; statement in the preface on the inclusion of names that exist in Manchu, 40; on *tama* (sole, Ch. *xiedi yu* 鞋底魚), 81, 303; on *uya* (Ch. *haiyan yu* 海鰋魚), 81; on the waxing and waning of fish brains, 67–68

—fauna chapters (*šošohon*)—"Large birds" (*gasha cecike*): four sections of, 31; *kūrcan* (hooded crane) in, 223; legendary jay (Ma. *isha*; Ch. *songya* 松鴉) in the *Catalog of Birds* related to the paradise flycatcher, 248–49; number of entries with translations in relation to the total number of entries in its section and to the number of translations overall, 77–79, *78*, *79*; ostriches as *to gi* (*tuoji* 駝雞 [camel chickens]) in, 39, 254–55; *tugu* (rabbit falcon) in, 294

—fauna chapters (*šošohon*)—"Livestock and domestic animals" (*ulha ujima*): "Bovines" (*ihan*) section, 32; "Domestic animals" (*eiten ujima*) section, 32; "domesticated geese, ducks, and other animals" identified as "domestic animals" (*ujima*), 67; "Horses and livestock" (*morin ulha*) section, 32; "male sheep (*honin*) and male goats (*niman*) called *buka*" (*hahahonin, haha niman be gemu buka sembi*), 251; *taiha* and *yolo*—as well as *beserei* (a mix between a *taiha* and a common house dog), 203

—fauna chapters (*šošohon*)—names of bugs: number of entries with translations in relation to the total number of entries in its section and to the number of translations overall, 77–79, *78*, *79*; seventy-five names listed in two sections, 32

—flora chapters (*šošohon*)—"Flowers" (*ilha*): defined as a general category in the revised *Mirror*, 330, *330*; *jilha* defined as "the innermost part in the center of a flower" (*yaya ilha-i dulimbai niyaman be, jilha sembi*), 287; names of fourteen plants included in, 31; number of entries with translations in relation to the total number of entries in its section, 77, *78*; as a reference for Manchu translations of Chinese words in Grand Council documents, 287; wooden peony (Ma. *modan*; Ch. *mudan* 牡丹), 71–72

—flora chapters (*šošohon*)—"Foodstuffs" (*jetere jaka*): "Cooked grains and meats" (*buda yali*) as the first section in the chapter titled "Foodstuffs" (*jetere jaka*), 29; *honggoco* moved to the "Cooked grains and means" section in the *Expanded and Emended Mirror*, 301

—flora chapters (*šošohon*)—"Foodstuffs" (*jetere jaka*)—"Fruits" section: on the Chinese *uli* (Ch. *huahong* 花紅 [the red of the flower]), 164–65; on *ilhamuke* (flower water), 144; *jancuhūn hengke* (sweet melon) defined with a reference to the *Change Classic*, 70; *jancuhūn hengke* (sweet melon) described in terms of its taste, 61; on *misu hūsiha* (Ch. *mi-wu-shi-ha* 米孫烏什哈, "cherry") also found in *Shengjing tongzhi*, 140–41, 147, 173, 221; *nikan hengke* (Chinese melon) described in terms of its taste, 61; overview, 29; *solho hengke* (Korean muskmelon) described in terms of its taste, 61; *ulana* as an alternative name for *mamugiya* in, 163; *una* ("red color, sweet taste" [*boco fulgiyan, amtan jancuhūn*]) defined with a reference to the *Poetry Classic*, 186; *yengge* described as "black-colored" and "astringently flavored" fruit that

looked like small grapes (*boco sahaliyan, mucu de adalikan, amtan fekcuhun*) in, 140, 181
—flora chapters (*šošohon*)—"Foodstuffs" (*jetere jaka*)—"Vegetables and dishes" (*sogi booha*): *gio ura* (lit. "rear end of Siberian roe deer") mushroom, 142–43; *giyangdu* (cowpea, perhaps from early Mandarin *kjaŋtəw* 豇豆), 285–86; *guwalase* compared to a *hengke* (gourd), 60, 284–85; *huhucu* (Adenophera [bellflower], Ch. *mingzhe cai* 茗拤菜), 76, 304n84; *nasan hengke* (salting gourd, or pickling gherkin), 69–70; number of entries with translations in relation to the total number of entries in its section and to the number of translations overall, 77–79, 78, 79; as the second section in, 29. See also mushrooms (Ma. *Megu;* Ch. *mogu* 蘑菇), in Kangxi Mirror, "Vegetables and dishes" (*sogi booha*)
—flora chapters (*šošohon*)—"Grasses" (*orho-ihacin*): *ayan hiyan* described in, 244; on the Chinese toon tree (Ma. *cūn moo*), 35; defined as a general category in the revised *Mirror*, 330; first page of, 30; four kinds of *foyo* (Ula grass) introduced in, 143; *gabtama* (nettles), 35, 159–60, 161n35, 167; *ulhū* (defined as a plant that "grows in places covered in water..."), 216; *ulhū* (elephant grasss; Ch. *wei* 葦), 216
—flora chapters (*šošohon*)—"Herbs": division into two segments, 29, 31; on *gaha hengke* (small red snake gourd, lit. crow tub), 61, 62, 72–73; *kailari orho* (oriental motherwort; Ch. *yimu cao* [益母草]), 79; number of entries with translations in relation to the total number of entries in its section and to the number of translations overall, 77–80, 78, 79; *okjiha* (sweet flag) in, 291; *onggoro orho* (day lily; lit. forgetting herb), 80
—flora chapters (*šošohon*)—"Trees" (*moo*): defined as a general category in the revised *Mirror*, 330; division into five segments, 31; *fursun* defined as sawdust, 271n113; *jak moo* (jay [boundary?] tree) described in, 136; on larch (Ma *isi;* Ch. *luoye song* 落葉松), 37–38; lemmata containing the elements *fodoho* and *burga* for willow, 62; number of entries with translations in relation to the total number of entries in its section and to the number of translations overall, 77–79, 78, 79; wild peach tree (*hasuran* or *karkalan*) identified as *kara hūna* (*monggo kara hūna sembi*), 35
—scholarly tradition referenced in: erudition derived from the *leishu* tradition evidenced by, 75; a late-Ming encyclopedia quoted in the definition of the wooden peony (*mudan*), 71–72; phrasing of the section on "Dragons and snakes" traced to *Da Dai liji*, 65; *Poetry Classic* cited on *xuancao* 諼草 (forgetting herb), 80; quote from the *Shuo gua* 說卦 commentary on the trigram *gen* 艮 cited in the definition of *jancuhūn hengke* (sweet melon) in the Kangxi *Mirror*, 70; quotes appended to definitions, used to anchor a Manchu word to the tradition of classical learning, 71–72, 164–65
—terms and words glossed in: arrangement of words according to real-world physical similarity rather than similarity in name, 62; *buka* (castrated male sheep) and *kūca* (intact male sheep [*aktala-hakū haha honin*]), 251; *dondon* defined as a "small butterfly" (*ajige gefehe*), 32; *fafaha* fruit described in (compared with Yang Bin's description of *fa-fo-ha* 法佛哈), 140; *gaha hengke* (small red

—terms and words glossed in (*continued*)
snake gourd, lit. crow tub), 61, 62, 72–73; jay (Ma. *isha*; Ch. *songya* 松鴉), 248–49; *miantiao yu* (noodle fish), 296; *misu hūsiha* "cherry" also found in *Shengjing tongzhi*, 140, 141, 147, 173, 221; water buffalo (*mukei ihan*), 296; *yengge* described as "astringently flavored" fruit that looked like small grapes (*boco sahaliyan, mucu de adalikan, amtan fekcuhun*), 140, 181. See also *yasa jerkišembi* (dazzling eyes)

Mirror of the Manchu Language, Manchu-Mongol edition (1717): *edeng* in, 47; headwords in Mongolian, 228; *mamugiya* translated *asulan-a*, 163n39; plurilinguality of, 232, 233; Schlesinger on its textual scholarship, 22

Mongolia and Mongolian words: *ayan hiyan* (Ch. *yunxiang*; Mo. *sülü* [incense]), 244; blackish (Ma. *sengkiri hiyan*; Mo. *qarabur*), 244; *buka* (bull), 251n60; cognate for the word *tojin* (peacock) in Manchu, 234; diplomatic encounter between Tulišen and Torghut Mongol leader Ayuka Khan (1646–1724), 166; inclusion of Mongolian words in Gao Shiqi's wordlist, 125n7; information on the natural environment in Mongolia in the Kangxi *Mirror*, 33; Jean-François Gerbillon's Mongolian travelogue, 123n1, 133; Kangxi's observations of Mongolian seasonal migration customs, 176; Kangxi's visit near Ulaan Khad, 162, 165; Manchu neologisms ending in -*tu*, as loans from Mongolian, 236; *ob* ("squirrel" or "bat"), 184; the *qin-da-han* 秦達罕 (Ma. *cindahan*; Mo. *činday-a*[*n*]) related to hares (*tu* 兔) in *Kangxi ji xia gewu bian*, 178; wild peach tree (*hasuran* or *karkalan*) identified as *kara hūna* (*monggo kara hūna sembi*) in the Kangxi *Mirror*, 35; willow (Ma. *fodoho* and *burga*, from Mo. *uda* and *buryasu*[*n*]), 62; Zhang Penghe's Mongolian travelogue, 123–24, 133, 188. See also flora—golden lotus; *Mirror of the Manchu Language*, Manchu-Mongol edition (1717); Qian Liangze 錢良擇 (b. 1645), Mongolian travelogue

moose. *See* deer and moose

Mosca, Matthew W., 7; on Chinese-language court scholarship on Inner Asia, 128, 128n12

Mount Wutai: Kangxi's tours to, 152, 153, 154, 194, 307; "terrestrial golden lotus" (*han jin lian* 旱金蓮) associated with, 155, 156

multilingualism: plurilingual distinguished from, 2n2

mushrooms (Ma. *megu*; Ch. *mogu* 蘑菇), in Kangxi *Mirror*, "Vegetables and dishes" (*sogi booha*): *coko megu* (chicken mushroom), 142n51; *gio ura* (lit. "rear end of Siberian roe deer") mushroom, 142–43; *hailan megu* (elm mushroom), 142n51; *kūwara megu* (encircling mushroom), 142n51; *sanca megu* (wood ear mushroom), 142n51; *sisi megu* (hazelnut mushroom), 142n51

mythical animals and strange beasts (*yi shou* 異獸): *fulgiyentu* (Ch. *xiushi* 嗅石 [stone smeller]), 309; *kandatu* used to translate the name of a legendary Chinese bovine, 264; *qilin* 麒麟 (or *kilin*, unicorn), 72, 295–96; *sabintu* (unicorn), 236, 295–96, 297

Nappi, Carla S., 13n23, 14

Needham, Joseph and Lu Gwei-djen: close attention to lexicographical and encyclopedic sources, 14, 116n61; scant treatment of the Qing in their survey with of Chinese botany, 11n19

Neo-Confucianism. *See* Learning of the Way

Ningguta: Fang Gongqian's exile to, 137; *uli* (*ou lizi* 歐李子) found in Ningguta and Aigun, 165–66

Ningguta zhi 寧古塔志 (Treatise on Ningguta) (Fang Gongqian 方拱乾): crawfish (*hasima*) included in, 138–39; Manchu terms used only for certain place names and official titles, 138, 138n36; Ningguta described as a desolate place, 137–38

Nivkh language and the Nivkh people: ɔzŋ ("master," or "orca," associated with the "Master of the Waters"), 50; loan words from, in the Kangxi *Mirror*, 126; the name *kalimu* as an alternative name for *boo nimaha* (whale) in the *Mirror* traced to Nivkh, 43

Ode to Mukden by Qianlong: *cangji* 鶬 (bird) translated as *tsang gasha*, 223; difference in lexical register between the Manchu and Chinese versions of the *Ode to Mukden* noted by Rémusat, 212, 214; neologisms *fujurun* (ode), 206, 209, 210, 210n5; neologism *šutucin* (preface) introduced in, 206; *nunggele* tree rendered as *duan* 椴 in, 171, 225; printing of (QL 8), 206; Qianlong's reform of the Manchu lexicon evident in the 1747 edition, 227, 230; *Shengjing tongzhi* as the main source for the names of plants and animals included in it, 200; *si* 麚 (Siberian deer), 218; timing of the second edition of, 206–7, 226–28; two-stage revision process of, 195, 207, 213; use of Chinese transcriptions instead of Manchu words, 200, 202; on *xilin* 細鱗 (narrow-scaled fish), 222. *See also* Amiot, Joseph-Marie—translation of the *Ode to Mukden*
—coining of Manchu neologisms for plant and animal names evidenced in, 4; *bianxu* 萹蓄 (knotweed) used for *biyan hioi*, 224; *di orho* changed to *darhūwa orho* (*darhūwa* grass; Ch. lucao 蘆草 [reed canary grass, lady's-laces]), 216; lists of floral and fauna as its focus, 214; *miyahūtu* used instead of *gibuhū*, 219; *niomošon* used to translate *xilin* 細鱗 (narrow-scaled fish), 222–23, 281; *sangguji* used for *haizha* 海蛇 (sea jellyfish) in the second edition, 218, 236; *tugu* 兔鶻 (rabbit falcon), 294; *tulin cecike* (black-naped oriole), 248

Omont, Henri, 26n15

Ortai 鄂爾泰 (1677–1745): Xi-qing 西清 (fl. 1806) identified as his great-grandson, 167

Ottoman Empire, insight into the dynamics of knowledge and state power in afforded by the global scope of historical inquiry, 11

Ouyang Xiu 歐陽修 (1007–1072), on *ouli* 歐李, 167

Parrenin, Dominique (1665–1741), on Manchu words for dogs, 203–4, 314n6

Peiwen yunfu 佩文韻府 (Storehouse of rhymes from the [Studio] of Literary Admiration), 86

Pentaglot Mirror of the Manchu Language: inadvertent creation of new words for newly included languages, 236; *kīšmīš* (or *kishmish*) given as the equivalent of *kišimiši* (seedless green grape) in, 238; lemmata from *Guang Qunfang pu* (Expanded catalog of myriad flowers), carried over to the *Pentaglot Mirror*, 305–8; the need for the study of words for aquatic creatures suggested by Doerfer, 40, 233n7, 236, 236n19; *yolo* used for (big eagle) and *yolo* used for "Tibetan dog" (Ch. *Zanggou* 藏狗), 314

Persian: cognate for the word *tojin* (peacock) in Manchu, 234
Peterson, Willard J., 118, 127n9
pharmacy and pharmacopeia: *bencao* pharmacopeias as the predominant genre of natural erudition since the Tang, 104–5; *changpu* and *cangzhu* distinguised in the *bencao* pharmacopeia tradition, 393; classification based on utility not abstracted physical characteristics, 59; ginseng associated with *jia* trees in a "Koryŏ" (Gaoli 高麗) ditty, 170; historical study of pharmacy and Manchu language studies in late imperial China, 3; *honzōgaku* 本草学 adopted and reinvented by Japanese scholars, 12; *Jiuhuang bencao* 救荒本草 (Materia medica for famine relief) (Zhu Su 朱橚), 167; loss of the encyclopedic scope of *bencao* 本草 after the 1700s, 12; Tibeto-Mongolian materia medica of 'Jam-dpal-rdo-rje, 322; woody berry shrub (*mumei* 木莓) named *pupan* 普盤 related to a woody berry shrub in *bencao* by Kangxi, 185. See also *Bencao gangmu* 本草綱目 (Systematic materia medica) (Li Shizhen)
Piya 埤雅 (Supplement to *Approaching perfection*) (Lu Dian 陸佃), 131–32, 179
plurilingualism: of Kangxi's Manchu-Mongolian *Mirror*, 232, 233; multilingualism distinguished from, 2n2; Qianlong's goal of revealing the "essence of sound and rhymes" (Ma. *jilgan mudan-i fulehe*; Ch. *sheng lü zhi yuan* 聲律之元) of any language, 276–77; of scholarship at the Qianlong court, 231, 332–33; of Takahashi's study of the Manchu *Mirrors*, 329
Poetry Classic: dried spadix (Ma. *ibaga hiyabun*; Ch. *changpu* 菖蒲) in, 291; *fu* 賦 repeated and explained as one of six Principles of Poetry (*liu yi* 六義),

197; *hūwara* used to translate Ch. *li* 鱧, 226; Manchu translation published in 1655 (*Han-i araha Ši ging bithe*), 24–25; names of birds, beast, and plants in, 25, 69; *nasan hengke* (salting gourd, pickling gherkin) in, 70; new words in the revised translation of 1768, not included in the new *Mirror*, 239–40; quote on *qi* willows 南山有杞 in, used to describe "Chinese" *uli* in the Kangxi *Mirror*, 164; *una* used to translate *qi* 杞 (goji), 186; on the *xuancao* 諼草 (forgetting herb), 80
pre-conquest period Manchu scholarship: knowledge of the natural environment evident in the Kangxi *Mirror*, 33; translation of ancient works "associated with a rising power on the eve of its conquest," 23–24; unsuccessful pre-conquest petitioning for a Manchu translation of the *Four Books*, 23; *Wanbao quanshu* 萬寶全書 (Complete book of a myriad treasures), 24
Prince Cheng. See Yinchi 胤祉 (in Chinese) or In c'y (in Manchu) (Prince Cheng, 1677–1732)

Qian Daxin 錢大昕 (1728–1804), 108
Qian Liangze 錢良擇 (b. 1645), Mongolian travelogue: bird called *elan* (likely a crested lark) encountered by, 134–35; Khalkha lands described as inhospitable, 133–34; Mongolian word *dabaya(n)* (modern Mo. *dabaa*) transcribed as *daba* 打八, 151n9; natural-historical observations in, 133; unidentified tree described by, 137, 135–36
Qianlong emperor (r. 1736–1796): *cao lizhi* 草荔支 (grass lychee) in his poetry composed between 1758 and 1759 (QL 13–24), 268–69; Chinese poem about the *qishimishi* 奇石蜜食

(marvelous stone honeyed food), 238; construction of universal rulership, 6; immature locusts identified as born from fish and shrimp eggs, 255–56. See also *Ode to Mukden* by Qianlong
—Manchu language reform: field officials' suggestions submitted to the throne, 237; *ihasi* (rhinoceros) introduced (on March 11, 1749), 237, 322; impact of Tibeto-Mongol learning, 322; linguistic parity of Manchu, especially vis-à-vis the classical Chinese lexicon, as a primary motivation, 237–38, 240; neologisms added during, 4, 16–17, 234–35; new words for government institutions compiled by Yunggui 永貴 (1706–1783), 238–39; revision of names for plants (sugarcane, apple, Chinese crabapple, pomegranate) during, 237. See also *Expanded and Emended Manchu Mirror* (*Yuzhi zengding Qingwen jian* 御製增訂清文鑑) (1772); "unification of writing" (Ma. *hergen be emu ombime*; Ch. *tongwen* 同文)

Qiju zhuce 起居注冊 (Register of rise and repose): neologism *sebseheri* mentioned in, 257; the *Ode* mentioned in, 206

Qinding Huangchao tongzhi 欽定皇朝通志 (Imperially authorized Comprehensive treatises of our august dynasty), 319n14

Qingwen buhui 清文補彙 (*Manju gisun be niyeceme isabuha bithe* [Manchu collected, supplemented]) (Ihing 宜興): inclusion of lemmata from the Manchu-Chinese *Mirror* included in, 284; *yuanyang* (Mandarin duck) equated with lama duck (*lama ya*), 303

Qingwen huishu 清文彙書 (*Manju isabuha bithe*) (Li Yanji): acceptations of the verb *jerkišembi* from the *Mirror*, 45n67; *boo nimaha* defined as *jingni* 鯨鯢, 43n64; *boo nimaha* identified with *shenyu* (a legendary creature with the appearance of a snake and horns like a dragon), 47–48; on the *edeng*, 48, 48n78, 51, 52n90, 53; on the *fen shu* 鼢鼠 (or zokor, dung rodent), 329n42; "fly broom" (*cangying zhouzi* 蠅拂箒子) and "fly whisk" (*ying fuzi* 蠅拂子) in, 262; *kandahan* listed in, 264; on *koojiha* (fish called *latun* [cured piglet]), 303, 304; *ucika* defined as "the front paddle[s] on a fish's belly" 魚肚子前划水, 44n67; *wolan* 窩攔 identified as a name for *wenderhen* and *wolan que*, 135

Qunfang pu (Catalog of myriad flowers) (Wang Xiangjin): on clover (*muxu* 苜蓿), 104; Er ru ting 二如亭 (Pavilion of Double Resemblance) used in its title, 89, 91–92; initial composition by Wang Xiangjin (1615–1625), 89; structure of entries in, 104, *106*; taboo character *xuan* 玄 included at the beginning of multiple volumes, 91n10; Wang Xiangjin's reliance on "old farmers" and "old gardeners," 91–92; wide scope of, 95
—section on "Fragrant traces of past luminaries": exclusion from the *Guang Qunfang pu*, 98; short biographies of worthy gentlemen collected in, 97–98, *98*, 98n30; on Wang's reliance on "shaved heads and long beards" (*pingtou changxu* 平頭長鬚), 92, 92n14
—section on "Grains," 99

Rémusat, Jean-Pierre Abel (1788–1832): background of, 324; difference in lexical register between the Manchu and Chinese versions of the *Ode to Mukden* noted by, 212, 214; *kaji* (Ch. *chu*) used to translate *hoošari moo* (paper mulberry), 325, 327; the *Mirrors* viewed as encyclopedias by, 323–25,

Rémusat, Jean-Pierre Abel (*continued*) 327; natural historical card catalog assembled from Manchu *Mirrors* and Japanese sources, 325, 325n29, *326*; notebook for Manchu and Mongolian words confusingly titled "Table alphabétique de l'Encyclopédie japonaise," 235n29; research on the Manchu-Mongol and Manchu-Chinese *Mirrors* as example of research on Asian plants in which Manchu *Mirrors* played a role, 324–25, 327

Rogaski, Ruth, 9, 124, 157; on "landscapes of exile," 138

Ross, Edward Denison Ross (1871–1940), 236

Rozycki, William: on the etymology of the word *edeng*, 50; on Mo. *jay*, 136n31; on the name *kalimu* as an alternative name for *boo nimaha* (whale) in the *Mirror*, 43

Ruiying tu 瑞應圖 (Illustrated compendium of auspicious responses) (Sun Rouzhi 孫柔之), 72

Russia: anecdotes about animals from brought to Kangxi, 182–83; insight into the dynamics of knowledge and state power in Imperial Russia afforded by the global scope of historical inquiry, 11; peace settlement sought by the Qing court with (1688), 133; Seuberlich's Russian secondary and tertiary education in Harbin, 314

Sangge's Manchu-Chinese dictionary. See *Man-Han leishu* (Manchu and Chinese classified writings) (Sangge 桑額)

Schäfer, Dagmar, 12n22, 92n15, 96n26

Schiebinger, Londa, 10

Schlesinger, Jonathan, 132n19, 184n92; on the integration of Beijing and Manchuria by the mid-Qing, 33; on the "natural fringes of Qing rule," 8, 21; on *niomošon* left untranslated in Tulišen's travelogue, 222; on words used for the natural world as a challenge for lexicographers, 21–22, 126

Sela, Ori, 13n22, 108

Seuberlich, Wolfgang (1906–1985): background of, 314; words for plants and animals in the Manchu *Mirrors* discussed with Walter Fuchs, 313–14, 316, 331

1772 *Mirror*. See *Expanded and Emended Manchu Mirror* (*Yuzhi zengding Qingwen jian* 御製增訂清文鑑) (1772)

Shandong region: correspondence with the ancient state of Qi, 248; Qianlong's visit to, 271; Wang Xiangjin on the crops unique to, 104

Shanhai jing 山海經 (Classic of mountains and seas): mermaid with a "human face and fish body without legs" described in, 56; mythical trees in, 101; topically arranged vocabulary in, 64–65

Shen Qiliang 沈啟亮: *jolo buhū* (doe deer) in his *Sishu yaolan* 四書要覽 (*Sye šu oyonggo tuwara bithe* [Essential readings from the *Four Books*]) (Shen Qiliang), 261n80. See also *Da Qing quanshu* 大清全書 (Complete book of the Great Qing) (Shen Qiliang 沈啟亮)

Shen Yongmao 申用懋 (sometimes written as *mao* 楙) (1560–1638), preface written for the *Qunfang pu*, 90

Shengjing tongzhi 盛京通志 (Comprehensive gazetteer of Mukden): as an early attempt to describe Manchurian nature, 127; aquatic creatures identified as strange and unknown to locals, 130–31; *cangji* 鶬雞 bird in, 223; on *chou lizi* 稠梨子 (dense pear), 180, 180–81n82; *Da Ming yitongzhi* used as a source for interviewing "locals" (*turen* 土人) about plants, 130; *elan* 阿藍 described as a generic name for small sparrows, 134–35; flora and fauna named using local languages but associated with older Chinese literature, 131–32; *huanggu yu* listed in the 1684 edition of, 301n70; *lan hushou* 攔虎獸 (*lan* tiger beast) included in, 294; local product list showing *fafaha* and *misu usiha*, 141, 141; *lu* 蘆 *wei* 葦, and *di* 荻 in, 215–16; as the main source for the names of plants and animals included in the *Ode*, 200; map of Manchuria in, 128, 129; *mi lu* (*mi* deer) defined in, 143; *mi-sun-wu-shi-ha* 米孫烏什哈 listed as a product from Ningguta but not translated as *misu hūsiha*, 141, 147, 221; *nunggila* (read: *nunggile*) *moo* and *nunggele moo* with the translation *jiamu* 椵木 (*jia* tree), 171; *pupan* described as a "vine-grown" (*man sheng* 蔓生), 184; the sequence "tiger, leopard, black bear, brown bear" in, 64; on the *si* 麂 ("a kind of muntjac deer [*jun*] with long fur and dog fee, hide suitable for shoes") (or *pao* 麃), 218–19, 220n91; a source for Yang Bin's description of "peach flower water," 145; the spelling of Khingan in the Kangxi *Mirror* possibly linked to, 37n46; *tugu* 兔鶻 used for falcon (*gu*) and *yagu* 鴉鶻 used for crow falcon, 294n49; on *xilin yu* 細鱗魚 (narrow-scaled fish), 222

Shengzu Ren huangdi yuzhi wenji 聖祖仁皇帝御製文集 (Collection of imperially authored prose by the Sagacious Progenitor, Emperor Ren): on the *chake* 查克, 137n35; *Kangxi ji xia gewu bian* as part of, 175–76; Kangxi's assertion that American cochineal presented to him by the Jesuits was the same as insect-derived red pigment imported since Tang and Song times, 184–85; Kangxi's proposal of a tripartite classification of berries into vine-grown ones, upright-growth shrub berries, and herbaceous berries that simply crawl on the ground, 185; on *mo-men-tuo-wa* 摩門橐窪, 183n87; woody berry shrub (*mumei* 木莓) named *pupan* 普盤 related to a woody berry shrub in *bencao* by Kangxi, 185. See also *Kangxi ji xia gewu bian* 康熙幾暇格物編 (Collection of the investigation of things during times of leisure from the myriad affairs [of state] in the Kangxi period)

Shenyi jing 神異經 (Classic of divine marvels): *endurin soro* (divine jujube) described in, 306; on the *xi* rodent, 183, 183n87

Shinbunkan meibutsu goshō 清文鑑名物語抄 (Content words excerpted from the *Expanded and Emended Mirror of the Manchu Language*), 328

Shunzhi reign (1644–1661): books printed by the Imperial Household Department during, 24–25

Shuowen jiezi 說文解字 (Explaining the graphs and unraveling the written words): *bao* 苞 identified as a kind of grass, 289; Chinese characters used in, 318; inclusion in Needham and Lu's survey of Chinese botany, 14; used to gloss entries in the *Guang Qunfang pu*, 105

Sibe, accusative case particle *be*, 45n67

Siebert, Martina, 13n23, 120

Siku quanshu 四庫全書 (Complete books of the four repositories): Fang Shiji's account of Manchuria included in, 139n42; hierarchical arrangement of text determined by their socio-scholarly function, 59; incorporation of the 1772 *Mirror* and the 1780 trilingual *Mirror*, 318, 321; office (*guan*) opened for (1773), 317–18; preface to the *Expanded and Emended Mirror of the Manchu Language* found in the trilingual *Mirror* in, 276n4, 281n15;

Siku quanshu 四庫全書 (*continued*) the professionalization of scholarship resulting from evidential learning movement marked by, 321; *Summaries* (*tiyao* 提要) of, 318, 318n12, 320

Sinor, Denis, 315

snakes. *See* dragons and snakes

Song dynasty: jottings (*biji* 筆記) on the "Great efficacious bean" (*Daling dou* 大靈豆) as a fantastical famine food, 113; Song-dynasty poem and Song pharmacopoeia cited by Kangxi to identify *shanzha* 山楂 (haw) as *wenpu* 榅桲 (which closely resembles Ma. *umpu* in sound), 187

Sŏng Paek-in 成百仁 (1933–2018), 283

Stary, Giovanni: Gao Shiqi's wordlist studied by, 125, 125n7; on the *Ode to Mukden*, 211, 213; Qianlong's *irgebun* studied by, 210

strange beasts (*yi shou* 異獸). *See* mythical animals and strange beasts (*yi shou* 異獸)

supplement of the 1772 bilingual *Mirror*. *See* Expanded and Emended Manchu Mirror (*Yuzhi zengding Qingwen jian* 御製增訂清文鑑) (1772)—supplement (*bubian* 補編)

Taiping yulan 太平御覽 (Imperial digest [from the era of rousing the state through] great tranquility): on the ancient idea that the brains of fish grow with the moon, 68; *Bencao gangmu* published several centuries after it, 74; as a source for the Kangxi *Mirror*, 66, 68

Takahashi Kageyasu 高橋景保 (1785–1829): background of, 327; the discrepancy between *xi shu* and *fen shu* in the *Expanded and Emended Mirror* noted by, 329, 329n42; *Yakugo shō* 譯語抄 (Excerpted translations), 328; *Zōtei Manbun shūin* 增訂滿文輯韻 (Compiled Manchu rhymes, expanded and emended) by, 330, 330n45

Tao Hongjing 陶弘景 (456–536), on Koguryŏ ginseng, 170–71n56

Tibet: impact of Qianlong's Manchu language reform on Tibeto-Mongol learning, 322; Tibeto-Mongolian materia medica of 'Jam-dpal-rdo-rje, 322; *yolo* used for "Tibetan dog" (Ch. *Zanggou* 藏狗) in the *Pentaglot Mirror*, 314

trees: Chinese toon tree (Ma. *cūn moo*; *chunmu* 椿木), 35; *douban nan* 豆瓣楠 (bean-lobe nan), 291; *kaji* (Ch. *chu* 楮) used to translate *hoošari moo* (paper mulberry) by Rémusat, 325, 327; larch (Ma *isi*; Ch. *luoye song* 落葉松), 37–38, 174n64; mythical trees in *Zhuangzi* 莊子, 101, 113; "night-glow wood" (*yeguang mu* 夜光木) in *Guang Qunfang pu*, 157; *uli* (*qi* 杞) translated by Karlgren as a kind of willow, 164–65n42; willow (Ma. *fodoho* and *burga*), 62. *See also Mirror of the Manchu Language* (Kangxi Mirror) (1708)—flora chapters (*šošohon*)—"Trees" (*moo*)

—*jiamu* 椵木 (*jia* tree); broad-leafed Manchurian tree: ginseng associated with *jia* trees in a "Koryŏ" (Gaoli 高麗) ditty, 170; ginseng associated with *jia* trees in the *Guang Qunfang pu*, 171; *nunggila* (read: *nunggile*) *moo* and *nunggele moo* with the translation *jiamu* 椵木 (*jia* tree) in the *Shengjing tongzhi*, 171; tree called *duan* encountered by Yang Bin in Ningguta, 170, 171n60; tree identified as Ch. *duanmu* 椴木 and Ma. *nunggele* (Tilia) in Amiot's translation of the *Ode to Mukden*, 171, 225; white *jia* tree described by Wang Hao, 168; white *jia* tree in the *Guang Qunfang pu*, 168; Zha Shenxing's association of it with *duan* 椴, 169; Zha Shenxing's description of it and its uses, 168–69

Tsai Ming-che 蔡名哲: on the influence of Chinese culture on definitions in the Kangxi *Mirror*, 58; on the Manchu legend about avaricious jays, 249; on the *nukyak gasha* (Malabar pied hornbill), 293; on the role of quotes in the standardization of Manchu-Chinese terminology, 69

Tu Long 屠隆 (1542–1605), 98n30

Tulišen 圖麗琛 (1667–1741): on *isi* in his 1723 travelogue, 37; *ouli* 歐黎 ("Ula crabapple" [*Wula nai* 烏喇柰]) identified correctly by, 166

Tungusic language: loan words from, in the Kangxi *Mirror*, 36, 126; *nieniye* (snake hibernation) as alternate word for Ma. *meihe bulunambi*, 36; spread of the Manchu language to areas where other Tungusic languages were spoken, 34; the word *kenggin* as a loan word from, 55; *yengge* (wild cherry, or bird cherry), 180

Turki: *kišmīš* (or *kishmish*) given as the equivalent of *kišimiši* (seedless green grape) in the *Pentaglot Mirror*, 238, 238n26; Ross on Turki names in the *Pentaglot Mirror*, 236

Uehara Hisashi 上原久 (1908–1997), 327–28

"unification of writing" (Ma. *hergen be emu ombime*; Ch. *tongwen* 同文): Qianlong's ideological commitment to the ideal of, 273, 275–78; transformation of lexicography by, 16–17, 276, 278–80

Vedal, Nathan, 14, 119

vegetables: "Beans" (*dou* 豆, alternatively "Pulses"), 113; the "Great efficacious bean" (*Daling dou* 大靈豆) identified as a fantastical famine food, 113. See also *Mirror of the Manchu Language* (Kangxi *Mirror*) (1708)—flora chapters (*šošohon*)—"Foodstuffs" (*jetere jaka*)—"Vegetables and dishes" (*sogi booha*)

Verbiest, Ferdinand (Nan Huairen 南懷仁, 1623–1688): map of the world with drawings of animals (1674), 41; sea creature presented to Kangxi by Koreans identified by, 42. See also *Kunyu tushuo* 坤輿圖說 (Illustrated explanation of the world) (Verbiest)

Wakan sansai zue 和漢三才圖會 (Illustrated dictionary of the three realms in Japanese and Chinese), 325

Wakeman, Frederic, 3n3

Walravens, Hartmut: criticism of Hauer's dictionary, 315n9; *edeng* defined as "Sea-beaver, sea-otter, Enhydra marina," 49n82; on the etymology of the term *sumaltu* (for the opossum), 247n47; on Verbiest's "swordfish," 53n93

Wang Hao 汪灝 (b. 1651): background of, 93–94; compared to gentlemanly observers to experiments in seventeenth-century England, 188n110; as a member of the *Guang Qunfang pu* editorial team, 94, 119; sentencing of his household to servitude under the Eight Banners, 119

—*Suiluan jien* 隨鑾紀恩 (An account of grace during my trip accompanying His Majesty's carriages): diary kept during Kangxi's 1703 imperial tour and seasonal hunt, 150; on looking out for scorpion grass that was as "poisonous as a bee sting," 160; on luminescent wood shown to him, by Kangxi, 157; "Ula crabapple" described and eaten by, 162–63, 167; on a wild boar with "steel hooks" (*gang gou* 鋼鉤) shot by the emperor, 151;

—*Suiluan jien* 隨鑾紀恩 (*continued*) "yellow flower" encountered while entering "Mongol territory," 150–51

Wang Long 汪瀧 (d. 1742), 94

Wang Xiangjin 王象晋 (1561–1653, *jinshi* 1604): reinstated in office at the beginning of the Chongzhen reign, 89. See also *Qunfang pu* (Catalog of myriad flowers) (Wang Xiangjin)

Wang Zhao 王釗, 152

Wang Zuwang 王祖望, 261, 261n80, 298n63

Wu Siyu 吳思雨 and Wang Le 王乐, 299n65

Wu, Yulian, 7

Wu Zhenchen 吳振臣: flora and fauna described in his memoir of Ningguta, 139; *yi-er-ha-mu-ke* 衣而哈目克 fruit compared to a small *yangmei* (bayberry), 58

Wuti Qingwenjian. See *Pentaglot Mirror of the Manchu Language*

Xiang Xuan 項旋, 172–73

Xi-qing 西清 (fl. 1806), *ouli* 歐李 identified as the same as *yuli* 郁李 (bush cherry), 167

Yan Zinan 颜子楠, on the "routine production" aspect of court culture, 272–73

Yanagisawa Akira 柳澤明: jay (Ma. *isha*; Ch. *songya* 松鴉) identified with *garrulus glandarius*, 249n52; on the "southern" quail, 283n19; 361 specimens of birds in Qianlong's Grand Council's collection asserted by, 247n46

Yang Bin 楊賓 (1650–1720): on *fa-fo-ha* 法佛哈 and *mi-sun-wu-shi-ha* 米孫烏什哈, 140–41; on *mi-sun-wu-shi-ha* 米孫烏什哈, 140–41, 147, 221; *Shengjing tongzhi* used as a source for his description of "peach flower water,"

145; travelogue about crossing the "Willow Palisade" (*liubian* 柳邊), 139; tree called *duan* encountered in Ningguta, 170, 171n60

yasa jerkišembi (dazzling eyes): acceptations for *jerkišembi* in the *Mirror*, 45n67; in the description of the *edeng*, 44–45; interpreted as referring to the shining properties of a steel-like bone (by Zakharov), 49n81; in Li Yanji's *Qingwen huishu*, 47, 47n71; rendered as *nidü jirgelümüi* without a case particle following "eye," 47n71

Yinchi 胤祉 (in Chinese) or In c'y (in Manchu) (Prince Cheng, 1677–1732): as Chen Menglei's patron, 114; identified as Kangxi's third son, 13; on the need to compile a "comprehensive compilation" (*huibian* 彙編), 114–15

Yin Han Qingwen jian 音漢清文鑑 (Mirror of the Manchu language translated into Chinese) (Mingdo): *guwalase* translated as *kegua* 客瓜 in, 272, 272n115; *ilhamuke* translated as *di shen* 地椹 (ground mulberry fruit), 265; *mi-sun-wu-shi-ha* 米孫烏什哈 identified with *wuweizi* in, 140n47

Yi xue san guan Qingwen jian 一學三貫清文鑑 (Juntu 屯圖): on *cuisheng mu* 催生木 (hasted-growth wood), 137; on *edeng*, 46–47n69; *guwalase* translated as *kegua* 客瓜 in, 272n115

Yongzheng emperor (r. 1723–1735, Yinzhen): *Gujin tushu jicheng* published by, 115–16; Jiang Tingxi assigned as the editor of the *Guang Qunfang pu*, 173, 174; provincial gazetteers ordered to be updated, 128n11; succession of, 115, 172

Yu Minzhong 于敏中 (1714–1779), 266, 267–68, 268n109

Yuan Mei 袁枚 (1716–1798), on gazetteers and encyclopedias used for writing *fu*, 199

Yuanjian leihan 淵鑑類函 (Categorized boxes of the Profound Mirror Studio): additional "categories" (*lei* 類) and "entries" (*mu* 目) added to the Ming original it was based on, 110, 110–11n50; on "Beans" (*dou* 豆, alternatively "Pulses"), 113; completion in 1701, 86, 110; Kangxi's preface bestowed on (1710), 117–18; *leishu* conventions followed by, 112; lexicographical encyclopedism of, 116–17; new entries and sources not seen in the Ming original marked with the character "added" (*zeng* 增), 111, *111*; search for *yi* 異 (unusual/ strange), 112–13; *Tang lei han* 唐類函 (Categorized boxes of Tang sources) compiled by Yu Anqi as the original work it was based on, 110n48; *xi shu* 磎鼠 (*xi* rodent) described in, 183

Yunggui 永貴 (1706–1783), 238–39

Yuzhi zengding Qingwen jian 御製增訂 清文鑑. See Expanded and Emended Manchu Mirror (*Yuzhi zengding Qingwen jian* 御製增訂清文鑑) (1772)

Zach, Erwin Ritter von (1872–1942), on neologisms created at the Qianlong court, 234–35

Zakharov, Ivan Il'ich (1817–1885), the *edeng* identified as "sawfish" (пила-рыба), 49–50, 49n81, 53

Zha Shenxing 查慎行 (1650–1727, *jinshi* 1703): compared to gentlemanly observers to experiments in seventeenth-century England, 188n110
— "*Peilie biji*" 陪獵筆記 (Jottings during my trip accompanying the imperial hunt): diary kept during Kangxi's 1703 imperial tour and seasonal hunt, 150; *jia* tree described in, 168–69; on practicing fishing in front of the heir apparent's tent, 162, 162n37; "Ula crabapple" described and eaten by, 162–63
— *Renhai ji* 人海記 (My account in the sea of humankind): anecdotes gathered during his sojourn in the imperial capital—including on Kangxi's imperial hunting trips, 150; the characters *jia* 椵 and *duan* 椴 used to gloss a Manchurian tree that was "not found in the central state [of China]," 169, 182; on luminescent wood shown to him, by Kangxi, 157; on scorpion grass that was "more poisonous than a bee sting," 160

Zhang Hong 张虹, Cheng Dakun 程大鯤, and Tong Yonggong 佟永功, 237

Zhang Penghe 張鵬翮 (1649–1725), Mongolian travelogue, 123–24, 133, 188

Zhang Yishao 張逸少 (fl. 1687–1716), 94, 94n22

Zheng Qiao 鄭樵 (1104–1162), *Tongzhi* 通志 (Comprehensive treatises), 319n14

Zhi-kuan 志寬 and Pei-kuan 培寬, *Duiyin jizi* 對音輯字 (*Dui yen ji dzi bithe* [Collection of characters for transcription]), 315

Zhou li 周禮 (Rites of Zhou): *fu* 賦 listed as one of six Principles of Poetry (*liu yi* 六義), 197; *shen* 蜃 (oysters) in, 47

Zhu Guosheng 朱國盛 (*jinshi* 1610), preface written for the *Qunfang pu*, 90

Zhuangzi 莊子, mythical trees in, 101, 113

Zou Zhenhuan 鄒振環, 13–14, 56

Harvard-Yenching Institute Monographs
(most recent titles)

97. *Materializing Magic Power: Chinese Popular Religion in Villages and Cities,* by Wei-Ping Lin
98. *Traces of Grand Peace: Classics and State Activism in Imperial China,* by Jaeyoon Song
99. *Fiction's Family: Zhan Xi, Zhan Kai, and the Business of Women in Late-Qing China,* by Ellen Widmer
100. *Chinese History: A New Manual, Fourth Edition,* by Endymion Wilkinson
101. *After the Prosperous Age: State and Elites in Early Nineteenth-Century Suzhou,* by Seunghyun Han
102. *Celestial Masters: History and Ritual in Early Daoist Communities,* by Terry F. Kleeman
103. *Transgressive Typologies: Constructions of Gender and Power in Early Tang China,* by Rebecca Doran
104. *Li Mengyang, the North-South Divide, and Literati Learning in Ming China,* by Chang Woei Ong
105. *Bannermen Tales (Zidishu): Manchu Storytelling and Cultural Hybridity in the Qing Dynasty,* by Elena Suet-Ying Chiu
106. *Upriver Journeys: Diaspora and Empire in Southern China, 1570–1850,* by Steven B. Miles
107. *Ancestors, Kings, and the Dao,* by Constance A. Cook
108. *The Halberd at Red Cliff: Jian'an and the Three Kingdoms,* by Xiaofei Tian
109. *Speaking of Profit: Bao Shichen and Reform in Nineteenth-Century China,* by William T. Rowe
110. *Building for Oil: Daqing and the Formation of the Chinese Socialist State,* by Hou Li
111. *Reading Philosophy, Writing Poetry: Intertextual Modes of Making Meaning in Early Medieval China,* by Wendy Swartz
112. *Writing for Print: Publishing and the Making of Textual Authority in Late Imperial China,* by Suyoung Son
113. *Shen Gua's Empiricism,* by Ya Zuo
114. *Just a Song: Chinese Lyrics from the Eleventh and Early Twelfth Centuries,* by Stephen Owen
115. *Shrines to Living Men in the Ming Political Cosmos,* by Sarah Schneewind
116. *In the Wake of the Mongols: The Making of a New Social Order in North China, 1200–1600,* by Jinping Wang
117. *Opera, Society, and Politics in Modern China,* by Hsiao-t'i Li
118. *Imperiled Destinies: The Daoist Quest for Deliverance in Medieval China,* by Franciscus Verellen
119. *Ethnic Chrysalis: China's Orochen People and the Legacy of Qing Borderland Administration,* by Loretta E. Kim
120. *The Paradox of Being: Truth, Identity, and Images in Daoism,* by Poul Andersen

121. *Feeling the Past in Seventeenth-Century China*, by Xiaoqiao Ling
122. *The Chinese Dreamscape, 300 BCE–800 CE*, by Robert Ford Campany
123. *Structures of the Earth: Metageographies of Early Medieval China*, by D. Jonathan Felt
124. *Anecdote, Network, Gossip, Performance: Essays on the* Shishuo xinyu, by Jack W. Chen
125. *Testing the Literary: Prose and the Aesthetic in Early Modern China*, by Alexander Des Forges
126. *Du Fu Transforms: Tradition and Ethics amid Societal Collapse*, by Lucas Rambo Bender
127. *Chinese History: A New Manual (Enlarged Sixth Edition)*, Vol. 1, by Endymion Wilkinson
128. *Chinese History: A New Manual (Enlarged Sixth Edition)*, Vol. 2, by Endymion Wilkinson
129. *Wang Anshi and Song Poetic Culture*, by Xiaoshan Yang
130. *Localizing Learning: The Literati Enterprise in Wuzhou, 1100–1600*, by Peter K. Bol
131. *Making the Gods Speak: The Ritual Production of Revelation in Chinese Religious History*, by Vincent Goossaert
132. *Lineages Embedded in Temple Networks: Daoism and Local Society in Ming China*, by Richard G. Wang
133. *Rival Partners: How Taiwanese Entrepreneurs and Guangdong Officials Forged the China Development Model*, by Wu Jieh-min; translated by Stacy Mosher
134. *Saying All That Can Be Said: The Art of Describing Sex in* Jin Ping Mei, by Keith McMahon
135. *Genealogy and Status: Hereditary Office Holding and Kinship in North China under Mongol Rule*, by Tomoyasu Iiyama
136. *The Threshold: The Rhetoric of Historiography in Early Medieval China*, by Zeb Raft
137. *Literary History in and beyond China: Reading Text and World*, edited by Sarah M. Allen, Jack W. Chen, and Xiaofei Tian
138. *Dreaming and Self-Cultivation in China, 300 BCE–800 CE*, by Robert Ford Campany
139. *The Painting Master's Shame: Liang Shicheng and the* Xuanhe Catalogue of Paintings, by Amy McNair
140. *The Cornucopian Stage: Performing Commerce in Early Modern China*, by Ariel Fox
141. *The Collapse of Heaven: The Taiping Civil War and Chinese Literature and Culture, 1850–1880*, by Huan Jin
142. *Elegies for Empire: A Poetics of Memory in the Late Work of Du Fu*, by Gregory M. Patterson
143. *The Manchu Mirrors and the Knowledge of Plants and Animals in High Qing China*, by He Bian and Mårten Söderblom Saarela